AF360716

The Molecular and Cellular Mechanisms of Inflammation and Tissue Regeneration

The Molecular and Cellular Mechanisms of Inflammation and Tissue Regeneration

Editor

Krisztina Nikovics

MDPI • Basel • Beijing • Wuhan • Barcelona • Belgrade • Manchester • Tokyo • Cluj • Tianjin

Editor
Krisztina Nikovics
Department of Platforms and
Technology Research, French
Armed Forces Biomedical
Research Institute,
Brétigny-sur-Orge, France

Editorial Office
MDPI
St. Alban-Anlage 66
4052 Basel, Switzerland

This is a reprint of articles from the Special Issue published online in the open access journal *Biomedicines* (ISSN 2227-9059) (available at: http://www.mdpi.com).

For citation purposes, cite each article independently as indicated on the article page online and as indicated below:

LastName, A.A.; LastName, B.B.; LastName, C.C. Article Title. *Journal Name* **Year**, *Volume Number*, Page Range.

ISBN 978-3-0365-7898-9 (Hbk)
ISBN 978-3-0365-7899-6 (PDF)

Cover image courtesy of Krisztina Nikovics.

Contents

About the Editor

Krisztina Nikovics

Krisztina Nikovics received her Ph.D. degree in Biology from the University of Szeged (Hungary) in 2001. She works as a project leader at the Biological Research Institute of the French Armed Forces. Her research focuses on molecular communication, i.e. in situ analysis of cytokine–receptor interactions, as well on the in situ identification of macrophages using different "in situ hybridization" methods.

Editorial

Molecular and Cellular Mechanisms of Inflammation and Tissue Regeneration

Anne-Laure Favier and Krisztina Nikovics *

Imagery Unit, Department of Platforms and Technology Research,
French Armed Forces Biomedical Research Institute, 91223 Bretigny sur Orge, France;
anne-laure.favier@intradef.gouv.fr
* Correspondence: krisztina.nikovics@def.gouv.fr; Tel.: +33-(0)17-8651-3331

Citation: Favier, A.-L.; Nikovics, K. Molecular and Cellular Mechanisms of Inflammation and Tissue Regeneration. *Biomedicines* **2023**, *11*, 1416. https://doi.org/10.3390/biomedicines11051416

Received: 4 May 2023
Accepted: 9 May 2023
Published: 10 May 2023

Over the past 70 years, significant progress has been made in understanding the molecular and cellular mechanisms of inflammation and tissue regeneration. This has become most evident in the last 20 years when revolutionary techniques in biology have exponentially increased the number of publications in the field. Following tissue damage caused by infection, mechanical or toxic injury, or autoimmune diseases, the healing process involves a series of highly regulated molecular and cellular processes that lead to the restoration of tissue homeostasis. If any step in the regeneration process is dysfunctional, chronic inflammation, tissue fibrosis or tumor formation can develop [1].

There are three overlapping steps in wound healing: (1) coagulation and inflammation, then (2) the proliferation and formation of new tissue, and (3) finally the tissue remodeling [2]. The initial phase of acute wound healing is the coagulation and the formation of a temporary wound matrix. This phase begins immediately after the injury and is completed within a few hours [3]. Inflammation is crucial to the clean-up-repair process. Early inhibition of inflammation can hinder regeneration processes [4]. Inflammation is associated with the activation of the innate immune system. At the site of inflammation, neutrophils appear first, followed by monocytes, which may differentiate into macrophages. The main function of macrophages and immune cells is to remove cell debris and microorganisms. These cells, in addition to the functions mentioned below, play an essential role in preparing the next phase by coordinating cellular processes [5]. The second phase starts with the division of the cells. This process allows damaged and lost structures to be replaced. Granulation tissue formed by the extracellular matrix (ECM) and new blood vessels generated by angiogenesis fills the lesion. This process takes 2–10 days [6]. In the final phase, the blood vessels regress, the inflammation resolves, and the granulation tissue become functional tissue. In this phase, the ECM transforms from a temporary ECM to a permanent collagen matrix. This phase starts 2–3 weeks after injury and can last up to years if tissue regeneration is inadequate [7].

The role of immune cells, in particular macrophages, in the inflammatory process has long been known, but it has recently been discovered that these immune cells also play an important role in the second and third stages of tissue regeneration [8–11]. They are involved in the activation of stem/progenitor cells, the clearance of damaged tissues, the remodeling of the extracellular matrix, and the promotion of angiogenesis [12]. Macrophages and molecular actors of tissue regeneration continuously produce soluble mediators and extracellular vesicles that stimulate fibroblasts, immune cells, and stem/progenitor cells [12–14]. Soluble mediators, together with extracellular vesicles (which may contain various bioactive molecules such as nucleic acids, proteins, lipids, and sugars), play an important role in cell-to-cell communication at all steps of regeneration (Figure 1) [15,16].

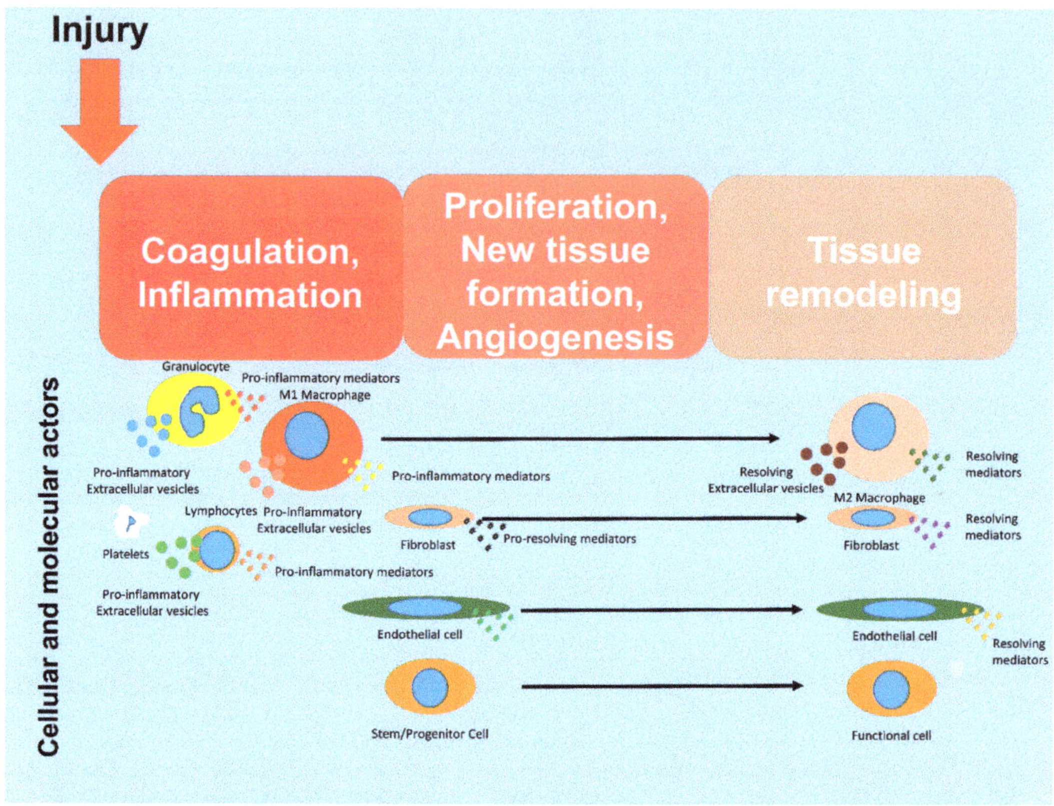

Figure 1. Molecular and cellular actors during tissue regeneration. Regeneration occurs after three schematical steps: coagulation/inflammation, profiferation/angiogenesis and finally tissue remodeling.

Several interesting findings and overviews were shortly mentioned to illustrate molecular and cellular actors involved in inflammation and tissue regeneration.

Rangarajan and co-workers (2022) reviewed the effects of different pro-resolving mediators on atherosclerosis [17]. Atherosclerosis is a chronic inflammation that can last for several years. The healing of this disease can be greatly facilitated by the presence of various pro-resolving mediators such as resolvin, lipoxin, maresin, and protectin. The study of the complex temporal and functional relationships of these molecules offers a new approach to the treatment of tissue regeneration.

Durand and co-authors (2022) investigated the osteogenic effects of two different biomaterials (metakaolin and polymethylmethacrylate) in the Masquelet-induced membrane in a rat model animal [18]. They showed that the nature of the biomaterials influences the immune microenvironment and macrophage responses, which strongly affects bone regeneration. The more intense osteogenic effect of metakaolin was attributed to a higher number of M1- and M2-like macrophages and to a more intense expression of transforming growth factor-β (TGF-β) and bone morphogenetic protein-2 (BMP-2).

Spinal cord injury often causes paralysi and currently no therapy is available. Following injury, nerve and glial cells die as a result of inflammation. In addition, activation of microglial cells and infiltration of macrophages and lymphocyte cells can be observed in the injured tissue. Wu and colleagues (2022) showed that the bile acid tauroursodeoxycholic acid (TUDCA) was able to inhibit inflammation and promote the temporary recovery of motor function in rats [19]. Unfortunately, these positive effects are not noticeable in the long term, so the authors do not recommend TUDCA for the treatment of spinal cord injury.

Toll-like receptor 8 (TLR8), localizes in the endosome, is able to bind to single-stranded RNAs of viral and bacterial origin. In human monocytes and macrophages, pro-inflammatory cytokines and type I interferons are secreted in response to the binding. Nilsen and colleagues (2022) investigated this mechanism [20]. Their work showed that the Toll-interleukin 1 receptor domain adaptor protein (TIRAP) played an important role in regulation and was involved in signal transduction as an adaptor protein. This protein promotes the translocation of a transcription factor, interferon regulatory factor 5 (IRF5), into the nucleus and the production of IRF5-dependent cytokines.

Alumina nanoparticles (Al_2O_3 NPs) are one of the most frequently produced particles in the world. Dekali and colleagues (2022) reviewed the current knowledge on the effects of Al_2O_3 NPs on animal health [21]. Al_2O_3 NPs have been shown to significantly increase the risk of inflammation, pulmonary fibrosis, reduced lung function, and increased risk of lung cancer in the lung parenchyma of various animals. In contrast, in vitro studies on the effects of Al_2O_3 NPs on lung cells have been rather controversial.

Recently, there has been renewed interest in analyzing the mechanism of action of chemical warfare agents derived from organophosphorus materials. François and colleagues (2022) analyzed the long-term effects of 4′-nitrophenyl isopropyl methyl phosphonate (NIMP), a sarin surrogate [22]. The experimental mice received two sub-lethal doses of NIMP and were studied for 6 months. Six months after exposure, inflammation of the gut, anxiety-like behavior, and significant changes in leukocyte count was observed.

The development of inflammatory and tissue regeneration disorders is largely due to an imbalance between reactive oxygen species (ROS) and endogenous antioxidants. Perez and colleagues (2022) reviewed the relationship between the effects of antioxidants on neuronal remyelination and muscle regeneration [23]. In summary, it was pointed out that if the remyelination of damaged nerve cells is inadequate, the regeneration of muscle cells will not be efficient. In addition, antioxidants have been shown to facilitate the remyelination of nerve cells, which promotes muscle cell regeneration.

Hart and Nakamura (2022) reviewed the most important cell therapy methods for the regeneration of damaged musculoskeletal tissue [24]. The cell therapies that have been used over the past 30 years and have greatly improved the regeneration of musculoskeletal tissues have been Platelet-rich Plasma (PRP) or mesenchymal stem cell (MSC) therapies. Factors in the PRP or secreted by the MSC activate the regenerative capacity of endogenous cells. To ensure optimal regeneration, the first step is to reduce inflammation and then introduce properly activated cells into the damaged tissue. Cell therapy was less effective when the cells were injected into the bloodstream, as the cells were probably not able to reach the site of injury.

Failure of tissue regeneration can have serious clinical consequences, such as tumor formation because proteins and cells involved in signaling mechanisms that are important for healing are also involved in cancer metastasis. In their review, Lopez and colleagues (2022) compared the relationship between wound healing and metastasis [2]. They have underlined the role of oxidative stress in these processes. It is very important to reduce oxidative stress, for example by limiting its production, using scavenger agents, and increasing the antioxidant capacity of the cells. These treatments offer interesting therapeutic options that can promote proper tissue regeneration and prevent metastasis.

In our laboratory, we have investigated a method to identify mRNA transcripts in cryosections of undecalcified rat bone. In collaboration with the Institute of Plant Sciences of Paris-Saclay (IPS2), in situ hybridization and hybridization chain reaction (in situ-HCR) was adaptedto better understand gene expression in in situ bone tissue [25]. Our objective was to study a section of a whole rat femur. The muscle significantly delayed the penetration of the decalcifying solution (EDTA) into the bone, so paraffin embedding was not applicable. We have developed an improved version of the CryoJane tape transfer system for making bone tissue sections. This technique was used to investigate the expression of different genes in bone tissue by in situ-HCR.

As the phenotype of in vivo macrophages is still poorly understood, we used the above techniques to identify in vivo macrophages in regenerating bone [26]. There is no cell surface marker available to identify macrophage subtypes that can distinguish between different subtypes. Another approach to determining macrophage subtypes is to identified the patern of cytokines they express. However, since cytokines are usually secreted, immunostaining techniques are not suitable for detecting subtypes. We have shown that in situ-HCR hybridization is one of the most suitable methods to detect cytokines in order to characterize macrophage subtypes. This technique is based on the detection of messenger RNA (mRNA) of targeted genes.

Understanding the mechanisms of inflammation and tissue regeneration is of great scientific and clinical importance. However, the molecular and cellular mechanisms through which they exert their effects are still largely unknown. The study of these processes is essential for the development of therapeutic strategies aimed at tissue repair. We hope that readers of this special issue of Biomedicines will enjoy reading the excellent papers of many of the leading scientists in the field. We hope also that future generations of researchers will be inspired by the topics published in this special issue to further improve our knowledges in this field.

Author Contributions: K.N. and A.-L.F. wrote the editorial. All authors have read and agreed to the published version of the manuscript.

Funding: Work was supported by the Délégation Générale de l'Armement (DGA) (PDH2-NRBC-4-NR-4306 and PDH-SAN-1-217/206).

Conflicts of Interest: The authors declare no conflict of interest.

References

1. Park, M.H.; Lee, E.D.; Chae, W.-J. Macrophages and Wnts in Tissue Injury and Repair. *Cells* **2022**, *11*, 3592. [CrossRef] [PubMed]
2. Lopez, T.; Wendremaire, M.; Lagarde, J.; Duquet, O.; Alibert, L.; Paquette, B.; Garrido, C.; Lirussi, F. Wound Healing versus Metastasis: Role of Oxidative Stress. *Biomedicines* **2022**, *10*, 2784. [CrossRef] [PubMed]
3. Reinke, J.M.; Sorg, H. Wound Repair and Regeneration. *Eur. Surg. Res.* **2012**, *49*, 35–43. [CrossRef] [PubMed]
4. Cooke, J.P. Inflammation and Its Role in Regeneration and Repair. *Circ. Res.* **2019**, *124*, 1166–1168. [CrossRef]
5. Gurtner, G.C.; Werner, S.; Barrandon, Y.; Longaker, M.T. Wound Repair and Regeneration. *Nature* **2008**, *453*, 314–321. [CrossRef]
6. Oishi, Y.; Manabe, I. Macrophages in Inflammation, Repair and Regeneration. *Int. Immunol.* **2018**, *30*, 511–528. [CrossRef]
7. Duffield, J.S.; Forbes, S.J.; Constandinou, C.M.; Clay, S.; Partolina, M.; Vuthoori, S.; Wu, S.; Lang, R.; Iredale, J.P. Selective Depletion of Macrophages Reveals Distinct, Opposing Roles during Liver Injury and Repair. *J. Clin. Investig.* **2005**, *115*, 56–65. [CrossRef]
8. Murray, P.J.; Allen, J.E.; Biswas, S.K.; Fisher, E.A.; Gilroy, D.W.; Goerdt, S.; Gordon, S.; Hamilton, J.A.; Ivashkiv, L.B.; Lawrence, T.; et al. Macrophage Activation and Polarization: Nomenclature and Experimental Guidelines. *Immunity* **2014**, *41*, 14–20. [CrossRef]
9. Sica, A.; Mantovani, A. Macrophage Plasticity and Polarization: In Vivo Veritas. *J. Clin. Investig.* **2012**, *122*, 787–795. [CrossRef]
10. DeNardo, D.G.; Ruffell, B. Macrophages as Regulators of Tumour Immunity and Immunotherapy. *Nat. Rev. Immunol.* **2019**, *19*, 369–382. [CrossRef]
11. Kaur, S.; Raggatt, L.J.; Batoon, L.; Hume, D.A.; Levesque, J.-P.; Pettit, A.R. Role of Bone Marrow Macrophages in Controlling Homeostasis and Repair in Bone and Bone Marrow Niches. *Semin. Cell Dev. Biol.* **2017**, *61*, 12–21. [CrossRef] [PubMed]
12. Abdelaziz, M.H.; Abdelwahab, S.F.; Wan, J.; Cai, W.; Huixuan, W.; Jianjun, C.; Kumar, K.D.; Vasudevan, A.; Sadek, A.; Su, Z.; et al. Alternatively Activated Macrophages; a Double-Edged Sword in Allergic Asthma. *J. Transl. Med.* **2020**, *18*, 58. [CrossRef] [PubMed]
13. Murray, P.J. Macrophage Polarization. *Annu. Rev. Physiol.* **2017**, *79*, 541–566. [CrossRef] [PubMed]
14. Théry, C.; Witwer, K.W.; Aikawa, E.; Alcaraz, M.J.; Anderson, J.D.; Andriantsitohaina, R.; Antoniou, A.; Arab, T.; Archer, F.; Atkin-Smith, G.K.; et al. Minimal Information for Studies of Extracellular Vesicles 2018 (MISEV2018): A Position Statement of the International Society for Extracellular Vesicles and Update of the MISEV2014 Guidelines. *J. Extracell. Vesicles* **2018**, *7*, 1535750. [CrossRef]
15. Altan-Bonnet, G.; Mukherjee, R. Cytokine-Mediated Communication: A Quantitative Appraisal of Immune Complexity. *Nat. Rev. Immunol.* **2019**, *19*, 205–217. [CrossRef]
16. Zheng, D.; Ruan, H.; Chen, W.; Zhang, Y.; Cui, W.; Chen, H.; Shen, H. Advances in Extracellular Vesicle Functionalization Strategies for Tissue Regeneration. *Bioact. Mater.* **2023**, *25*, 500–526. [CrossRef]

17. Rangarajan, S.; Orujyan, D.; Rangchaikul, P.; Radwan, M.M. Critical Role of Inflammation and Specialized Pro-Resolving Mediators in the Pathogenesis of Atherosclerosis. *Biomedicines* **2022**, *10*, 2829. [CrossRef]
18. Durand, M.; Oger, M.; Nikovics, K.; Venant, J.; Guillope, A.-C.; Jouve, E.; Barbier, L.; Bégot, L.; Poirier, F.; Rousseau, C.; et al. Influence of the Immune Microenvironment Provided by Implanted Biomaterials on the Biological Properties of Masquelet-Induced Membranes in Rats: Metakaolin as an Alternative Spacer. *Biomedicines* **2022**, *10*, 3017. [CrossRef]
19. Wu, S.; García-Rama, C.; Romero-Ramírez, L.; de Munter, J.P.J.M.; Wolters, E.C.; Kramer, B.W.; Mey, J. Tauroursodeoxycholic Acid Reduces Neuroinflammation but Does Not Support Long Term Functional Recovery of Rats with Spinal Cord Injury. *Biomedicines* **2022**, *10*, 1501. [CrossRef]
20. Nilsen, K.E.; Skjesol, A.; Frengen Kojen, J.; Espevik, T.; Stenvik, J.; Yurchenko, M. TIRAP/Mal Positively Regulates TLR8-Mediated Signaling via IRF5 in Human Cells. *Biomedicines* **2022**, *10*, 1476. [CrossRef]
21. Dekali, S.; Bourgois, A.; François, S. Critical Review on Toxicological Mechanisms Triggered by Inhalation of Alumina Nanoparticles on to the Lungs. *Biomedicines* **2022**, *10*, 2664. [CrossRef] [PubMed]
22. François, S.; Mondot, S.; Gerard, Q.; Bel, R.; Knoertzer, J.; Berriche, A.; Cavallero, S.; Baati, R.; Orset, C.; Dal Bo, G.; et al. Long-Term Anxiety-like Behavior and Microbiota Changes Induced in Mice by Sublethal Doses of Acute Sarin Surrogate Exposure. *Biomedicines* **2022**, *10*, 1167. [CrossRef] [PubMed]
23. Cabezas Perez, R.J.; Ávila Rodríguez, M.F.; Rosero Salazar, D.H. Exogenous Antioxidants in Remyelination and Skeletal Muscle Recovery. *Biomedicines* **2022**, *10*, 2557. [CrossRef]
24. Hart, D.A.; Nakamura, N. Creating an Optimal In Vivo Environment to Enhance Outcomes Using Cell Therapy to Repair/Regenerate Injured Tissues of the Musculoskeletal System. *Biomedicines* **2022**, *10*, 1570. [CrossRef] [PubMed]
25. Nikovics, K.; Castellarin, C.; Holy, X.; Durand, M.; Morin, H.; Bendahmane, A.; Favier, A. In Situ Gene Expression in Native Cryofixed Bone Tissue. *Biomedicines* **2022**, *10*, 484. [CrossRef]
26. Nikovics, K.; Durand, M.; Castellarin, C.; Burger, J.; Sicherre, E.; Collombet, J.-M.; Oger, M.; Holy, X.; Favier, A.-L. Macrophages Characterization in an Injured Bone Tissue. *Biomedicines* **2022**, *10*, 1385. [CrossRef] [PubMed]

 *biomedicines*

Article

Influence of the Immune Microenvironment Provided by Implanted Biomaterials on the Biological Properties of Masquelet-Induced Membranes in Rats: Metakaolin as an Alternative Spacer

Marjorie Durand [1,*], Myriam Oger [2], Krisztina Nikovics [2], Julien Venant [1,3], Anne-Cecile Guillope [1], Eugénie Jouve [1], Laure Barbier [4], Laurent Bégot [2], Florence Poirier [3], Catherine Rousseau [4], Olivier Pitois [5], Laurent Mathieu [1,6], Anne-Laure Favier [2], Didier Lutomski [3] and Jean-Marc Collombet [1]

[1] Osteo-Articulary Biotherapy Unit, Department of Medical and Surgical Assistance to the Armed Forces, French Armed Forces Biomedical Research Institute, 91223 Brétigny-sur-Orge, France
[2] Imaging Unit, Department of Platforms and Technology Research, French Armed Forces Biomedical Research Institute, 91223 Brétigny-sur-Orge, France
[3] Tissue Engineering Research Unit-URIT, Sorbonne Paris Nord University, 93000 Bobigny, France
[4] Molecular Biology Unit, Department of Platforms and Technology Research, French Armed Forces Biomedical Research Institute, 91223 Brétigny-sur-Orge, France
[5] Laboratoire Navier, Gustave Eiffel University, Ecole des Ponts ParisTech, CNRS, 77447 Marne-la-Vallée, France
[6] Department of Surgery, Ecole du Val-de-Grace, French Military Health Service Academy, 1 Place Alphonse Laveran, 75005 Paris, France
* Correspondence: durand.irba@orange.fr

Citation: Durand, M.; Oger, M.; Nikovics, K.; Venant, J.; Guillope, A.-C.; Jouve, E.; Barbier, L.; Bégot, L.; Poirier, F.; Rousseau, C.; et al. Influence of the Immune Microenvironment Provided by Implanted Biomaterials on the Biological Properties of Masquelet-Induced Membranes in Rats: Metakaolin as an Alternative Spacer. *Biomedicines* 2022, 10, 3017. https://doi.org/10.3390/biomedicines10123017

Academic Editor: Matthias Wiens

Received: 29 September 2022
Accepted: 17 November 2022
Published: 23 November 2022

Publisher's Note: MDPI stays neutral with regard to jurisdictional claims in published maps and institutional affiliations.

Abstract: Macrophages play a key role in the inflammatory phase of wound repair and foreign body reactions—two important processes in the Masquelet-induced membrane technique for extremity reconstruction. The macrophage response depends largely on the nature of the biomaterials implanted. However, little is known about the influence of the macrophage microenvironment on the osteogenic properties of the induced membrane or subsequent bone regeneration. We used metakaolin, an immunogenic material, as an alternative spacer to standard polymethylmethacrylate (PMMA) in a Masquelet model in rats. Four weeks after implantation, the PMMA- and metakaolin-induced membranes were harvested, and their osteogenic properties and macrophage microenvironments were investigated by histology, immunohistochemistry, mass spectroscopy and gene expression analysis. The metakaolin spacer induced membranes with higher levels of two potent pro-osteogenic factors, transforming growth factor-β (TGF-β) and bone morphogenic protein-2 (BMP-2). These alternative membranes thus had greater osteogenic activity, which was accompanied by a significant expansion of the total macrophage population, including both the M1-like and M2-like subtypes. Microcomputed tomographic analysis showed that metakaolin-induced membranes supported bone regeneration more effectively than PMMA-induced membranes through better callus properties (+58%), although this difference was not significant. This study provides the first evidence of the influence of the immune microenvironment on the osteogenic properties of the induced membranes.

Keywords: Masquelet-induced membrane; macrophages; PMMA; metakaolin

1. Introduction

In the face of large bone defects, surgery is required to restore the shape and function of the bone. The induced membrane technique (IMT), also known as the Masquelet technique, is a widely used two-stage surgical procedure. This technique is unique in preparing the bed graft by molding a polymethylmethacrylate (PMMA) spacer to fill the bone defect [1]. The implantation of this spacer leads to the formation of an induced membrane (IM)—granulation tissue surrounding the spacer. In the second step, the spacer is removed while preserving the integrity of the IM. A standard autologous bone graft is then implanted into the IM cavity to repair the bone.

The IM, which acts as a biologically privileged membrane at the site of the defect, is the key element in this procedure. Preclinical and clinical studies have highlighted the various roles of the IM. It prevents graft resorption and muscle invasion of the defective bone by acting as a barrier membrane. It also creates an osteogenic and osteoinductive environment by secreting many growth factors and cytokines. These factors include bone morphogenic protein-2 (BMP-2), interleukin-6 (IL-6), transforming growth factor-β (TGF-β), vascular endothelial growth factor A (VEGF-A), von Willebrand factor (vWF) and metalloproteinase-9 (MMP-9) [2–6]. The IM has also been shown to be highly vascularized and to serve as a source of bone progenitor cells.

Biologically, the IM results from a foreign-body reaction (FBR). Upon implantation, all biomaterials elicit a FBR, a natural immunoinflammatory process that isolates the implant from the rest of the body in a collagenous capsule [7]. Macrophages are plastic cells that play a key role in the FBR by orchestrating the inflammatory environment around the implanted biomaterial. Indeed, macrophages can adopt diverse functional phenotypes upon activation, ranging from M1 (pro-inflammatory) to the M2 (pro-healing) profiles. Interestingly, macrophage activation depends on the shape and surface properties of the biomaterial (chemistry, porosity, wettability, roughness and stiffness) [8].

We hypothesized that changing the chemical composition of the spacer generating the IM in the Masquelet technique would modify the immune microenvironment in which the FBR occurred, thereby altering the osteogenic properties of the IM and potentially enhancing bone regeneration. We tested this hypothesis by replacing the PMMA of standard spacers with an alternative biomaterial, a metakaolin-based geopolymer (Davidovits) [9]. This polymer is synthesized by an alkaline activator solution's reaction (geopolymerization) with metakaolin particles. Chemically, metakaolin is a dehydroxylated form of the clay mineral kaolinite, an aluminosilicate material. Aluminosilicates and their derivatives are known to have immunostimulatory effects due to induction of macrophage activation [10]. Metakaolin was, therefore, chosen for this study based on its immunogenicity, and its innocuity relative to other clay minerals, such as bentonite [11]. Wiemann et al. [12] recently showed that the intratracheal instillation of kaolin in rats induced transient macrophage-based hypercellularity in rat lungs, with no signs of inflammation or structural change in the lung parenchyma, whereas bentonite instillation leads to a very intense lung inflammation with changes to the structure of the lung epithelium. Metakaolin is listed in the US Pharmacopeia, suggesting that its transfer into clinical practice might be facilitated in terms of the requirements for medical device regulation. Kaolinite and its chemical derivatives have been widely used in the pharmaceutical domain for decades as well-characterized pharmaceutical excipients: diluents, binders, disintegrants, pelleting agents, granulating agents, amorphizing agents, film-coating additives or even drug carriers [13,14]. They are also used as active pharmaceutical ingredients in hemostatic wound dressings, dermatological protectors, gastrointestinal protectors and antidiarrheal agents.

We used a validated Masquelet model in rats to assess the osteogenic properties of metakaolin-generated IM with histological and immunohistochemical methods. We first analyzed the distribution of M1-like and M2-like macrophage populations within the IM. Finally, we determined the impact of metakaolin spacers as an alternative to PMMA on bone-healing outcomes.

2. Materials and Methods

2.1. Animals

Animal procedures were approved by the appropriate institutional animal care and use committee (protocol 65 DEF_IGSSA_SP). Interventions were performed at an accredited animal facility. Male Sprague Dawley rats (Charles River, France) were housed individually in cages with controlled temperature and lighting conditions, and food and water supplied ad libitum. The rats were eight weeks old (mean weight of 200 g) when they underwent the first surgical procedure. In cases of postoperative complications, such as deep infection or bone fixation failure, the animals were excluded from the study and euthanized. Animals were killed by the intraperitoneal injection of sodium pentobarbital (150 mg/kg) at the age of 12 weeks (for IM analysis) or 22 weeks (for bone repair assessment).

2.2. Surgical Procedures

IMT surgery (steps 1 and 2) was performed as previously described [15]. The first stage of surgery was performed under general anesthesia induced by the intraperitoneal administration of a ketamine/medetomidine mixture (60 and 0.42 mg/kg, respectively). Rats were placed in the prone position and an incision was made through the skin and muscle to expose the right femur. A mini external fixator (RatExFix RISystem, Davos, Switzerland) was screwed onto the anterolateral surface of the femur shaft, and a Gigli wire saw was used to create a 6 mm segmental defect. The bone defect was filled with either hand-made PMMA or metakaolin spacers ($n = 5$/group). Four weeks later, the animals underwent the second stage of graft surgery or were killed for stage 1 membrane studies. This time point was chosen based on the results of our previous model validation study [15]. For stage 2 of the IMT surgery, rats ($n = 5$/group) were anesthetized with isoflurane (1.5 to 2% isoflurane in 1 to 1.5 L of O_2/min). An incision was carefully made in the membrane for spacer removal. The defect was then filled with a morselized corticocancellous allograft harvested from the distal femur of littermates killed on the same day. On three consecutive days after each surgical procedure, the animals received subcutaneous injections of a cephalosporin antibiotic (10 mg/kg enrofloxacin) and an opioid painkiller (0.05 mg/kg buprenorphine, twice daily). Unprotected weight-bearing activity was allowed immediately after surgery. The animals were weighed daily, and animal facility staff also evaluated their behavior, pain, normal movements and the appearance of the wound every day. Radiographic follow-up evaluations were performed every two weeks to check for incorrect spacer positioning and implantation failure. The animals were killed after 10 weeks for bone-healing assessment.

2.3. Spacers

PMMA (Palacos R + G, Heraeus, Hanau, Germany) spacers were made by hand under sterile conditions before surgery. They were macroscopically smooth and cylindrical. Metakaolin spacers were prepared in advance, as follows. Activated metakaolin paste was prepared by mixing sodium silicate (activating solution) with metakaolin particles. The activating solution was prepared by mixing NaOH solution (mass concentration Cw = 0.35) with a solution containing Na_2O (Cw = 0.08) and SiO_2 (Cw = 0.27) provided by MERCK KGaA, and water. The metakaolin particles (Argical M 1200S) were provided by AGS Minéraux (Clérac, France). The Brunauer, Emmett and Teller (BET) specific surface area of these particles was 19 m^2/g and their mass mean diameter was about 2 μm. The chemical composition of the resulting paste was characterized by the following ratios: Si/Al = 1.71 (molar), Na_2O/Al_2O_3 = 1.01 (molar) and solid/liquid = 1.50 (mass ratio). The paste was poured into cylindrical PMMA molds, all of the same diameter (4 mm), but with three different lengths: 5.5, 6 and 6.5 mm. The opening was covered and the molds were left at room temperature for 48 h. Geopolymerization resulted in very slight shrinkage, facilitating the removal of the metakaolin spacers from the molds. Prior to animal implantation, spacers were exposed to steam sterilization accomplished in an autoclave (20 min, +121 °C).

2.4. Bone Turnover Assessment

Serum samples were used to assess bone turnover markers, both markers of formation (procollagen-1 N terminal telopeptide or P1NP) synthesized by osteoblasts, and markers of resorption (tartrate-resistant alkaline phosphatase C or TRAP-C) released by osteoclasts during bone matrix remodeling. The levels of these markers were determined by ELISA. For P1NP assessments, we used the Rat/Mouse PINP EIA^{TM} assay kit (ref. AC-33F1, IDS Inc., El Segundo, CA, USA). The rat TRAPTM (TRAcP-5b) ELISA kit (ref SBTR102, IDS Inc.) was used for TRAP-C assays. Duplicate determinations were performed for each sample, and the two results were then averaged. The P1NP/TRAP-C ratio was calculated to express bone turnover four weeks after creating bone defects.

2.5. Histology

Membrane fragments were fixed in a 4% paraformaldehyde solution for embedding in paraffin. Sections (5 μm) were cut and stained with hematoxylin-eosin-saffron (HES) or prepared for BMP-2 immunostaining and CD68 CD206 immunofluorescence analysis. A pathological histologist examined all HES-stained sections. All immuno-stained sections on glass slides were digitized with a Nanozoomer S60 slide scanner (Hamamatsu) to quantify whole-slide images. Scanning resolution at 20× magnification was 0.46 μm/px. Virtual slide images were saved in 16-bit raw format for immunofluorescence analysis and RGB TIFF format for sections with standard staining. All image processing was performed with Fiji software [16]. To quantify IM cellularity, sections were stained with DAPI to visualize the cell nuclei. A region of interest (ROI) was drawn manually to exclude muscle fibers from the areas analyzed. The DAPI image was thresholded with the Triangle algorithm to select the brightest objects. Each object was then isolated to segment clusters of nuclei based on the local maxima of the initial image (with the segmented particles option). A "logical and" was used between the first threshold and the segmented particles.

2.6. BMP-2 Immunostaining

We assessed the expression of BMP-2, a potent osteogenic growth factor, by performing immunohistochemical analyses on paraffin-embedded IM sections, as previously described [15]. The rabbit polyclonal antibody specific for BMP-2 (Bioworld 90141) was used at a dilution of 1:200. The ready-to-use ImmPRESS HRP Anti-Rabbit IgG detection kit (Vector, MP-7451) was incubated with the slides for 30 min, and hematoxylin counterstaining was then performed.

2.7. Real-Time PCR Analysis

Membrane tissues for molecular biology analysis were collected and stored in RNA later® (Ambion, Austin, TX, USA). Samples were kept at +4 °C for 24 h and then stored at −20 °C until homogenization in guanidium-based lysis buffer with a TissueLyser II (RLT buffer, Qiagen, 20 Hz, 2 min, two 3 mm-carbide beads). According to the manufacturer's recommendations, total RNA was extracted with the Nucleospin RNA XS kit (Macherey Nagel, France) but with an additional proteinase K digestion step (Qiagen, Les Ulis, France). RNA was eluted in 15 μL of RNase-free water. The quantity and purity of the RNA were determined with an Agilent TapeStation 4200 automated electrophoresis system, with RNA screen tape and reagents (Agilent Technologies, Santa Clara, CA, USA), according to the manufacturer's instructions. The total RNA concentration of each sample was expressed in nanograms per microliter. RNA quality was assessed by determining the RNA integrity number (RIN) on a scale of 1 (completely degraded RNA) to 10 (intact RNA), as described by Schroeder and collaborators [17]. The mean RIN value was 7.4 for the PMMA group and 7.6 for the metakaolin group. A real-time PCR study was carried out as described in the MIQE guidelines [18].

Based on the manufacturer's instructions, the first-strand cDNA was generated by reverse transcription with the EuroScript reverse transcriptase on 400 ng total RNA (Eurogentec #RT-RTCK-03, Seraing, Belgium). RNA integrity and reverse transcription yields were confirmed with the 5′/3′ integrity assay and *Rplp0* selected primers (supplementary data, Table S1) [19]. Primers were designed and optimized with MacVector® 3.5 software (Accelrys, San Diego, CA, USA) to prevent dimerization, self-priming and melting temperature. Primers binding to flanking introns were selected to exclude genomic DNA amplification and were assessed for specificity to rats with the Blast nucleotide algorithm. Oligonucleotide primers were synthesized by Eurogentec (Sereing, Belgium). Real-time qPCR was performed with a LightCycler® 480 instrument (Roche Applied Science, Mannheim, Germany) with SybrGreen I Mastermix (Roche Applied Science). Quantification of mRNA was measured using the comparative threshold method [20] with efficiency correction estimated from a standard curve. The qPCR primers used for the three reference genes (ribosomal protein lateral stalk subunit P0 (*Rplp0*), peptidylprolyl isomerase A (*Ppia*), hypoxanthine phosphoribosyltransferase 1 (*Hprt*)) and the five target genes (transforming growth factor beta 2 (*TGFβ2*), interleukin-6 (*IL-6*), interleukin-1-beta (*IL-1β*), insulin-like growth factor (*IGF1*) and vascular endothelial growth factor A (*VEGF-A*) are listed in Supplemental Table S1, along with the optimized concentration and annealing temperature for each primer. Normalization was assessed with geNorm software. A geometric mean for the three internally validated reference genes (*Rplp0*, *Ppia* and *Hprt*) was calculated [21]. The pairwise variation of these three genes was 0.119, which is below the threshold (0.15), requiring the inclusion of an additional normalization gene.

2.8. Immunofluorescence Assays and Macrophage Quantification

Immunofluorescence analysis was performed to study the phenotypic profiles of the macrophages in the IM. "M1-like macrophages" were defined as CD68-positive cells, whereas "M2-like macrophages" were defined as cells positive for both CD68 and CD206, as previously described [22]. Cells negative for CD68 but positive for CD206 were defined as muscle satellite cells [22,23].

Sections were permeabilized by incubation for 15 min with 0.5% Triton X100 (v/v) buffered with PBS. Non-specific binding sites were blocked by incubation with Emerald Antibody Diluent (Sigma 936B-08) for 1 h. The sections were then incubated overnight at +4 °C with the primary mouse anti-CD68 (BIO-RAD MCA341GA, Hercules, CA, USA) antibody at a dilution of 1:100 and the primary rabbit anti-CD206 (Sigma HPA045134) antibody at a dilution of 1:100. They were washed in PBS and incubated with an anti-rabbit green fluorescent Alexa Fluor 488 (A-21206, Thermo Fisher Scientific) secondary antibody and an anti-mouse red fluorescent Alexa Fluor 568 (A10037, Thermo Fisher Scientific) secondary antibody, both at a dilution of 1:1000, for two hours at room temperature. Finally, sections were washed in PBS for 20 min and mounted in Fluoroshield mounting medium with DAPI (Abcam, Cambridge, UK, ab104139). Fluorescence was detected under an epifluorescence microscope DM6000 (Leica, Wetzlar, Germany) equipped with monochrome and color digital cameras. Macrophages were quantified on whole-slide images with FIJI software. The M2-like cells displayed double labeling (green + red), whereas M1-like macrophages displayed only red labeling. A "zone of influence" was defined around each nucleus, with nuclei segmented for cellularity measurement as seeds. On Alexa Fluor 488-labeled images, the Otsu method set a double threshold for the previously drawn ROI. On the Alexa Fluor 568-labeled images, an Otsu threshold was determined within the same ROI. A geodesic reconstruction of the cells was performed with each type of immunofluorescence labeling used as a seed and the "zone of influence" of the nuclei as a mask. These analyses yielded the number of stained cells/total number of cells expressed as a percentage.

2.9. Liquid Chromatography–Tandem Mass Spectrometry (LC-MS/MS)

Proteins secreted by IM fragments were identified by mass spectrometry. No labeling/tagging techniques were used in our LC-MS/MS study. Therefore, we could not determine the abundance of the secreted proteins. Instead, we aimed to identify all the secreted proteins and their molecular networks and compare protein secretion frequencies between the two batches. The proteins secreted by IMs were purified by an organic solvent-based protein precipitation method. Briefly, nine volumes of ice-cold acetone-methanol (8:1) were added to one sample volume, and the resulting mixture was incubated overnight at −20 °C. The samples were then centrifuged at 10,000× *g* for 30 min, and the protein pellet was dissolved in 40 μL of 2X Laemmli buffer (Biorad).

Protein samples were briefly subjected to SDS-PAGE (8% acrylamide gel, 8 × 8 cm) until the sample had completely penetrated the gel. Following in-gel fixation (ethanol 30% v/v, acetic acid 7% v/v) for 1 h and protein staining with Coomassie Brilliant Blue, each band was excised manually and cut into small pieces with a scalpel. Gel pieces were dehydrated by incubation in 100 µL acetonitrile for 15 min and rehydrated by incubation with 100 µL 25 mM NH_4HCO_3 for 10 min. This operation was repeated twice. After final dehydration in 100 µL acetonitrile, gel pieces were covered with 100 µL 10 mM DTT in 25 mM NH_4HCO_3 and incubated at +56 °C for 45 min. The supernatant was removed, and 100 µL of 55 mM iodoacetamide in 25 mM NH_4HCO_3 was added. The mixture was left in the dark at room temperature for 30 min and the supernatant was then removed. The gel pieces were covered with 100 µL 25 mM NH_4HCO_3 for 10 min and dehydrated by incubation with 100 µL acetonitrile for 15 min. The volume of the dehydrated gel was evaluated and three volumes of trypsin (12 ng/µL) in 25 mM NH_4HCO_3 (freshly diluted) were added. The digestion was allowed to proceed at +35 °C overnight. Peptides were finally extracted from the gel pieces by incubation in 60% acetonitrile/5% HCOOH for 1 h. The supernatant was collected, the volume of each peptide sample was reduced to 15 µL, and the peptides were analyzed by mass spectrometry.

Peptide samples were then analyzed with a QToF instrument (Xevo G2-XS QTof, Waters, Milford, MA, USA) coupled to a nano liquid chromatography apparatus (ACQUITY UPLC M-Class system, Waters) running with two buffers: 0.1% formic acid in water (A) and 0.1% formic acid in acetonitrile (B). We separated 3 µL of each sample on a C18 reverse-phase column (NanoE MZ HSS C18 T3, 1.7 µ 75 µm × 100 mm, Waters), with a linear gradient of 5% to 85% buffer B over 120 min at a flow rate of 300 nL min^{-1}. Peptide ions were analyzed with Masslynx v4.1, with the following data-independent acquisition steps (DIA): MS scan range: 50–2000 m/z, scan time 0.5 s, ramp collision energy from 15 to 40 V. Proteins were identified with Progenesis QI for proteomics v3.0 (Waters) with the following parameters: enzymatic cleavage by trypsin with two missed cleavages allowed, carbamidomethylation for cysteine residues and potential oxidation for methionine residues. Only peptides with a score of at least 5 were considered. The Uniprot KB database (www.expasy.org (accessed on 1 October 2019)) and a custom-built contaminant database (trypsin, keratin, etc.) were used. The species of origin was restricted to the rat. The identified proteins were filtered to retain only those with a minimum of three fragments per peptide and one peptide per protein. Analysis was performed on n = 5 animals/group. A protein was considered differentially secreted if its detection frequency in a group differed from that of the other group by at least two animals.

2.10. MicroCT

Three-dimensional microcomputed tomography (µCT) was used to quantify bone regeneration 10 weeks after stage 2 of the Masquelet technique. The rats were killed, and the limb on which surgery was performed was collected, together with the surrounding soft tissues, and fixed by incubation in 10% phosphate-buffered formalin for two weeks. The area between the inner pins was scanned by microCT (Skyscan 1174, Bruker Micro-CT, Billerica, MA, USA) with a voltage source of 50 keV, a current of 745 mA and an isotropic resolution of 14.4 µm. Three-dimensional reconstruction was performed for all scans and analyzed with the same parameter setup (NRecon v.1.6 and CTAn v.1.11 software, SkyScan, Kontich, Belgium) to separate mineralized elements from the background, with the software histogram tool used to determine grayscale level threshold values. As a dedicated external fixator with a guide saw had been used to create the defect, it was possible to locate the 6 mm long defective region with precision (the distance between the two adjacent pins) and to identify it as the region of interest. The following data were collected within the region of interest: total defect volume (TV in mm^3) and bone volume (BV in mm^3) for the calculation of the bone volume ratio (BV/TV, as a %).

2.11. Statistical Analysis

All results are reported as means $\pm$ standard error of the mean (SEM). The Shapiro-Wilk test was used to determine whether the data followed a normal distribution. An F-test was performed to verify the assumption of equal variances. Two-tailed Student's t-tests were used for comparisons if the data met both these requirements (normal distribution and equal variances). If one or both the assumptions were not met, the PMMA and metakaolin groups were compared in non-parametric Mann–Whitney U-tests. Values of $p < 0.05$ were considered significant in all tests. Statistical analyses were performed with GraphPad Prism 5 statistical software (GraphPad Software Inc, La Jolla, CA, USA).

3. Results

3.1. Animals and Blood Parameters at the End of IMT Stage 1

All rats tolerated surgical procedures well and gained weight steadily from day 4 after stage 1 surgery onwards. Two animals (one PMMA and one metakaolin) were excluded from the analysis due to infection-related fixator failure. Given the inflammatory potential of the aluminosilicate present in metakaolin, we determined blood cell counts for the animals to assess systemic inflammation at the time of death. White blood cell counts and red blood cell parameters were similar between the PMMA and metakaolin groups (Figure 1A). Serum P1NP and TRAP-C concentrations and ratios were similar in the two groups, suggesting that bone remodeling activity four weeks after the creation of the bone defect was similar in the PMMA and metakaolin groups (Figure 1B).

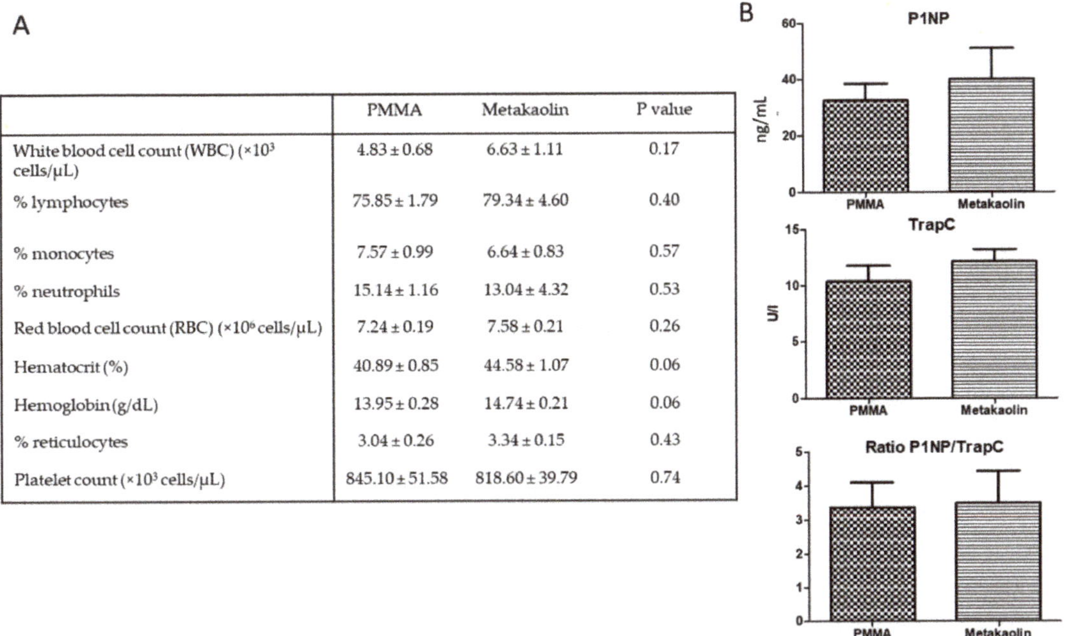

A

	PMMA	Metakaolin	P value
White blood cell count (WBC) ($\times 10^3$ cells/µL)	4.83 ± 0.68	6.63 ± 1.11	0.17
% lymphocytes	75.85 ± 1.79	79.34 ± 4.60	0.40
% monocytes	7.57 ± 0.99	6.64 ± 0.83	0.57
% neutrophils	15.14 ± 1.16	13.04 ± 4.32	0.53
Red blood cell count (RBC) ($\times 10^6$ cells/µL)	7.24 ± 0.19	7.58 ± 0.21	0.26
Hematocrit (%)	40.89 ± 0.85	44.58 ± 1.07	0.06
Hemoglobin (g/dL)	13.95 ± 0.28	14.74 ± 0.21	0.06
% reticulocytes	3.04 ± 0.26	3.34 ± 0.15	0.43
Platelet count ($\times 10^3$ cells/µL)	845.10 ± 51.58	818.60 ± 39.79	0.74

Figure 1. Panel (**A**) shows the hematological parameters of animals four weeks after spacer implantation. Blood was collected into EDTA-containing tubes when the animals were killed, and blood parameters were determined with an optical hematology analyzer (MS-9, Melet Schloesing) with rat-specific analysis software. Panel (**B**) shows the serum levels of bone turnover markers, as determined by ELISA, four weeks post-spacer implantation. Concentrations of a bone formation marker P1NP (**top**) and a bone resorption marker TRAP-C (**middle**) were determined, and turnover for bone remodeling was evaluated by calculating the P1NP/TRAP-C ratio (**bottom**).

3.2. Comparison of Biological Properties between Metakaolin- and PMMA-Induced Membranes

3.2.1. Membrane Architecture and Cellularity

We previously showed that bioactive IMs are organized as bilayered structures and have a rich cellular network. Figure 2 illustrates typical sections of PMMA-induced (Figure 2A) and metakaolin-induced (Figure 2B) membranes, with an inner layer in contact with the biomaterial, including fibroblasts, lymphocytes and macrophages. A thick outer layer principally consists of fibroblasts with a dense vascular network in contact with the muscle. The quantification of DAPI-stained nuclei showed cell density to be slightly higher in metakaolin-IMs than in PMMA-IMs, although this difference was not statistically significant (4553 ± 51 nuclei/mm^2 in the PMMA group versus 5882 ± 695 nuclei/mm^2 in the metakaolin group, $p = 0.15$).

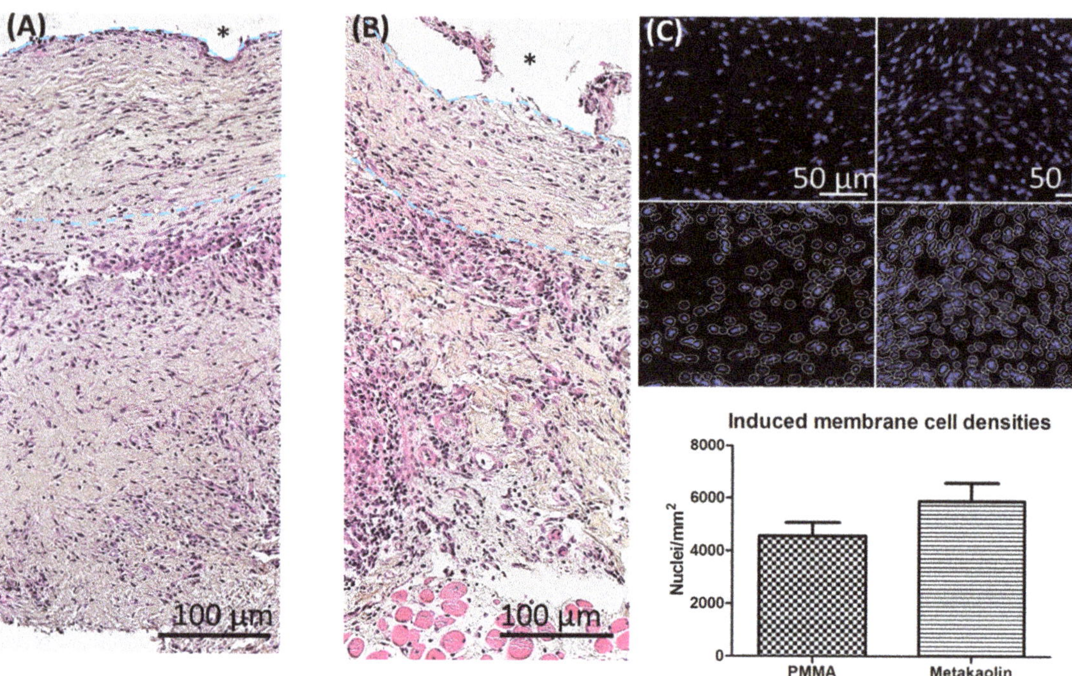

Figure 2. Representative hematoxylin-eosin-saffron-stained sections of (**A**) PMMA-IMs and (**B**) metakaolin-IMs showing their histological organization. * Indicates the site of the spacer before its removal. The right panel (**C**) illustrates the semi-automatic counting process of DAPI-stained nuclei and the comparison of cell density (the mean number of nuclei per mm^2 $\pm$ SEM) in PMMA-IMs and metakaolin-IMs.

3.2.2. Gene Expression within Membranes

We compared the expression of key inflammation-related genes involved in wound healing between PMMA-IMs and metakaolin-IMs. Real-time RT-PCR analysis (Figure 3) showed that the relative levels of insulin-like growth factor-1 (*IGF-1*), vascular endothelial growth factor (*VEGF*), interleukin-6 (*IL-6*) and interleukin-1-beta (*IL-1β*) expression was similar in PMMA-IMs and metakaolin-IMs. However, in metakaolin-IMs, transforming growth factor-β (*TGF-β*) mRNA levels were significantly upregulated (fold change = 2.74, $p = 0.016$), potentially enhancing bone healing and regeneration.

3.2.3. Secretion of Proteins by the IM and BMP-2 Expression within Membranes

IMs form a biological chamber containing secreted angiogenic and osteogenic factors around the bone defect. We, therefore, performed a descriptive mass spectrometric analysis to compare the secretome profiles of PMMA-IMs and metakaolin-IMs. We detected a total of 688 proteins in both groups (Figure 4A), 683 (99.3%) of which were not differentially secreted between PMMA-IMs and metakaolin-IMs (i.e., the frequency of secretion of these proteins was similar in the two groups). In contrast, the secretion frequency differed between the two groups for five proteins (0.72%): four were more frequently secreted by metakaolin-IMs, and one was more frequently secreted by PMMA-IMs. The four proteins more frequently secreted by metakaolin-IMs were identified as cysteine- and glycine-rich protein 3, the GON7 subunit of the KEOPS complex, carboxylic ester hydrolase and synaptogyrin. These proteins are involved in various metabolic pathways, including myogenesis and apoptosis. The protein most frequently secreted by PMMA-IMs was the neurotrophin tyrosine kinase receptor 1 TrkA L0 variant, which is involved in the MAPK pathway.

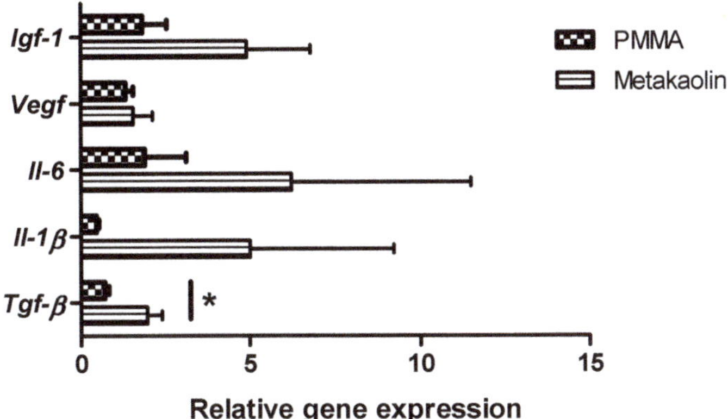

Figure 3. Relative levels of *Igf-1*, *Vegf*, *Il-6*, *Il-1β* and *Tgf-β* mRNA in four-week-old PMMA-IMs or metakaolin-IMs. Data are expressed as the mean ± SEM, * $p < 0.05$.

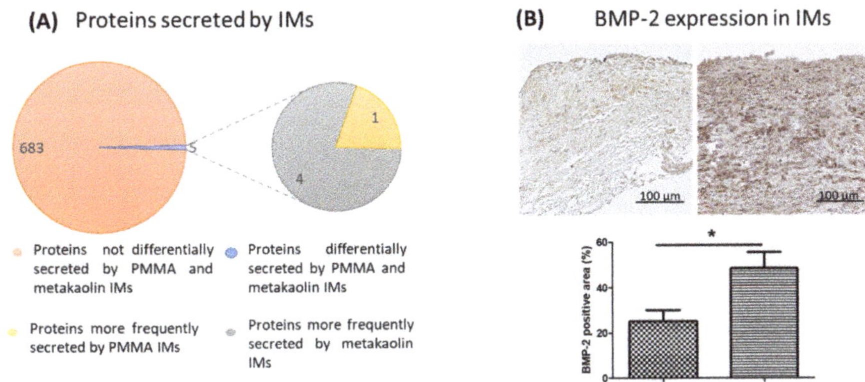

Figure 4. (**A**) Secretome profiles of PMMA-IMs and metakaolin-IMs. In total, 683 proteins with similar frequencies of secretion in the PMMA and metakaolin groups were identified by mass spectrometry. By contrast, five proteins were differentially secreted: four proteins were more frequently secreted by the metakaolin-IMs, and the other was more frequently secreted by PMMA-IMs. (**B**) Representative histological slide of in situ BMP-2 immunostaining in PMMA-IMs (left panel) and metakaolin-IMs (right panel). The diagram shows the percentage of the area of the collected membranes positive for BMP-2; * $p < 0.05$.

We also investigated the expression of the pro-osteogenic mediator BMP-2 within the membranes by immunohistochemistry (Figure 4B). BMP-2-expressing cells were uniformly distributed throughout the membranes, but BMP-2 staining was more intense in metakaolin-IMs than in PMMA-IMs. Furthermore, the percentage of the membrane area positive for BMP-2 was 1.9 times higher in metakaolin-IMs than in PMMA-IMs (25.25% ± 4.83% versus 48.41% ± 7.11%, $p = 0.0021$).

3.3. Macrophage Distribution in IMs

We characterized the macrophage populations in IMs by immunofluorescence analysis to detect both CD68 and CD206, with CD68 used as a phenotypic marker of the M1-like subtype and CD68+/CD206+ double labeling as a marker of the M2-like subtype (Figure 5A). CD68-/CD206+ cells were defined as satellite cells. Semi-automatic quantification revealed that the total macrophage population was significantly larger in metakaolin-IMs than in PMMA-IMs (25.77% ± 5.48% versus 48.11% ± 5.77%, $p = 0.02$; Figure 5B). This larger total macrophage population reflected a significant expansion of the M1-like population (20.01% ± 3.81% versus 36.30% ± 4.45%, $p = 0.02$) and a smaller, non-significant expansion of the M2-like subtype (5.75% ± 2.40% versus 11.81% ± 1.72%, $p = 0.07$).

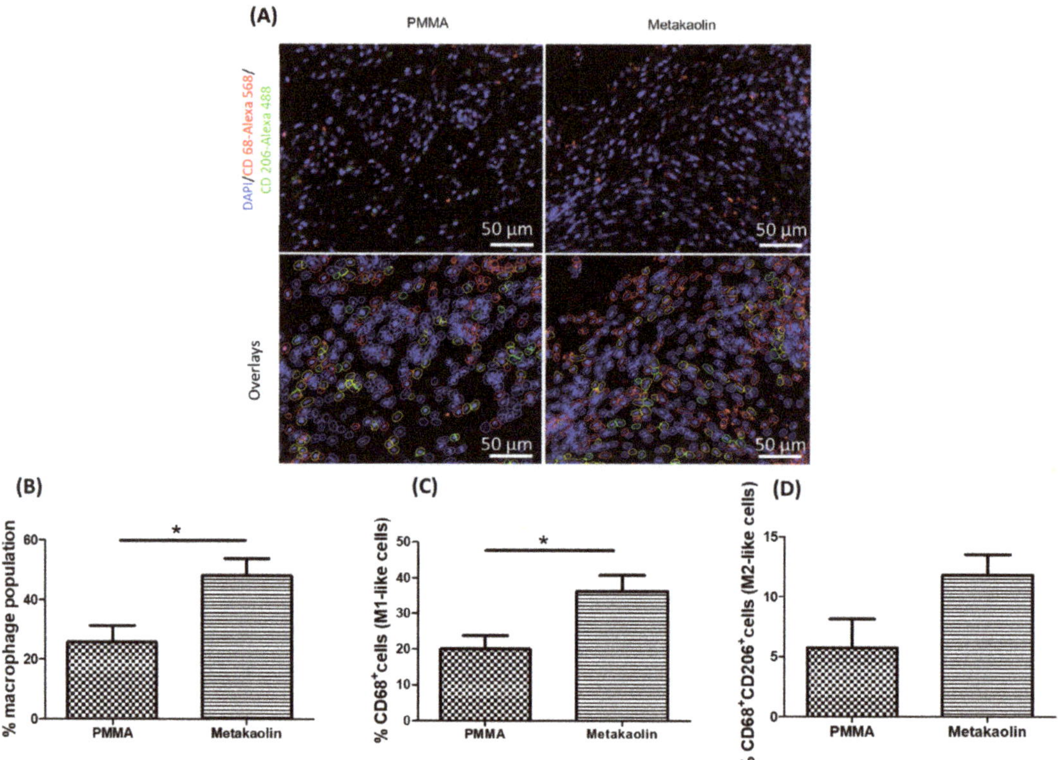

Figure 5. Identification and quantification of M1-like and M2-like macrophages in PMMA-IMs and metakaolin-IMs: (**A, top panel**) Representative immunolabeling with anti-CD68 (red) and anti-CD206 (green) antibodies and DAPI (blue) nuclear staining. (**A, bottom panel**) Illustration of semi-automatic macrophage quantification. Green objects represent satellite cells, and red and yellow objects correspond to M1-like and M2-like macrophages, respectively. Histograms show (**B**) the % total (M1-like + M2-like) macrophages in IMs, (**C**) the % M1-like macrophages and (**D**) M2-like macrophages. * $p < 0.05$.

3.4. Bone Healing after IMT Stage 2 Surgery

We compared the bone-healing properties of PMMA-IMs and metakaolin-IMs, by performing a quantitative analysis of callus volume within the osteotomy region 10 weeks after bone graft implantation in the IM cavities. We observed a small, non-significant difference in new bone volume within the defect, with a slightly greater volume in the metakaolin group (16.65% ± 3.59% versus 26.37% ± 6.5%, *p* = 0.22, Figure 6A) than in the PMMA group.

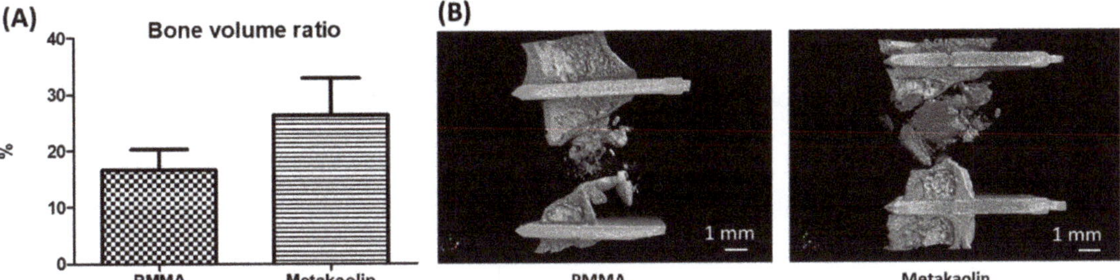

Figure 6. Bone formation was assessed by microCT. (**A**) The quantitative and comparative analysis showed no difference in bone volume between the metakaolin and PMMA groups. (**B**) Representative three-dimensional reconstructions of the region of interest in the two groups.

4. Discussion

In this study, we evaluated the use of a metakaolin spacer as an alternative to standard PMMA spacers for the induced membrane technique. This technique is increasingly used in orthopedic surgery to repair large bone defects in humans. We first analyzed the osteogenic, biological and inflammatory properties of IMs in rat bone defects treated with metakaolin or PMMA spacers. We then assessed bone repair efficiency in rats 10 weeks after the implantation of a morselized corticocancellous allograft into the IM cavity generated by the two types of spacers.

4.1. Metakaolin Modifies Several Osteogenic and Biological Parameters of IMs

IMs are well-organized bilayer encapsulation membranes resulting from a foreign body reaction to the implanted spacer [15,24,25]. In a previous investigation, we demonstrated the importance of both the cellularity and collagen density of IMs on their biological properties in humans. Indeed, patients in which the Masquelet technique was unsuccessful (absence of bone repair resulting in non-union) had IMs with 50% lower levels of cellularity and a much higher collagen density (fibrosis-like status membrane) than those in which this technique was successful [4]. Conversely, here, the replacement of the PMMA spacer with a metakaolin spacer tended to increase IM cellularity (+30%). The IM acts as a biological chamber, promoting bone graft vascularity and corticalization by the secretion of various cytokines and growth factors [3,5,26–28]. Mass spectrometry showed that the metakaolin-IM and PMMA-IM secretomes differed by only 0.72%, suggesting that the secreted protein profiles of metakaolin-IMs and PMMA-IMs differed very little. Our mass spectrometry proteomic analysis was purely descriptive. We did not, therefore, have precise data for protein secretion levels. However, evidence from other molecular and protein analyses suggests that the expression levels of several proteins are modified by metakaolin-IMs. We observed a non-significant trend towards higher *Igf-1*, *Il-6* and *Il-1β* transcript levels with the metakaolin spacer. Fischer et al. [29] showed that serum *Igf-1* levels were higher in patients successfully treated with the IMT than in patients presenting treatment failure.

We found that *Tgf-β* transcript levels and BMP-2 protein levels were significantly higher (2.7-fold and 1.9-fold increases, respectively) in metakaolin-IMs than in PMMA-IMs. BMP-2 is undoubtedly the most osteoinductive growth factor, promoting the migration, proliferation and osteoblastic differentiation of osteoprogenitor cells. TGF-β has dual activity in bone remodeling activity [30], acting as a chemoattractant for osteoprogenitor cells at bone lesion sites and stimulating bone formation (osteoprogenitor proliferation and active osteoblastic differentiation; collagen synthesis) while inhibiting bone resorption (inhibition of osteoclast proliferation and activity). Interestingly, BMP-2 and TGF-β belong to the same growth factor superfamily. They bind to serine/tyrosine kinase receptors, and this interaction activates the SMAD intracellular signaling transduction pathway, which is involved in various steps of the bone regeneration process during fracture healing [31]. Tang et al. [32] suggested that activation of the SMAD pathway by both BMP-2 and TGF-β might underlie the osteogenic effects mediated by IMs. Taken together, the increase in IM cellularity and higher levels of IGF-1, IL-6, IL-1β IM transcripts and BMP-2 protein are consistent with the theory that the membranes induced by metakaolin are more osteogenic than those induced by PMMA spacers.

4.2. Metakaolin Spacers Modulate the IM Immune Microenvironment

The nature of the spacer did not affect systemic inflammation, as estimated from white blood cell counts in our animals. However, the increases in *Igf-1*, *Il-6*, *Il-1β* and *Tgf-β* transcripts suggested that the metakaolin spacer modulated the local inflammatory response. Macrophages are one of the most abundant sources of cytokines [33]. In this context of biomaterial implantation in a bone lesion area, it is difficult to separate the local inflammation process induced by the bone lesion from that triggered by biomaterial implantation. Macrophages form a highly heterogeneous and plastic population of cells and are, therefore, of particular interest in the IMT context due to their involvement in both wound repair processes and the foreign body response [34,35]. After activation, tissue-resident and monocyte-derived macrophages are recruited to the inflammation site. Depending on local environment cues, they transiently gain and lose functions by undergoing major phenotypic changes. A consensus has emerged concerning a sequential macrophage polarization pattern in the bone-healing process [35–37]. Following the formation of the bone lesion, there is a rapid, massive infiltration of monocytes and undifferentiated M0 subtype macrophages at the fracture site. During the first few days after the injury, the polarization of macrophages to the M1 "pro-inflammatory" phenotype is driven by secreted inflammatory and chemoattractant mediators, such as IL-6, IL-1β, IFN-γ, TNFα and monocyte chemotactic protein 1 or MCP-1.

M1 macrophages remove the provisional fibrin matrix and necrotic cells by phagocytosis. By secreting TNFα, IL-1β, IL-6 and MCP-1, they support inflammation by recruiting additional immune cells, but they also initiate the recruitment of fibroblasts and osteoprogenitor cells to the lesion site. Later in inflammation/repair kinetics, under the influence of IL-4, IL-10 and IL-13 signaling, macrophage polarization switches to the M2 "anti-inflammatory" phenotype. The secretion of VEGF, matrix metalloproteinases (MMPs), BMP-2 and platelet-derived growth factor (PDGF) by M2 macrophages triggers both angiogenesis and bone tissue remodeling during the healing process [35,37]. Macrophages are also crucial regulators of the FBR [38]. Following the implantation of biomaterials, plasma components adsorb onto the surface of the material, promoting neutrophil inflammation and macrophage recruitment. The macrophage-driven secretion of TGF-β around the implant triggers the transdifferentiation of fibroblasts into myofibroblasts, thereby promoting myofibroblast collagen production, leading to encapsulation of the biomaterial. In addition, macrophages fuse to form foreign body giant cells (FBGCs). FBGCs are large multinucleated cells secreting cathepsin-K and reactive oxygen species to degrade the foreign body (in this case, the biomaterial).

Surprisingly, despite their key role in the FBR, little is known about the phenotypes of macrophages in vivo during this reaction. Conflicting reports have been published [39], probably because the characteristics of the biomaterial (including surface chemistry, porosity, stiffness, etc.) directly affect macrophage phenotype. However, there is a general consensus that both M1 and M2 macrophages are present throughout the FBR [40,41]. Moreover, higher levels of M2 macrophages than M1 macrophages surrounding implanted biomaterials are associated with more constructive remodeling [42,43]. For example, Zhu et al. [44] tested the capacity for orienting macrophage polarization of four scales of honeycomb-like titanium structures with honeycomb diameters ranging from 90 nm to 5 μm. Raw 264.7 macrophages cultured with the smoothest titanium structure had the highest M2-macrophage polarization rate, with the highest levels of CD206 expression (a specific marker of M2 macrophages) and IL-4, IL-10 and BMP-2. In vivo, the implantation of titanium rods with 90 nm honeycombs in rat tibia gave the best results for bone osteointegration [44]. Here, we compared macrophage polarization in IMs according to the nature of the spacer implant. In metakaolin-IMs, we observed a significant expansion of the total macrophage population. This observation is consistent with previous findings indicating that aluminosilicates stimulate the immune response by inducing the activation of macrophages. The expansion of the macrophage population in metakaolin-IMs is also consistent with the increase in key inflammation-related transcripts observed in the same membranes. More specifically, even though the expansion of the M2-like population was not significant ($p = 0.07$), both M1-like and M2-like macrophage subtypes increased markedly in frequency in metakaolin-IMs. To our knowledge, this study is the first in an IMT context to show a link between greater osteogenic properties of the induced membrane and a spacer-driven modulation of the phenotype and number of macrophages. Further studies are required to elucidate the mechanism underlying the balance between M1 and M2 macrophages in the induced membrane.

4.3. Metakaolin Slightly Improves Bone Repair Efficiency

Given the more osteogenic properties of the metakaolin-IMs, better bone regeneration was expected in this group. Unsurprisingly, we observed a trend towards better bone healing, as shown by the 1.58-fold increase in BV/TV when a metakaolin spacer was used to generate IM rather than a PMMA spacer. Other alternative biomaterials have been tested for the creation of IM mimetics [45] or the induction of IMs with enhanced osteogenic properties [46]. However, mixed results for bone repair outcomes have been obtained for these alternative spacers. Indeed, after four weeks of maturation in rats, smooth and rough titanium spacers generated thicker IMs than smooth and rough PMMA spacers but with similar histological structures and biochemical expression parameters [47]. The only difference observed concerned IL-6 protein levels in the IM, which were about 35% higher with rough spacers (both PMMA and titanium spacers) than with smooth spacers. Smooth PMMA spacers resulted in a more functional bone union than the other PMMA and titanium spacers tested [47]. Unfortunately, the authors did not investigate the immune microenvironment of the membranes, particularly the balance between M1 and M2 macrophages. Following on from the successful clinical use of polypropylene syringes as alternative spacers to PMMA cement to treat metacarpal bone lesions [48], we validated the potential of this biomaterial in a rat IMT model [15]. Polypropylene-induced membranes had a similar histologic organization, cell density and BMP-2 protein level to PMMA- IMs, and similar levels of serum bone turnover markers. In micro-CT analysis, bone regeneration capacities were similar in the polypropylene and PMMA groups of rats [15]. Our investigation highlights the value of polypropylene syringes as an alternative to PMMA cement for use as spacers in a military practice context and/or in low-medical resource environments. With a view to developing a modified IMT approach for efficient one-step surgery, Ma et al. [49] evaluated the osteogenic properties of calcium sulfate (CS)-induced membranes in rats. The histological characteristics of CS-IM and PMMA-IM were similar, except that the calcium sulfate spacer induced thicker membranes. Levels of the TGF-β1, BMP-2 and VEGF proteins were not significantly higher in CS-IMs at two, four, six and eight weeks post-implantation, whereas IL-6 protein levels were significantly higher in PMMA-IMs at two weeks post-implantation. Finally, CS-IMs promoted better endochondral ossification at the edges of the bone defect than PMMA-IMs at six and eight weeks post-implantation [49]. The authors concluded that calcium sulfate could replace PMMA as an alternative spacer in IMT. The results obtained with the metakaolin spacer in this study are equivalent to those obtained by Ma's research team for the calcium sulfate spacer.

5. Limitations, Conclusions and Future Directions

This study is the first to date to investigate a correlation between local inflammation/the immune microenvironment of the IM and osteogenic properties by comparing IMs generated with PMMA and metakaolin spacers in a preclinical rat model of IMT. Metakaolin induced a membrane with slightly better osteogenic properties than the PMMA spacer, improving bone-healing efficiency, albeit not significantly in our rat model. This significant success in bone repair was accompanied by an expansion of the macrophage population in the IM structure for both M1 and M2 subtype macrophages. This stronger local inflammation process was sustained by local overexpression of the osteogenic BMP-2 protein and several inflammatory cytokines, including TGF-β, IL-1β and IL-6.

This study had several limitations. The number of rats included was relatively small, which may have contributed to the high standard deviation in the RT-PCR analysis. Furthermore, our study included only male rats. Sex-specific differences in bone-healing outcomes remain underinvestigated, but most studies in the field have suggested that being female is a significant risk factor for compromised bone healing [50]. This influence of sex on fracture healing may be related to the smaller numbers of mesenchymal stromal cells (MSCs) in female bone marrow [51].

Given the presence of MSCs in the induced membrane [4], it would be desirable to investigate sex-specific differences in bone-healing outcomes, particularly in the IMT context. Another limitation of the study concerns the in situ characterization of polarized macrophages using CD68 and CD206 immunofluorescence. The CD68 protein is one of the most common monocyte/macrophage markers [52], whereas the CD206 protein is mostly expressed by M2 macrophages [53]. The co-expression of CD68 and CD206 is generally considered to indicate an M2-like phenotype. In this study, the expression of CD68 alone was considered to indicate a M1-like macrophage phenotype. The use of a single marker for identifying the M1-like population is questionable. Since a weak CD68 expression can be detected in some non-hematopoietic cells (mesenchymal stem cells, fibroblast, endothelial and tumor cells) [54], we assumed that the M1-like population is overestimated in our study. Indeed, the in situ detection of polarized macrophages is technically challenging [22,55], and none of the other discriminating markers we tested gave conclusive results. We acknowledged that the in situ CD68-based strategy for M1-like cell detection can be regarded as a "by default" identification of this population. Although there is no real direct evidence for the use of CD68 as a single marker for the M1-type population characterization in rats, this labeling approach is commonly described in the literature [22,56–58], thus providing robust indirect evidence to our conclusion.

In conclusion, metakaolin spacers would be a valuable biomaterial for replacing PMMA spacers in the Masquelet technique. One particularly interesting clinical application would be the healing of complicated bone defects. Indeed, this strategy would involve the manufacture of 3D printing molds in the shape of the injured bone areas based on the CT scans for the patients concerned. A metakaolin spacer could then be molded in a specific cast to obtain the appropriate shape before implantation into the bone defect. Finally, given the high absorbency of metakaolin, the metakaolin spacer could be impregnated with a large panel of antibiotics to eradicate potential bone infections that might lead to a failure of bone repair. Conversely, only heat-resistant antibiotics could be loaded onto PMMA spacers due to the exothermic nature of the PMMA polymerization reaction.

Supplementary Materials: The following supporting information can be downloaded at: https://www.mdpi.com/article/10.3390/biomedicines10123017/s1, Table S1: Primers sets, forward (F) and reverse (R), used for quantitative PCR assays. Product size and specific qPCR conditions are indicated.

Author Contributions: M.D. and J.-M.C. designed the experiments. M.D., E.J. and L.M. performed the surgery. M.O. and M.D. processed the images. E.J. and A.-C.G. provided technical support in histology, C.R. and L.B. (Laure Barbier) in molecular biology, L.B. (Laurent Bégot) in micro-CT analysis, K.N. in immunofluorescence and F.P. in QToF analysis. M.D. analyzed the results with input from L.B (Laure Barbier)., K.N., J.V., F.P., D.L., A.-L.F. and O.P. manufactured and provided the metakaolin spacer. M.D., J.-M.C. and J.V. wrote the manuscript with input from all the authors. All authors have read and agreed to the published version of the manuscript.

Funding: This research was funded by DGA (Délégation générale pour l'Armement) grants numbers SAN-1-217 and SAN-1-226.

Institutional Review Board Statement: The animal study protocol was approved by the Institutional Review Board of the Institut de Recherche Biomédicale des Armées (IRBA) (protocol 65 DEF_IGSSA_SP).

Informed Consent Statement: Not applicable.

Acknowledgments: The authors thank Stéphanie Yen-Nicolaÿ and Guillaume Ruellou from the Proteomics Facility of the IPSIT for assistance with proteomic analysis. We also thank the Région Île-de-France for providing this facility with support, Nathalie Guatto for paraffin embedding, and Michel Wasseff for the histological examination of IMs.

Conflicts of Interest: The authors declare no conflict of interest.

References

1. Masquelet, A.C.; Giannoudis, P.V. The Induced Membrane Technique for Treatment of Bone Defects: What Have I Learned? *Trauma. Case Rep.* **2021**, *36*, 100556. [CrossRef] [PubMed]
2. Zwetyenga, N.; Catros, S.; Emparanza, A.; Deminiere, C.; Siberchicot, F.; Fricain, J.-C. Mandibular Reconstruction Using Induced Membranes with Autologous Cancellous Bone Graft and HA-BetaTCP: Animal Model Study and Preliminary Results in Patients. *Int. J. Oral Maxillofac. Surg.* **2009**, *38*, 1289–1297. [CrossRef] [PubMed]
3. Christou, C.; Oliver, R.A.; Yu, Y.; Walsh, W.R. The Masquelet Technique for Membrane Induction and the Healing of Ovine Critical Sized Segmental Defects. *PLoS ONE* **2014**, *9*, e114122. [CrossRef] [PubMed]
4. Durand, M.; Barbier, L.; Mathieu, L.; Poyot, T.; Demoures, T.; Souraud, J.-B.; Masquelet, A.-C.; Collombet, J.-M. Towards Understanding Therapeutic Failures in Masquelet Surgery: First Evidence That Defective Induced Membrane Properties Are Associated with Clinical Failures. *J. Clin. Med.* **2020**, *9*, 450. [CrossRef]
5. Henrich, D.; Seebach, C.; Nau, C.; Basan, S.; Relja, B.; Wilhelm, K.; Schaible, A.; Frank, J.; Barker, J.; Marzi, I. Establishment and Characterization of the Masquelet Induced Membrane Technique in a Rat Femur Critical-Sized Defect Model. *J. Tissue Eng. Regen. Med.* **2016**, *10*, E382–E396. [CrossRef] [PubMed]
6. Xie, J.; Wang, W.; Fan, X.; Li, H.; Wang, H.; Liao, R.; Hu, Y.; Zeng, M. Masquelet Technique: Effects of Vancomycin Concentration on Quality of the Induced Membrane. *Injury* **2022**, *53*, 868–877. [CrossRef] [PubMed]
7. Anderson, J.M.; Rodriguez, A.; Chang, D.T. Foreign Body Reaction to Biomaterials. *Semin. Immunol.* **2008**, *20*, 86–100. [CrossRef]
8. Klopfleisch, R. Macrophage Reaction against Biomaterials in the Mouse Model-Phenotypes, Functions and Markers. *Acta Biomater.* **2016**, *43*, 3–13. [CrossRef]
9. Davidovits, J. Geopolymers. *J. Therm. Anal.* **1991**, *37*, 1633–1656. [CrossRef]
10. Jung, M.; Shin, M.-K.; Jung, Y.-K.; Yoo, H.S. Modulation of Macrophage Activities in Proliferation, Lysosome, and Phagosome by the Nonspecific Immunostimulator, Mica. *PLoS ONE* **2015**, *10*, e0117838. [CrossRef]
11. Bowman, P.D.; Wang, X.; Meledeo, M.A.; Dubick, M.A.; Kheirabadi, B.S. Toxicity of Aluminum Silicates Used in Hemostatic Dressings toward Human Umbilical Veins Endothelial Cells, HeLa Cells, and RAW267.4 Mouse Macrophages. *J. Trauma.* **2011**, *71*, 727–732. [CrossRef] [PubMed]
12. Wiemann, M.; Vennemann, A.; Wohlleben, W. Lung Toxicity Analysis of Nano-Sized Kaolin and Bentonite: Missing Indications for a Common Grouping. *Nanomaterials* **2020**, *10*, 204. [CrossRef]
13. Khurana, I.S.; Kaur, S.; Kaur, H.; Khurana, R.K. Multifaceted Role of Clay Minerals in Pharmaceuticals. *Future Sci. OA* **2015**, *1*, FSO6. [CrossRef] [PubMed]
14. Awad, M.E.; López-Galindo, A.; Setti, M.; El-Rahmany, M.M.; Iborra, C.V. Kaolinite in Pharmaceutics and Biomedicine. *Int. J. Pharm.* **2017**, *533*, 34–48. [CrossRef] [PubMed]
15. Mathieu, L.; Murison, J.C.; de Rousiers, A.; de l'Escalopier, N.; Lutomski, D.; Collombet, J.-M.; Durand, M. The Masquelet Technique: Can Disposable Polypropylene Syringes Be an Alternative to Standard PMMA Spacers? A Rat Bone Defect Model. *Clin. Orthop. Relat. Res.* **2021**, *479*, 2737–2751. [CrossRef] [PubMed]
16. Schneider, C.A.; Rasband, W.S.; Eliceiri, K.W. NIH Image to ImageJ: 25 Years of Image Analysis. *Nat. Methods* **2012**, *9*, 671–675. [CrossRef]
17. Schroeder, A.; Mueller, O.; Stocker, S.; Salowsky, R.; Leiber, M.; Gassmann, M.; Lightfoot, S.; Menzel, W.; Granzow, M.; Ragg, T. The RIN: An RNA Integrity Number for Assigning Integrity Values to RNA Measurements. *BMC Mol. Biol.* **2006**, *7*, 3. [CrossRef]
18. Bustin, S.A.; Benes, V.; Garson, J.A.; Hellemans, J.; Huggett, J.; Kubista, M.; Mueller, R.; Nolan, T.; Pfaffl, M.W.; Shipley, G.L.; et al. The MIQE Guidelines: Minimum Information for Publication of Quantitative Real-Time PCR Experiments. *Clin. Chem.* **2009**, *55*, 611–622. [CrossRef]

19. Pugniere, P.; Banzet, S.; Chaillou, T.; Mouret, C.; Peinnequin, A. Pitfalls of Reverse Transcription Quantitative Polymerase Chain Reaction Standardization: Volume-Related Inhibitors of Reverse Transcription. *Anal Biochem* **2011**, *415*, 151–157. [CrossRef] [PubMed]

20. Livak, K.J.; Schmittgen, T.D. Analysis of Relative Gene Expression Data Using Real-Time Quantitative PCR and the 2(-Delta Delta C(T)) Method. *Methods* **2001**, *25*, 402–408. [CrossRef] [PubMed]

21. Vandesompele, J.; De Preter, K.; Pattyn, F.; Poppe, B.; Van Roy, N.; De Paepe, A.; Speleman, F. Accurate Normalization of Real-Time Quantitative RT-PCR Data by Geometric Averaging of Multiple Internal Control Genes. *Genome Biol.* **2002**, *3*, RESEARCH0034. [CrossRef] [PubMed]

22. Nikovics, K.; Durand, M.; Castellarin, C.; Burger, J.; Sicherre, E.; Collombet, J.-M.; Oger, M.; Holy, X.; Favier, A.-L. Macrophages Characterization in an Injured Bone Tissue. *Biomedicines* **2022**, *10*, 1385. [CrossRef]

23. Kosmac, K.; Peck, B.D.; Walton, R.G.; Mula, J.; Kern, P.A.; Bamman, M.M.; Dennis, R.A.; Jacobs, C.A.; Lattermann, C.; Johnson, D.L.; et al. Immunohistochemical Identification of Human Skeletal Muscle Macrophages. *Bio. Protoc.* **2018**, *8*, e2883. [CrossRef] [PubMed]

24. Gindraux, F.; Rondot, T.; de Billy, B.; Zwetyenga, N.; Fricain, J.-C.; Pagnon, A.; Obert, L. Similarities between Induced Membrane and Amniotic Membrane: Novelty for Bone Repair. *Placenta* **2017**, *59*, 116–123. [CrossRef]

25. Gouron, R.; Petit, L.; Boudot, C.; Six, I.; Brazier, M.; Kamel, S.; Mentaverri, R. Osteoclasts and Their Precursors Are Present in the Induced-Membrane during Bone Reconstruction Using the Masquelet Technique. *J. Tissue Eng Regen. Med.* **2017**, *11*, 382–389. [CrossRef] [PubMed]

26. Cuthbert, R.J.; Churchman, S.M.; Tan, H.B.; McGonagle, D.; Jones, E.; Giannoudis, P.V. Induced Periosteum a Complex Cellular Scaffold for the Treatment of Large Bone Defects. *Bone* **2013**, *57*, 484–492. [CrossRef]

27. Aho, O.-M.; Lehenkari, P.; Ristiniemi, J.; Lehtonen, S.; Risteli, J.; Leskelä, H.-V. The Mechanism of Action of Induced Membranes in Bone Repair. *J. Bone Joint Surg. Am.* **2013**, *95*, 597–604. [CrossRef]

28. Wang, W.; Zuo, R.; Long, H.; Wang, Y.; Zhang, Y.; Sun, C.; Luo, G.; Zhang, Y.; Li, C.; Zhou, Y.; et al. Advances in the Masquelet Technique: Myeloid-Derived Suppressor Cells Promote Angiogenesis in PMMA-Induced Membranes. *Acta Biomater.* **2020**, *108*, 223–236. [CrossRef]

29. Fischer, C.; Doll, J.; Tanner, M.; Bruckner, T.; Zimmermann, G.; Helbig, L.; Biglari, B.; Schmidmaier, G.; Moghaddam, A. Quantification of TGF-SS1, PDGF and IGF-1 Cytokine Expression after Fracture Treatment vs. Non-Union Therapy via Masquelet. *Injury* **2016**, *47*, 342–349. [CrossRef]

30. Crane, J.L.; Xian, L.; Cao, X. Role of TGF-β Signaling in Coupling Bone Remodeling. *Methods Mol. Biol.* **2016**, *1344*, 287–300. [CrossRef]

31. Tsiridis, E.; Upadhyay, N.; Giannoudis, P. Molecular Aspects of Fracture Healing: Which Are the Important Molecules? *Injury* **2007**, *38* (Suppl. S1), S11–S25. [CrossRef]

32. Tang, Q.; Tong, M.; Zheng, G.; Shen, L.; Shang, P.; Liu, H. Masquelet's Induced Membrane Promotes the Osteogenic Differentiation of Bone Marrow Mesenchymal Stem Cells by Activating the Smad and MAPK Pathways. *Am. J. Transl. Res.* **2018**, *10*, 1211–1219. [PubMed]

33. Arango Duque, G.; Descoteaux, A. Macrophage Cytokines: Involvement in Immunity and Infectious Diseases. *Front. Immunol.* **2014**, *5*, 491. [CrossRef]

34. Thomas, M.V.; Puleo, D.A. Infection, Inflammation, and Bone Regeneration. *J. Dent. Res.* **2011**, *90*, 1052–1061. [CrossRef] [PubMed]

35. Loi, F.; Córdova, L.A.; Pajarinen, J.; Lin, T.; Yao, Z.; Goodman, S.B. Inflammation, Fracture and Bone Repair. *Bone* **2016**, *86*, 119–130. [CrossRef]

36. Niu, Y.; Wang, Z.; Shi, Y.; Dong, L.; Wang, C. Modulating Macrophage Activities to Promote Endogenous Bone Regeneration: Biological Mechanisms and Engineering Approaches. *Bioact. Mater.* **2021**, *6*, 244–261. [CrossRef] [PubMed]

37. Maruyama, M.; Rhee, C.; Utsunomiya, T.; Zhang, N.; Ueno, M.; Yao, Z.; Goodman, S.B. Modulation of the Inflammatory Response and Bone Healing. *Front. Endocrinol.* **2020**, *11*, 386. [CrossRef]

38. Chung, L.; Maestas, D.R.; Housseau, F.; Elisseeff, J.H. Key Players in the Immune Response to Biomaterial Scaffolds for Regenerative Medicine. *Adv. Drug Deliv. Rev.* **2017**, *114*, 184–192. [CrossRef]

39. Martin, K.E.; García, A.J. Macrophage Phenotypes in Tissue Repair and the Foreign Body Response: Implications for Biomaterial-Based Regenerative Medicine Strategies. *Acta Biomater.* **2021**, *133*, 4–16. [CrossRef]

40. Kyriakides, T.R.; Kim, H.-J.; Zheng, C.; Harkins, L.; Tao, W.; Deschenes, E. Foreign Body Response to Synthetic Polymer Biomaterials and the Role of Adaptive Immunity. *Biomed. Mater.* **2022**, *17*, 022007. [CrossRef]

41. Stahl, A.; Hao, D.; Barrera, J.; Henn, D.; Lin, S.; Moeinzadeh, S.; Kim, S.; Maloney, W.; Gurtner, G.; Wang, A.; et al. A Bioactive Compliant Vascular Graft Modulates Macrophage Polarization and Maintains Patency with Robust Vascular Remodeling. *Bioact. Mater.* **2023**, *19*, 167–178. [CrossRef] [PubMed]

42. Brown, B.N.; Valentin, J.E.; Stewart-Akers, A.M.; McCabe, G.P.; Badylak, S.F. Macrophage Phenotype and Remodeling Outcomes in Response to Biologic Scaffolds with and without a Cellular Component. *Biomaterials* **2009**, *30*, 1482–1491. [CrossRef] [PubMed]

43. Yu, T.; Wang, W.; Nassiri, S.; Kwan, T.; Dang, C.; Liu, W.; Spiller, K.L. Temporal and Spatial Distribution of Macrophage Phenotype Markers in the Foreign Body Response to Glutaraldehyde-Crosslinked Gelatin Hydrogels. *J. Biomater. Sci. Polym. Ed.* **2016**, *27*, 721–742. [CrossRef] [PubMed]

44. Zhu, Y.; Liang, H.; Liu, X.; Wu, J.; Yang, C.; Wong, T.M.; Kwan, K.Y.H.; Cheung, K.M.C.; Wu, S.; Yeung, K.W.K. Regulation of Macrophage Polarization through Surface Topography Design to Facilitate Implant-to-Bone Osteointegration. *Sci. Adv.* **2021**, *7*, eabf6654. [CrossRef] [PubMed]

45. Ganguly, P.; Jones, E.; Panagiotopoulou, V.; Jha, A.; Blanchy, M.; Antimisiaris, S.; Anton, M.; Dhuiège, B.; Marotta, M.; Marjanovic, N.; et al. Electrospun and 3D Printed Polymeric Materials for One-Stage Critical-Size Long Bone Defect Regeneration Inspired by the Masquelet Technique: Recent Advances. *Injury* **2022**, *53*, S2–S12. [CrossRef]

46. Liodakis, E.; Giannoudis, V.P.; Sehmisch, S.; Jha, A.; Giannoudis, P.V. Bone Defect Treatment: Does the Type and Properties of the Spacer Affect the Induction of Masquelet Membrane? Evidence Today. *Eur. J. Trauma. Emerg. Surg.* **2022**, 1–22. [CrossRef]

47. Toth, Z.; Roi, M.; Evans, E.; Watson, J.T.; Nicolaou, D.; McBride-Gagyi, S. Masquelet Technique: Effects of Spacer Material and Micro-Topography on Factor Expression and Bone Regeneration. *Ann. Biomed. Eng.* **2019**, *47*, 174–189. [CrossRef]

48. Murison, J.-C.; Pfister, G.; Amar, S.; Rigal, S.; Mathieu, L. Metacarpal Bone Reconstruction by a Cementless Induced Membrane Technique. *Hand Surg. Rehabil.* **2019**, *38*, 83–86. [CrossRef]

49. Ma, Y.-F.; Jiang, N.; Zhang, X.; Qin, C.-H.; Wang, L.; Hu, Y.-J.; Lin, Q.-R.; Yu, B.; Wang, B.-W. Calcium Sulfate Induced versus PMMA-Induced Membrane in a Critical-Sized Femoral Defect in a Rat Model. *Sci. Rep.* **2018**, *8*, 637. [CrossRef]

50. Haffner-Luntzer, M.; Fischer, V.; Ignatius, A. Differences in Fracture Healing Between Female and Male C57BL/6J Mice. *Front. Physiol.* **2021**, *12*, 712494. [CrossRef]

51. Strube, P.; Mehta, M.; Baerenwaldt, A.; Trippens, J.; Wilson, C.J.; Ode, A.; Perka, C.; Duda, G.N.; Kasper, G. Sex-Specific Compromised Bone Healing in Female Rats Might Be Associated with a Decrease in Mesenchymal Stem Cell Quantity. *Bone* **2009**, *45*, 1065–1072. [CrossRef]

52. Betjes, M.G.; Haks, M.C.; Tuk, C.W.; Beelen, R.H. Monoclonal Antibody EBM11 (Anti-CD68) Discriminates between Dendritic Cells and Macrophages after Short-Term Culture. *Immunobiology* **1991**, *183*, 79–87. [CrossRef] [PubMed]

53. Atri, C.; Guerfali, F.Z.; Laouini, D. Role of Human Macrophage Polarization in Inflammation during Infectious Diseases. *Int. J. Mol. Sci.* **2018**, *19*, 1801. [CrossRef]

54. Chistiakov, D.A.; Killingsworth, M.C.; Myasoedova, V.A.; Orekhov, A.N.; Bobryshev, Y.V. CD68/Macrosialin: Not Just a Histochemical Marker. *Lab. Investig.* **2017**, *97*, 4–13. [CrossRef] [PubMed]

55. Barros, M.H.M.; Hauck, F.; Dreyer, J.H.; Kempkes, B.; Niedobitek, G. Macrophage Polarisation: An Immunohistochemical Approach for Identifying M1 and M2 Macrophages. *PLoS ONE* **2013**, *8*, e80908. [CrossRef]

56. Hashimoto, A.; Karim, M.R.; Kuramochi, M.; Izawa, T.; Kuwamura, M.; Yamate, J. Characterization of Macrophages and Myofibroblasts Appearing in Dibutyltin Dichloride-Induced Rat Pancreatic Fibrosis. *Toxicol. Pathol.* **2020**, *48*, 509–523. [CrossRef]

57. Tsuji, Y.; Kuramochi, M.; Golbar, H.M.; Izawa, T.; Kuwamura, M.; Yamate, J. Acetaminophen-Induced Rat Hepatotoxicity Based on M1/M2-Macrophage Polarization, in Possible Relation to Damage-Associated Molecular Patterns and Autophagy. *Int. J. Mol. Sci.* **2020**, *21*, 8998. [CrossRef] [PubMed]

58. Nakagawa, M.; Karim, M.R.; Izawa, T.; Kuwamura, M.; Yamate, J. Immunophenotypical Characterization of M1/M2 Macrophages and Lymphocytes in Cisplatin-Induced Rat Progressive Renal Fibrosis. *Cells* **2021**, *10*, 257. [CrossRef]

 *biomedicines*

Article

Tauroursodeoxycholic Acid Reduces Neuroinflammation but Does Not Support Long Term Functional Recovery of Rats with Spinal Cord Injury

Siyu Wu [1,2], Concepción García-Rama [1], Lorenzo Romero-Ramírez [1], Johannes P. J. M. de Munter [3], Erik Ch. Wolters [3], Boris W. Kramer [2] and Jörg Mey [1,2,*]

[1]   Hospital Nacional de Parapléjicos, 45071 Toledo, Spain; wusy1029@gmail.com (S.W.); cgrama@externas.sescam.jccm.es (C.G.-R.); lromeroramirez@sescam.jccm.es (L.R.-R.)
[2]   School of Mental Health and Neuroscience and EURON Graduate School of Neuroscience, Maastricht University, 6229 ER Maastricht, The Netherlands; b.kramer@maastrichtuniversity.nl
[3]   Neuroplast BV, 6167 RD Geleen, The Netherlands; h.demunter@neuroplast.com (J.P.J.M.d.M.); e.wolters@neuroplast.com (E.C.W.)
*   Correspondence: jmey@sescam.jccm.es or joerg.mey@maastrichtuniversity.nl

Abstract: The bile acid tauroursodeoxycholic acid (TUDCA) reduces cell death under oxidative stress and inflammation. Implants of bone marrow-derived stromal cells (bmSC) are currently under investigation in clinical trials of spinal cord injury (SCI). Since cell death of injected bmSC limits the efficacy of this treatment, the cytoprotective effect of TUDCA may enhance its benefit. We therefore studied the therapeutic effect of TUDCA and its use as a combinatorial treatment with human bmSC in a rat model of SCI. A spinal cord contusion injury was induced at thoracic level T9. Treatment consisted of i.p. injections of TUDCA alone or in combination with one injection of human bmSC into the *cisterna magna*. The recovery of motor functions was assessed during a surveillance period of six weeks. Biochemical and histological analysis of spinal cord tissue confirmed the anti-inflammatory activity of TUDCA. Treatment improved the recovery of autonomic bladder control and had a positive effect on motor functions in the subacute phase, however, benefits were only transient, such that no significant differences between vehicle and TUDCA-treated animals were observed 1–6 weeks after the lesion. Combinatorial treatment with TUDCA and bmSC failed to have an additional effect compared to treatment with bmSC only. Our data do not support the use of TUDCA as a treatment of SCI.

Keywords: bile acid; spinal cord injury; bone marrow-derived stromal cells; rat; neuroinflammation

Citation: Wu, S.; García-Rama, C.; Romero-Ramírez, L.; de Munter, J.P.J.M.; Wolters, E.C.; Kramer, B.W.; Mey, J. Tauroursodeoxycholic Acid Reduces Neuroinflammation but Does Not Support Long Term Functional Recovery of Rats with Spinal Cord Injury. *Biomedicines* **2022**, *10*, 1501. https://doi.org/10.3390/biomedicines10071501

Academic Editor: Krisztina Nikovics

Received: 4 April 2022
Accepted: 23 June 2022
Published: 25 June 2022

Publisher's Note: MDPI stays neutral with regard to jurisdictional claims in published maps and institutional affiliations.

1. Introduction

Lesions of the spinal cord due to traumatic injury, tumors, or vascular ischemia frequently cause paralysis and the loss of autonomic functions. No disease modifying therapy for this pathology is currently available. The severe consequences of spinal cord injury (SCI) are in a large part due to a secondary inflammatory reaction, which is borne by local microglia cells, macrophages, and lymphocytes that infiltrate the lesion area. This reaction, in conjunction with increasing vascular permeability, causes cell death of neurons and glia cells [1]. For this reason, anti-inflammatory treatment is a therapeutic strategy for SCI and has been used in the clinic [2]. Unfortunately, the approved treatment option, methylprednisolone, is often ineffective and causes severe side effects such as a higher incidence of sepsis, gastrointestinal hemorrhage, or pulmonary embolism [3]. Therefore, new therapies for SCI are needed, and one of the most promising lines of research with this aim consists of the application of stem cells [4–6]. We have recently tested the safety and therapeutic benefits of human bone marrow-derived stromal cells (bmSC), which are prepared solely by negative selection without expansion in vitro (Neuroplast BV, patent WO2015/059300A1).

In rat models of SCI, intrathecal infusion of these cells reduced chronic inflammation and neural degeneration and provided a benefit on the functional level [7,8]. However, as in most other studies [5], the beneficial effect was limited. In comparison with control treatments, the recovery of sensory-motor function improved by 1.5 points (BBB scale) [9], and even at nine weeks after SCI, most rats did not recover beyond BBB 9-11 [8]. One reason for the limited effect of bmSC is that implanted cells die or are actively eliminated in the acute phase. We reasoned that a combinatorial therapy of bmSC implantation with additional cytoprotective measures would be advantageous to support the integration of implanted bmSC.

A recent strategy of cytoprotection for neuropathologies is based on bile acids, which have long been used in traditional Chinese medicine [10,11]. These are amphipathic molecules synthesized from cholesterol. Their biological effects are mediated via the Takeda G protein-coupled receptor-5 (TGR5) [12,13] as well as nuclear receptors farnesol X receptor, pregnane receptor, and liver X receptors [14,15]. While bile acids play important roles in lipid metabolism [16], they also have anti-inflammatory and cytoprotective effects by suppressing NFκB signaling, which make them interesting candidates for the treatment of neuropathologies [17,18]. One particular bile acid, tauroursodeoxycholic acid (TUDCA), has been tested in animal models for Parkinson's disease [19], multiple sclerosis [20], and a clinical trial of amyotrophic lateral sclerosis [21]. In rodent models of SCI, TUDCA reduced cellular apoptosis [22–26]. With rats, improvements on the functional level were observed within the first 5 days after injury, i.e., in the subacute phase, but it is not clear whether these treatments have a lasting effect after SCI.

Our objectives in the present study were (1) to assess the effects of TUDCA on neuro-inflammation and astrogliosis after SCI, (2) to clarify the long term therapeutic effects of TUDCA treatment on the recovery of sensory-motor function, and (3) to test whether a combinatorial therapy of TUDCA and bmSC transplantation provides additional benefit.

2. Materials and Methods

2.1. Experimental Animals and Study Design

Experimental protocols, surgical procedures, and post-operational care were reviewed by the ethics committee for animal care of the Hospital Nacional de Paraplejicos (163CEEA/ 2017) and approved by the Consejería de Agricultura y Ganadería de Castilla-la Mancha (ref. 210498, following EU directive 2010/63/EU). We used male Wistar rats (*Rattus norwegicus*), six to eight weeks of age, which had been bred in the animal facility of the hospital. Until the day of surgery, animals were kept in pairs and, subsequently, in individual cages. Standard housing conditions consisted of a 12 h light/dark cycle, 40–60% humidity, temperature of 22 °C with ad libitum access to food and water. A total of 75 animals entered the study (Figure 1a; sample size calculation for BBB scores at 6 weeks based on $\alpha = 0.05$, $\beta = 0.2$, $d = 3$, $SD = 2$, attrition 10%). In addition, the spinal cords of non-injured rats were processed for comparison of histological results. Since rats had the same sex, and a similar age and body weight, no randomization was performed to allocate them in the study. The order of treatment was mixed to ensure that all treatment groups were served throughout the entire period when surgery was performed. Rats were excluded when the SCI was considered invalid. Since the primary outcome of the study was sensory-motor function, the exclusion criteria were a BBB score above 2 at 24 h after SCI or a force/time plot of the impactor device that indicated that bone was hit.

2.2. Surgical Procedures and Postoperative Treatment

Spinal cord contusion injury and injections of bmSC were performed as described previously [8]. In short, anesthesia consisted of 2.5% isoflurane/97.5% oxygen at 0.5 L/min for SCI. For the injections of bmSC, we used one i.p. injection of ketamine 50 mg/kg combined with xylacine 5 mg/kg. Fifteen minutes before surgery, rats received one s.c. injection of buprenorfine 0.05 mg/kg to reduce pain. Corneal dehydration was prevented with ophthalmic ointment (Lubrithal, Dechra, Barcelona, Spain). With ketamine anesthesia,

0.04 mg/kg of atropin was given. Following laminectomy at thoracic level T9, a spinal cord contusion of 2 N (200 Kdyn, zero dwell time) was inflicted with the Infinite Horizon (IH) spinal cord impactor. We checked the procedure visually (hematoma) and by monitoring the IH displacement/time and force/time plots. To normalize biochemical data at 4 days post operation (dpo), a control group was operated on using the laminectomy procedure without SCI. Immediately after surgery, all animals received 2 × 2.5 mL of isotonic saline s.c. to prevent dehydration and antibiotic treatment with marbofloxacin 5 mg/kg (10 mg/mL s.c. Marbocyl, Alcobendas, Spain). Surgery and behavioral assessments were performed between 09:00 and 14:00.

The transplantation of bmSC was done 2 h after SCI. After the anesthetized animals were positioned in a stereotactic frame, the atlanto-occipital membrane was exposed and penetrated with a pointed scalpel blade. A catheter was then inserted and the cell suspension was slowly infused with a syringe pump (100 µL/3 min) into the cisterna magna. While a rat was being prepared by one researcher, a second person removed one batch of bmSC from storage in liquid nitrogen, thawed and washed the cells with saline, and resuspended them in 110 µL of saline. From this, 10 µL was removed for cytometric counting of cell numbers and determination of cell viability. On average, one injection of 100 µL contained 2.6 million viable cells.

Postoperative care, including analgesic and antibiotic treatment, was done as previously described [8]. The bladders were checked and voided manually every 12 h until the rats were urinating spontaneously. The volume of retained and manually expelled urine per 12 h was recorded. Euthanasia at the end of the study was induced by an i.p. injection of 100 mg/kg of sodium pentobarbital (Dolethal, Madrid, Spain).

2.3. Experimental Groups

Animals were assigned to seven experimental groups, six of which received the same SCI but differed in the treatment procedure (Figure 1a). One group (sham) had T9 laminectomy but underwent no contusion injury. Group *SCI-control* received two i.p. injections of saline, the first immediately after SCI (t0) and the second 24 h later (1 dpo). Group *SCI-T200* was treated with two i.p. injections of TUDCA 100 mg/kg body weight at t0 and 1 dpo. Group *SCI-T600* had two i.p. injections of TUDCA 300 mg/kg at t0 and 1 dpo. Group *SCI-T1500* had five injections of TUDCA 300 mg/kg at t0, 1 dpo, 2 dpo, 4 d po, and 6 dpo. Group *SCI-bmSC* received human bmSC implants at t0 + 2 h. Group *SCI-bmSC + T* had bmSC at t0 + 2 h and, in addition, two injections of TUDCA 100 mg/kg at t0 and 1 dpo. During the following six weeks, the investigators who performed the behavioral evaluation were blind with regard to the experimental condition of the individual animals. Nine rats from each of the laminectomy (sham), SCI-control, and SCI-T600 groups were sacrificed at 4 dpo for biochemical and histological analysis. Additional spinal cord sections from non-injured rats were used for comparing histological data.

2.4. Preparation of bmSC

Human bmSC for SCI treatment were prepared by negative selection eliminating erythrocytes with Ficoll density gradient centrifugation and, subsequently, B-cells (CD20), T-cells (CD3), monocytes (CD14) and natural killer cells (CD56) using antibody-based cell sorting with magnetic beads under GMP conditions. Cells were not expanded by cultivation (Neuroplast BV, patent WO2015/059300A1, Geleen, The Netherlands). All procedures for the collection of human bone marrow were approved by the ethics committee of Maastricht University Medical Center (METC 13-2-032). The viability and cell type composition of each batch were analyzed with flow cytometry (CD34, CD271, CD90, CD105, CD73). For the present study, bmSC were prepared at the Neuroplast facility in Geleen, NL, cryoprotected with DMSO, frozen in liquid nitrogen, shipped on dry ice to Toledo, Spain, and then stored in liquid nitrogen until use. Cell viability (exclusion of 7-amino-actinomycin D, cytometry) was again determined after thawing, i.e., immediately before application in vivo.

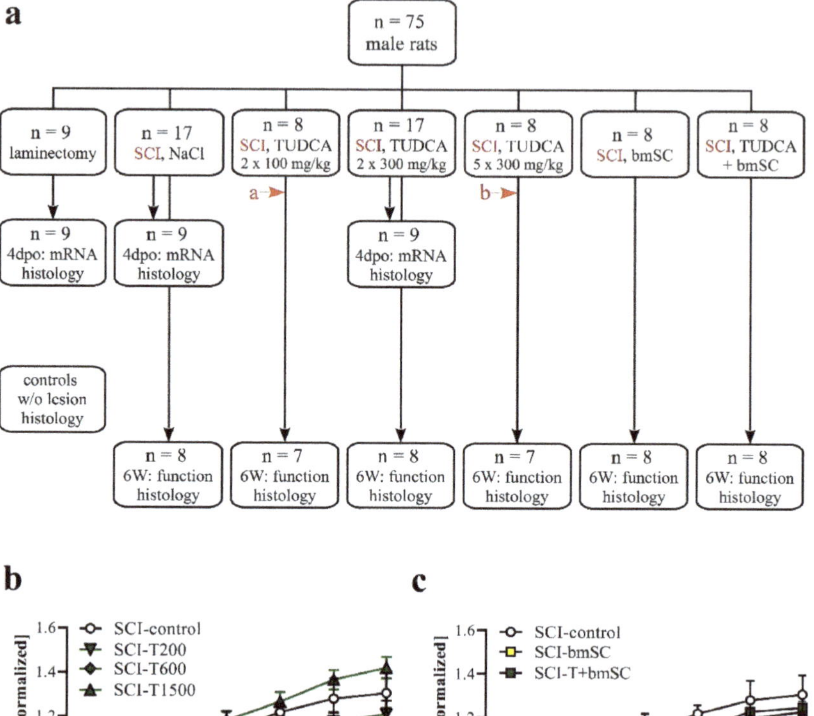

Figure 1. Experimental plan. (**a**) Animals were pseudo-randomly assigned to the different treatment groups. Functional analysis was carried out during a recovery period of six weeks after SCI with the following treatment conditions: (1) two injections of saline at the time of surgery (t0) and 24 h later, (2) two injections of 100 mg/kg TUDCA at t0, 24 h; (3) two injections of 300 mg/kg TUDCA at t0, 24 h; (4) five injections of 300 mg/kg TUDCA at t0, 24 h, 2 dpo, 4 dpo and 6 dpo; (5) one injection of bmSC at t0 + 2 h; (6) combinatorial treatment with bmSC and two injections of 100 mg/kg TUDCA. In addition, three treatment groups (laminectomy only, SCI-control, and SCI with 2 × 300 mg/kg TUDCA) were evaluated at 4 dpo with biochemistry and IF. In histology at 6 W, spinal cord tissue was also compared to tissue from non-lesioned rats. Two animals had to be excluded from the study because open field evaluation at 1 dpo suggested an incomplete lesion (red arrows a: BBB = 7, b: BBB = 8). (**b,c**) Changes in body weight following SCI: Animals in all treatment groups suffered from weight loss during the first 4 dpo and subsequently recovered [means +/− SD; see text for statistical evaluation].

2.5. Bile Acid Treatment

Tauroursodeoxycholic acid (Calbiochem CAS 14605-22-2, Millipore, Madrid, Spain) was dissolved at 150 mg/mL in 0.9% saline immediately before the intraperitoneal injection or stored for no longer than 24 h at 4 °C.

2.6. Evaluation of Locomotor Functions

This was the primary outcome of the study. Recovery of limb movements was evaluated in the open field using the Basso/Beattie/Bresnahan (BBB) locomotor function scale [9]. This was done before surgery (baseline), at 1 dpo, 2 dpo, 3 dpo, 4 dpo, 7 dpo, and subsequently once per week until six weeks after SCI. At the beginning, we established a criterion of BBB ≤ 2 at 1 dpo for inclusion in the study because a higher score was considered to indicate incomplete SCI. Scoring was performed independently by two investigators who were blind with respect to the treatment of the individual animals. Following assessment, both investigators discussed their evaluation, and in cases where different scores were given, the average of both was recorded.

A second assessment was made using the Rotarod test (Ugo Basile SRL, Gemonio, Italy). In this task, rats are positioned on a slowly rotating rod, which obliges them to use their hind legs in order to keep their balance [27]. In six training sessions of 5 min each, at three, two, and one days before SCI, all rats learned this task at a constant speed of 5 rpm of the rotating rod. During tests, which were administered at 7 dpo and then once per week, the rotation speed was accelerated from 5 rpm to 15 rpm over a period of 5 min. The readout in this assay was the time that the rats were able to stay on the rotating rod before falling off (two repetitions, separated by a break of ≥ 15 min). At 4 dpo, we confirmed that none of the SCI rats to be included in the study showed weight supported steps. Considering the high variability of this assay, we applied an additional performance assessment using the percentage of animals in each group that were able to maintain their balance for more than 30 s at 6 W after SCI.

2.7. Von Frey Test of Mechanical Allodynia/Hyperalgesia

Tactile allodynia/hyperalgesia were evaluated using a dynamic plantar aesthesiometer (von Frey test; Hugo Basile 37550, Gemonio, Italy). For each hind leg, a paw withdrawal threshold (PWT) was determined up to a maximum force of 50 g. This was done five times, with intervals of at least five minutes between tests. The lowest and highest values of these readings were excluded, and then the mean was calculated as PWT. This test was administered five weeks after SCI, when all animals were physically able to respond to the stimulation. On the basis of measurements before the lesion, we considered a PWT of below 20 g as an indication of neurogenic pain.

2.8. Quantification of Gene Expression

Animals for biochemical evaluation were sacrificed at 4 dpo with an overdose of sodium pentobarbital. After opening the thoracic cavity, a blood sample of 1 mL was taken from the heart, mixed with 100 µL 0.5 M EDTA to prevent coagulation, spun down, and the supernatant frozen (samples not intended for this study). This was followed by transcardial perfusion with phosphate buffered saline (PBS; 200 mL/rat), preparation of the brains and spinal cords. Spinal cord samples consisted of a 2 cm segment with the lesion site in the center. Tissues were homogenized mechanically with Trizol (Invitrogen, 15596018, Madrid, Spain) and the RNA extracted according to manufacturer's instructions. To remove genomic DNA, purified RNA was digested with DNase I (ThermoScientific, EN0521, Madrid, Spain). An aliquot corresponding to 0.5 µg of purified RNA was used for first-strand cDNA synthesis using Superscript III reverse transcriptase and oligo (dT) primers in a final volume of 40 µL (Invitrogen Life Technologies, K1632). A real-time quantification of cDNA was performed using a SYBR Green PCR assay. Each 15 µL SYBR green reaction mixture consisted of 1 µL cDNA, 7.5 µL SYBR Green PCR-mix (2×), 0.75 µL forward and reverse primers (10 pM) and 4.75 µL distilled water. PCR was performed with 5 min at 95 °C, followed by 40 cycles of 15 s at 95 °C, 60 s at 60 °C and a separate dissociation step for the melting curve. The specificity of the PCR product was confirmed by ascertaining a single melting peak in the temperature dissociation plots. All samples were run in triplicates and the level of expression of each gene was compared with the expression of acidic ribosomal phosphoprotein P0 (36B4). Amplification, detection of specific gene products and quantitative analysis were performed using an ABI 7500 sequence detection system (Applied Biosystems, Alcobendas, Spain). PCR efficiency was verified by dilution series (1, 1/3, 1/9, 1/27, 1/81, and 1/243) and relative mRNA levels were calculated using the comparative ΔCt method with normalization to 36B4. Gene identifiers, primer sequences, product sizes, and melting temperatures are listed in Table 1.

Table 1. Primer sequences used in quantitative RT-PCR.

Gene	Gene ID (NCBI Reference)	Primer Sequences	Product Tm [°C]	Product Size [bp]
36B4	AC130745.3	sense: TTCCCACTGGCTGAAAAGGT antisense: CGCAGCCGCAAATGC	60	60
IL-6	NM_012589	sense: TAGTCCTTCCTACCCCAATTTCC antisense: TTGGTCCTTAGCCACTCCTTC	60	76
CCL-2	NM_031530	sense: TGCTGTCTCAGCCAGATGCAGTTA antisense: TACAGCTTCTTTGGGACACCTGCT	64	131
Arg-1	NM_017134	sense: GCAGAGACCCAGAAGAATGGAAC antisense: CGGAGTGTTGATGTCAGTGTGAGC	62	144
IL-4Rα	NM_133380	sense: GATCTTCTGAGCCCGGTTGA antisense: CTCTCCGCTTGCTGCATT	59	59
IL-10	NM_012854	sense: GATGCCCCAGGCAGAGAA antisense: CCCAGGGAATTCAAATGCT	61	57
CD11b	NM_012711.1	sense: CTGCCTCAGGGATCCGTAAAG antisense: CCTCTGCCTCAGGAATGACATC	60	150
CD31	NM_031591	sense: GAGGTATCGAATGGGCAGAA antisense: GTGGAAGACCCGAGACTGAG	55	174
GFAP	NM_017009	sense: TGGCCACCAGTAACATGCAA antisense: CAGTTGGCGGCGATAGTCAT	60	134
CD20	NM_001107578	sense: TCTTGGGCATTCTGTCGGTG antisense: TCTACAACACCGGCTGTCAC	60	70
CD3ζ	NM_170789	sense: TCATACCCCAGCCCAGTTCT antisense: CGGCTCTGGGGACTTTACAA	60	148
FoxP3	NM_001108250	sense: TCATCACTGGCTTTCTGCGT antisense: GCTTTTAGCCTGAACCCCCT	60	95
TGR5	NM_177936	sense: AAAGGTGGCTACAAGTGCTTC antisense: TTCAAGTCCAAGTCAGTGCTG	58	103

See list for abbreviations, all gene sequences are from rat.

2.9. Tissue Preparation and Histological Staining

At four days or six weeks after SCI, rats were sacrificed with an overdose of sodium pentobarbital followed by transcardial perfusion with PBS and 4% paraformaldehyde/PBS (PFA). The spinal cords were prepared, post-fixed for 1 h, then stored at 4 °C in PFA for 1–3 days. For histological processing, 2 cm long spinal cord segments that included the lesion site in the center were dissected, dehydrated, embedded in paraffin, and cut in 3 µm parasagittal sections using a Leica RM2265 microtome. Sections were mounted on polylysine-coated glass slides (Superfrost Plus, Fisher Scientific, Madrid, Spain) and stored at room temperature (RT). Apoptotic cell nuclei were stained with terminal deoxynucleotidyl transferase dUTP nick end labeling (TUNEL) using the One-step fluorescence TUNEL apoptosis kit (Elabscience, E-CK-A325, Houston TX, USA) according to the manufacturer's protocol.

2.10. Immunofluorescence

For immunofluorescence (IF) staining, sections were rehydrated and incubated for 30 min at 90 °C (water bath) in 10 mM Na citrate/0.05% Tween 20, pH 6.0, for antigen retrieval. Standard procedure included blocking for 30 min at room temperature (RT) with 5% normal goat serum/0.05% Tween 20 in Tris-buffered saline (TBS-T), incubation with primary antibodies for 12 h at 4 °C in a humidified chamber, and 1 h of incubation with fluorescence-labeled secondary antibodies at RT. Nuclei were stained with 10 µg/mL Hoechst-33342 for 15 min at RT. Sections were cover slipped with ImmuMount (Thermoscientific). The following primary antibodies were used in double staining experiments: Polyclonal rabbit anti-GFAP (Sigma G9269; 1/500), polyclonal guinea pig anti-Iba1 (1/500; Synaptic systems 234004, Göttingen, Germany), monoclonal mouse anti CD68 (1/250; Serotec MCA341R1, Alcobendas, Spain). Secondary antibodies were labeled with fluorescent dyes: Goat anti-guinea pig IgG, Alexa-488 (Invitrogen A11073; 1/500), goat anti-rabbit

IgG, TRITC (1/500; Sigma T5268, Madrid, Spain), goat anti-mouse IgG, Alexa-594 (Invitrogen A11005; 1/500), goat anti-mouse IgG, Alexa-488 (1/500; Jackson 115-545003, Cambridge, UK).

2.11. Microscopy and Image Analysis

Immunohistochemical staining was evaluated using a Leica epifluorescence microscope (20×, 40× objective). Exposure conditions were kept constant for quantitative evaluation with GFAP, CD68, and Iba-1. Photographs were analyzed using Fuji Image-J, applying the same brightness/contrast adjustments and threshold values for each marker.

The intensity of immunoreactivity was measured as *integrated density* in regions of interest (ROI) in the ventral white matter at 8 mm distances anterior and posterior (ROI 0.3 mm^2) and within the lesion center (ROI 0.075 mm^2). Following background subtraction, signal intensities were normalized to values found in spinal cord sections from sham-operated rats (4 dpo) or rats without SCI (6 W survival). The number of CD68 positive cells was counted in the same regions. For the evaluation of apoptosis, we counted cell nuclei that were TUNEL positive and expressed the data as percentages of all nuclei in ROI of 1 mm^2, which were located in ventral white matter at 4 mm anterior and posterior of the lesion and in the lesion center.

A Sholl analysis was performed to quantify morphological changes of microglia cells. Iba-1-stained sections from spinal cord gray matter were photographed at an 8 mm distance from the site of injury (laminectomy or SCI +/− treatment) and visualized with Image-J. From three rats per treatment group, 15 cells were randomly selected and eight concentric rings superimposed over the cell nuclei (Fiji plugin *concentric circles*). The number of intersections of cell processes with each ring was counted manually, excluding extensions from neighboring cells.

2.12. Statistical Analysis

Unless stated otherwise in the figure legends, data are presented as mean values ± standard error of the mean (SEM). In box and whiskers plot data are shown with median, first and third quartiles and complete range. Statistical analysis, performed with GraphPad Prism software, consisted of one-factor or two-factor ANOVA, followed by post-hoc Dunnett's or Sidak's multiple comparison tests. In graphical data representation, statistical significance is indicated as follows: *, #: $p < 0.05$, **, ##: $p < 0.01$ and ***, ###: $p < 0.001$. Normal distribution of data within groups was assessed with a Kolmogorov-Smirnov test. Performance differences of treatment groups in von Frey and Rotarod tests were assessed using confidence intervals of proportions based on binomial calculation.

3. Results

3.1. Effects of SCI, TUDCA, bmSC and TUDCA/bmSC Combinatorial Treatment on the General Health Status and Body Weight of the Animals

The general health of the rats was not affected by either treatment with TUDCA alone or in combination with human bmSC, and no adverse effects such as sickness behavior or urinary infections were observed. Five weeks after SCI, two animals, one treated with 5 × 300 mg/mL TUDCA and one in the saline control group, had wounds on their flanks due to biting and were treated topically with antibiotic ointment. During the first four days following SCI surgery, the animals lost 9–11% of their weight, which they recovered with a weight gain of 9.5% in the second week (W1–2), subsequently 8.2% (W2–3), 7.3% (W3–4), 4.9% (W4–5), and 2.2% (W5–6). Changes in body weight over time were significant [two-factor ANOVA; effect of time: F (8, 374) = 370, $p < 0.001$; treatment: F (5, 374) = 37, $p < 0.001$], but post hoc tests showed no significant differences in weight loss or gain between the control group and TUDCA-treated (Figure 1b) or bmSC-treated rats (Figure 1c). Laminectomy without SCI caused an initial weight loss of 4%.

3.2. Expression of TGR5 after SCI

In the rat brain, astrocytes and neurons are immunoreactive for the bile acid receptor TGR5 [12]. We confirmed expression in the spinal cord with quantitative RT-PCR and found the level of TGR5 mRNA at 0.6 ± 0.2 ‰ (mean $\pm$ SD, n = 5) compared to the ribosomal gene 36B4, which was approximately the same as in the cerebral cortex (0.7 ± 0.2 ‰). It did not change significantly after SCI with or without treatment.

3.3. Faster Recovery of Bladder Control with TUDCA Treatment

Normal micturition requires coordinated activation of the detrusor muscle of the bladder and relaxation of the external urethral sphincter, which are controlled by spinal and supraspinal centers [28]. Since these connections were affected by the spinal cord contusion, the animals with SCI were unable to urinate spontaneously and needed manual assistance with bladder voiding during the first days after surgery. Laminectomy alone did not cause urinary retention. The volume of manually expelled urine was evaluated to assess the recovery of bladder control (Figure 2). One positive effect of TUDCA treatment, not reported before, was that recovery of this function occurred faster than in rats with saline injection (Figure 2a). The total volume of urine retained during the post-acute period was not significantly different between groups (Figure 2b), but the average duration of compromised bladder control was shorter for animals treated with TUDCA (Figure 2c). The recovery of the areflexive bladder in the group treated with additional bmSC was similar to that of the group treated with TUDCA only. Eventually, however, all rats recovered autonomic control of micturition.

3.4. Effect of TUDCA and bmSC Treatment on Allodynia/Hyperalgesia

Spinal cord injury can lead to neuropathic pain. This was assessed at five weeks after SCI by determining the withdrawal threshold (PWT) to mechanical stimulation of the hind paws using an automated von Frey test. Although we found no significant differences between experimental groups (Figure 2d; ANOVA, $F (6, 82) = 2.5$, $p < 0.05$, put post-hoc Dunnett's test vs. SCI n.s. for all groups), the number of rats that had a PWT below 20 g, which we considered indicative of allodynia or hyperalgesia, was higher after SCI than after the laminectomy operation. This measure of pain sensitivity appeared to be worse with high doses of TUDCA and lower with treatment of bmSC alone (Figure 2e), but variability and group size do not permit a conclusive interpretation.

3.5. Effect of TUDCA and bmSC Treatment on Recovery of Sensory-Motor Functions

One day after SCI, the rats' ability to use their hind legs was assessed in the open field. The SCI caused paralysis, as indicated by no or only slight movement of joints. Laminectomy without contusion resulted in temporary gait instability in some cases. Two TUDCA-treated SCI animals, which presented a BBB score above 2 at 1 dpo, were excluded from the analysis because we considered this an indication of a less severe SCI rather than an effect of treatment (indicated in Figure 1a). With time after injury, sensory-motor functions of the rats improved significantly, and different treatments were effective [two-factor ANOVA, time after SCI: $F (9, 470) = 121$, $p < 0.0001$ (pre-SCI data not included); treatment effect: $F (5, 470) = 3.1$, $p < 0.01$ (laminectomy group not included); interaction: $F (45, 470) = 0.7$, n.s.]. During the first 4 dpo, the majority of treated animals recovered the ability to move their hind legs. After treatment with 300 mg/kg TUDCA, this recovery occurred faster, such that BBB scores were significantly higher at 2 and 4 dpo compared to saline treatment (Figure 3a,b, subacute phase of these groups expanded in c,d). Improvements after two injections of 100 mg/kg TUDCA were not distinguishable from results after saline injections. Recovery of motor function improved further between one and three weeks after lesion, when all rats showed plantar stepping and weight support at least in stance but usually no coordination between fore and hind limbs (BBB 9–11). No improvement beyond this level was observed in the chronic phase (Figure 3a,b). Combinatorial therapy with TUDCA and bmSC or bmSC alone achieved higher BBB scores at 1–6 weeks (ΔBBB 0.7–1.9 higher than SCI-control, ΔBBB 0.5–1.2 higher than SCI-T200; $p < 0.05$), but there were no significant differences between treatment with bmSC and combinatorial treatment. The experiment confirmed previous results with bmSC injection only [8].

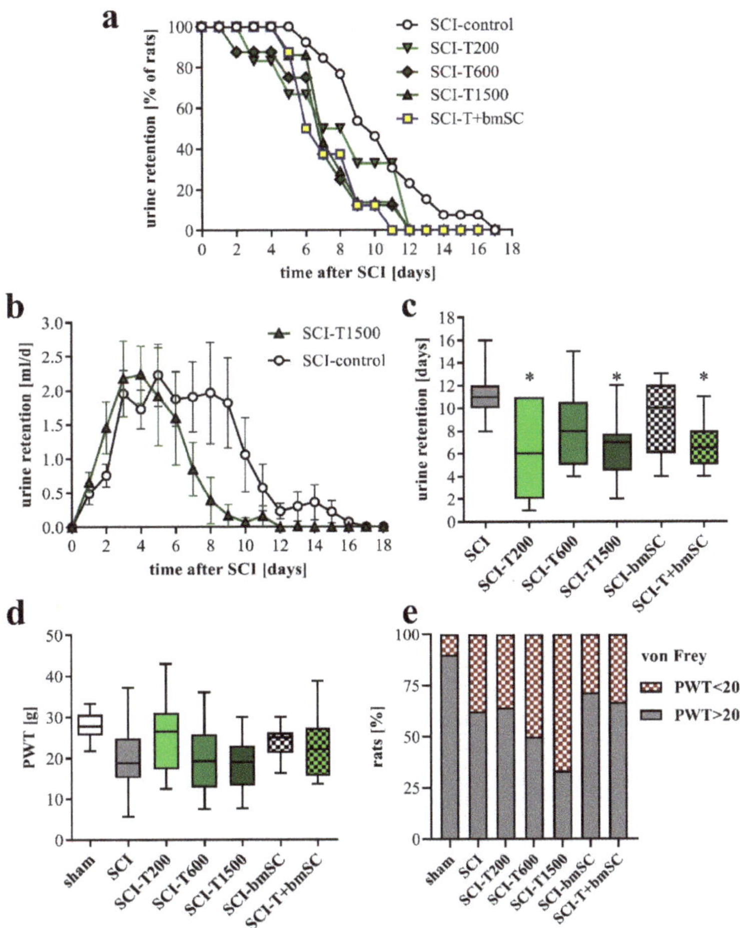

Figure 2. Autonomic functions and neuropathic pain. (**a**) Percentage of rats that required manual bladder voiding each day in the post treatment period. Bladders were voided manually every 12 h. (**b**) Average volume of retained urine per day and rat. Data shown here are for the SCI rats treated with saline and with five injections of TUDCA (mean ± SEM). (**c**) Time after SCI that passed until the animals no longer required manual voiding of the bladder [ANOVA, F (5, 43) = 2.7, $p < 0.05$; post hoc Dunnett's test vs. SCI control * $p < 0.05$]. (**d**) To assess neuropathic pain we measured the PWT to mechanical stimulation (von Frey test; median ± 25–75 percentile, range; ANOVA n.s.). (**e**) Percentage of rats that had a PWT of below 20 g, which was considered as an indication of allodynia. Experimental groups are abbreviated as follows. SCI-control: two injections of saline; SCI-T200: two injections of 100 mg/kg TUDCA at t0 and 24 h later; SCI-T600: two injections of 300 mg/kg TUDCA at t0, 24 h; SCI-T1500: five injections of 300 mg/kg TUDCA at t0, 24 h, 2 dpo, 4 dpo and 6 dpo; SCI-bmSC: one injection of bone marrow-derived stromal cells at t0 + 2 h; SCI-T + bmSC: combinatorial treatment with bmSC and two injections of 100 mg/kg TUDCA; sham: laminectomy only.

In addition to the assessment in the open field, rats were subjected to the Rotarod test, which measures their ability to maintain equilibrium on a rotating bar. Before SCI, all rats had been trained to perform this task for at least 300 s, and at 4 dpo, none of the animals that met the BBB inclusion criterion was able to do so. Spontaneous recovery caused a significant increase in Rotarod score during the first four weeks in all animals (Figure 4a,b). No further improvement occurred subsequently [ANOVA, effect of time after SCI: $F_{(3, 352)}$ = 22.0, $p < 0.001$; treatment effect: $F_{(5, 352)}$ = 1.5, n.s., interaction: $F_{(15, 352)}$ = 0.8, n.s.]. Some animals which showed weight supported plantar steps in the open field refused to hold on to the bar. In the absence of an independent criterion to distinguish between voluntary refusal and inability to perform the task, no data were excluded from the evaluation, resulting in a high variability in this assay. An additional assessment using the percentage of rats that at 6 W maintained themselves for longer than 30 s on the Rotarod showed no improvement at all following TUDCA treatment but revealed that more of the rats in the groups treated with bmSC or TUDCA + bmSC were able to do this (Figure 4c). With eight animals per group, the statistical power did not suffice for this effect to be significant.

3.6. Effect of TUDCA Treatment on Neuroinflammation

To a large degree, the devastating effects of SCI are due to the neuroinflammatory response of microglia and blood-derived macrophages. Since TUDCA is known for its anti-inflammatory effect on microglia, we confirmed this for our study by measuring the expression of marker genes in the spinal cord at 4 dpo. This experiment was done for the high dose of TUDCA. As in previous experiments, SCI caused mRNA upregulation of the inflammatory cytokine IL-6 and chemokine CCL-2 in spinal cord extracts. This appeared to be a local response and was not found in the cerebral cortex (Figure 5a,b). Treatment with TUDCA significantly reduced the expression of these and other (Table S1) inflammatory genes in the spinal cord. The lesion also caused alternative activation of microglia/macrophages, which was demonstrated by a highly significant increase in transcription of arginase-1 and IL-4R (Figure 5c,d). Contrary to our expectation, the bile acid treatment also reduced this response, indicating that it did not promote differentiation of an M2 phenotype. The cytokine IL-10 promotes alternative activation of macrophages and is produced by a subpopulation of these cells. At 4 dpo, we found a significant reduction of IL-10 transcripts in the spinal cord. This also occurred after TUDCA treatment but, compared to the SCI-controls, was no longer significant (Figure 5e).

The inflammatory activation of microglia or macrophages was also reflected by the expression of the complement receptor 3A (integrin αM, CD11b), which increased 7-fold after SCI. Similar to the other pro-inflammatory markers, its expression was significantly reduced after treatment with TUDCA (Figure 6a). Platelet endothelial cell adhesion molecule-1 (CD31) is found on endothelial cells, macrophages, and various lymphocytes. Its expression was detected in the control tissue, increased after SCI, and also was much reduced after TUDCA treatment (Figure 6b). Activation of astrocytes in neuropathologies is associated by an upregulation of the glial fibrillary acidic protein (GFAP). Transcripts of this gene, which was highly expressed in the spinal cord, increased 4-fold after SCI, and the response was inhibited by bile acid injections (Figure 6c). The validity of these effects on gene regulation can be appreciated by comparison with marker genes of other cell types, such as B lymphocytes (CD20; Figure 6d) and T lymphocytes (CD3ζ; Figure 6e). Indicators of these cells did not significantly change at 4 dpo with or without TUDCA injections. A marker gene of regulatory T-cells (FoxP3; Figure 6f) was expressed at a lower level after SCI and not influenced by treatment.

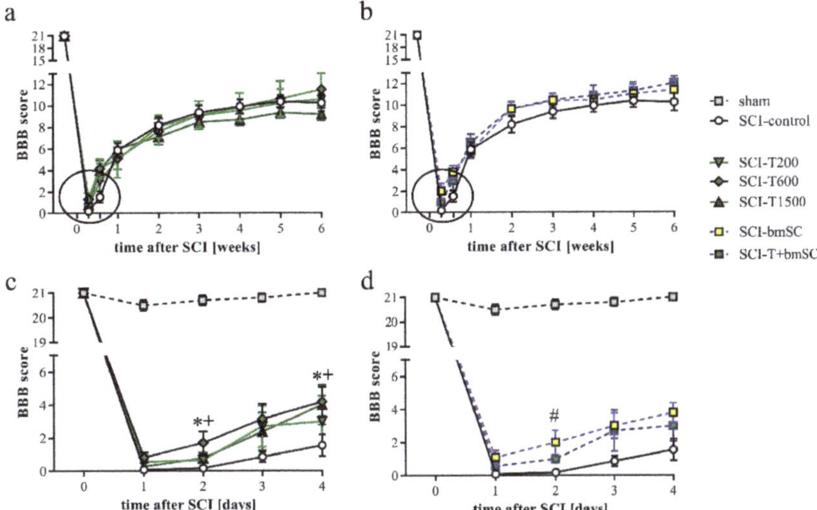

Figure 3. Recovery of motor functions after SCI assessed in the open field. Mean BBB scores were monitored before surgery, daily during the first 4 dpo after surgery, and then once per week during 6 weeks of evaluation. (**a**) Results for SCI-control, treated with saline, and the TUDCA treated groups. (**b**) Results for SCI-control and the SCI groups treated with bmSC and TUDCA + bmSC. (**c**,**d**) Changes in the acute/subacute phase; effects of treatment with 2×300 mg/kg TUDCA or with bmSC were significant at 2 dpo and/or 4 dpo [see main text for statistical evaluation; * (SCI-T600), + (SCI-T1500), # (SCT-bmSC) indicate $p < 0.05$; error bars indicate SEM]. Recovery of motor function improved in all treatment groups until a plateau was reached after three weeks; experimental groups are indicated as for Figure 2.

The lesion-induced activation of microglia and astrocytes was confirmed by microscopical inspection using double-staining IF for Iba-1, CD68, and GFAP (Figures 7 and 8). As expected from previous experiments, SCI disrupted the blood spinal cord barrier, caused an influx of hematogenous macrophages, and activated resident microglia and astrocytes. At 4 dpo and 6 weeks, under all treatment conditions, the center of the lesion was filled with cellular debris and Iba-1/CD68 positive cells. These made up $8.3 \pm 1.1\%$ of the cells in the center of the lesion (ROI 0.075 mm^2; Figure 7a–c). The majority of CD68 positive cells were likely to be of hematogenous origin, and these were absent in the spinal cords of animals that had received only the laminectomy. Already at 4 dpo, the lesion center was devoid of neurons (NeuN IR) and astrocytes (GFAP IR). After TUDCA treatment, $5.7 \pm 1.0\%$ of the cells in this area were macrophages, as identified by morphology and CD68 IR. With increasing distance from the lesion center, the number of CD68 positive cells decreased such that at 8 mm posterior and anterior distances, only a few macrophages were observed in the SCI animals. Therefore, distance to the lesion significantly affected the strength of the CD68 signal, and a treatment effect was only significant at the central location (see quantification below).

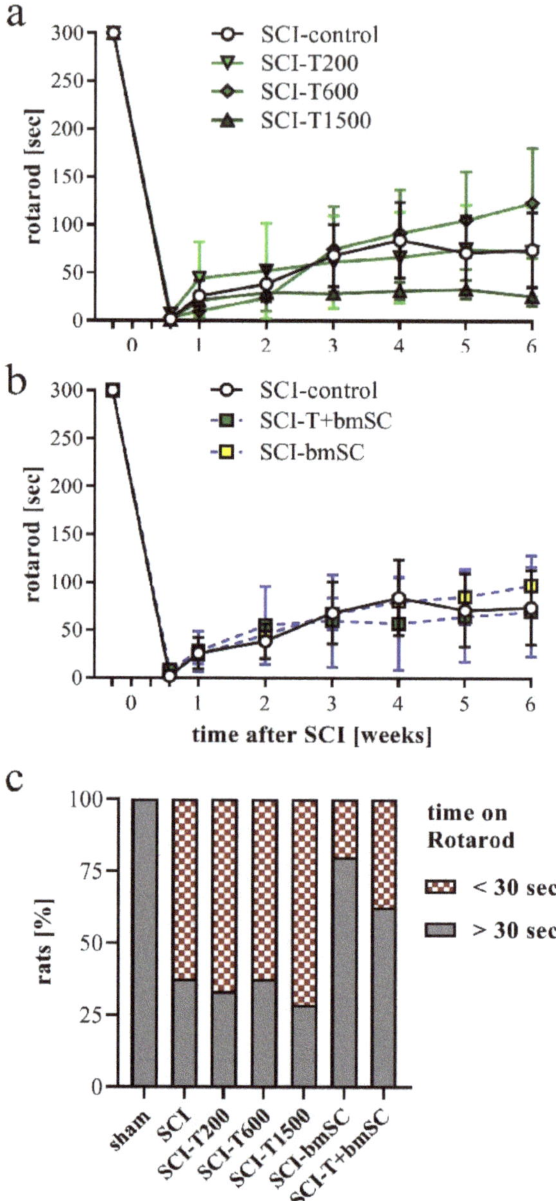

Figure 4. Recovery of motor functions after SCI assessed with the Rotarod assay. At 4 dpo, we confirmed that no SCI treated animals showed weight supported steps. Beginning at 7 dpo, some rats in all groups showed gradual improvement. (**a**,**b**) Time that rats were able to keep their balance on the rotating bar; mean +/− SEM, no significant differences between groups were detected. (**c**) Percentage of rats that performed the Rotarod task for more than 30 s at 6 W after SCI (n = 7–8 rats/group). Experimental groups are indicated as in Figure 2.

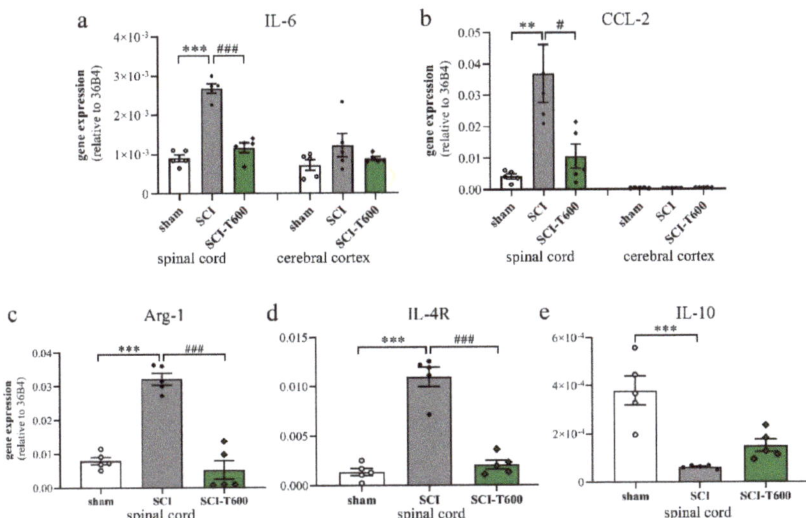

Figure 5. Expression of marker genes of inflammation. At 4 dpo, RNA extracts from spinal cord and cerebral cortex were analyzed with quantitative RT-PCR (treatment conditions abbreviated as in Figure 2). (**a**) Gene expression of IL-6 [ANOVA, spinal cord: F (2, 12) = 71.6, $p < 0.0001$; cortex: n.s.]; (**b**) CCL-2 [ANOVA, spinal cord: F (2, 12) = 8.9, $p < 0.01$; cortex: n.s.]; (**c**) arginase-1 [ANOVA, F (2, 12) = 56.0, $p < 0.0001$], (**d**) IL-4Rα [ANOVA, F (2, 12) = 66.0, $p < 0.0001$], (**e**) IL-10 [ANOVA, F (2, 12) = 18.7, $p < 0.001$]. Significant differences detected with post hoc Dunnett's tests ($p < 0.05$, $p < 0.001$, $p < 0.001$) are indicated with **, *** symbols for SCI vs. laminectomy and #, ### for SCI vs. SCI-T600.

3.7. Effect of TUDCA Treatment on Glial Activation after SCI

Increasing the IR of GFAP (Figure 8a–c,g–i) and Iba-1 (Figure 8d–i) indicated activation of astrocytes and microglia at 4 dpo. In the anterior and posterior of the SCI site, both signals were stronger than in animals that had only received the laminectomy. Contusion injury caused a 15-fold increase in Iba-1 IR in and close to the lesion center compared to levels in sham-operated animals. After treatment with 2 × 300 mg/kg TUDCA, Iba-1 and CD68 signals were significantly reduced in the lesion center (Figure 9a,b). Morphological changes of GFAP positive cells around the lesion site indicated the activation of astrocytes at 4 dpo (Figure 8j,k). Quantification of the GFAP IR confirmed this observation for the white matter posterior of the lesion center. Astrocyte activation was less pronounced (i.e., not significantly different from laminectomy treatment) after bile acid treatment (Figure 9c).

At 6 W after SCI, Iba-1 IR was no longer as elevated as at 4 dpo but still 3- to 4-fold higher than in non-lesioned tissue. No significant differences between TUDCA-, bmSC- or TUDCA/bmSC-treated and saline-treated animals were observed in the chronic phase (Figure 9d). At this time, a prominent glial scar had formed in all SCI animals. Quantification of GFAP IR in this area (Figure 9e) confirmed a strong increase from the acute to the chronic phase after SCI (note different scales of y-axis in panels Figure 9c,e). The scar area appeared to be reduced by bile acid/bmSC treatment, but the differences in GFAP intensity did not reach significance.

Responding to an inflammatory environment, microglia cells are known to respond with a morphological transformation from a branched appearance to an ameboid shape. We confirmed this using a Sholl analysis of Iba-1 IR cells in gray matter at an 8 mm distance from the lesion center (Figure 10a–c). The morphological change associated with SCI was highly significant. Microglia cells in SCI injured animals at 4 dpo had fewer branches than in rats with laminectomy. This change was less pronounced after TUDCA treatment ($p < 0.001$), corroborating the hypothesis that the bile acid affected microglial activation.

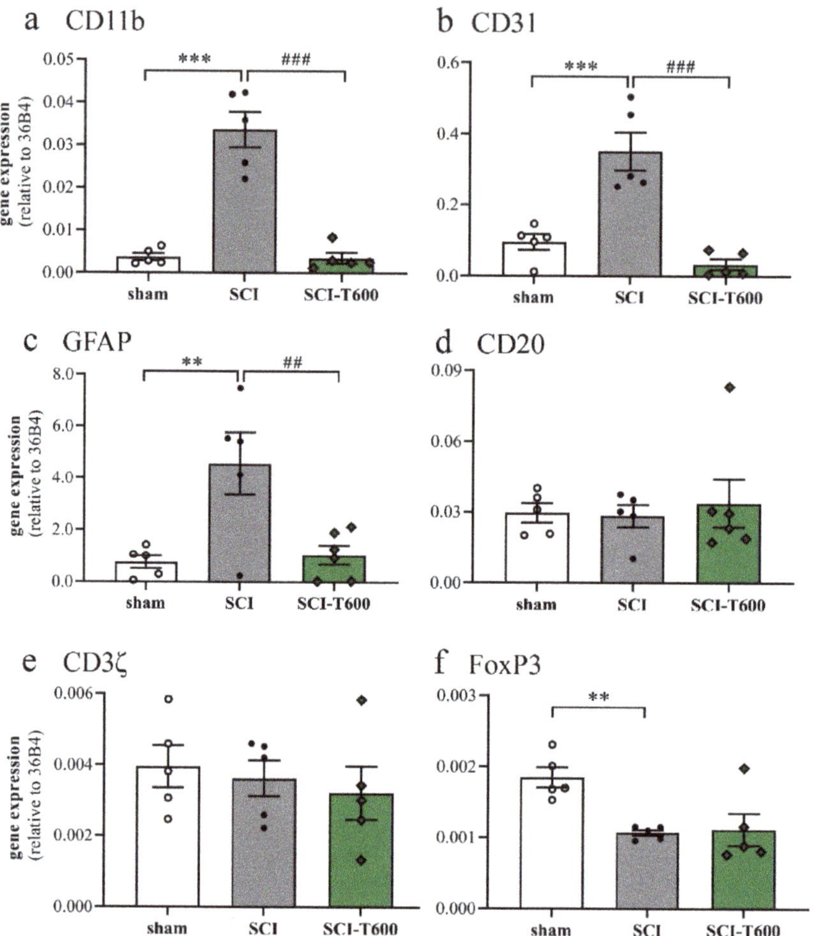

Figure 6. Expression of marker genes of cell type activation. At 4 dpo, RNA extracts from the spinal cord and cerebral cortex were analyzed with quantitative RT-PCR (treatment conditions abbreviated as in Figure 2). (**a**) Gene expression of CD11b, marker of microglia and macrophage [ANOVA, F (2, 12) = 46.3, $p < 0.0001$]; (**b**) CD31, endothelial cells, macrophages and lymphocytes [ANOVA, F (2, 12) = 23.9, $p < 0.0001$]; (**c**) GFAP, astrocytes [ANOVA, F (2, 12) = 8.6, $p < 0.01$]; (**d**) CD20, B-cells [ANOVA, F (2, 12) = 1.6, n.s.]; (**e**) CD3ζ, T-cells [ANOVA, F (2, 12) = 2.2, n.s.]; (**f**) FoxP3, regulatory T-cells [ANOVA, F (2, 12) = 7.7, $p < 0.01$]. Significant differences detected with post hoc Dunnett's tests ($p < 0.05$, $p < 0.001$, $p < 0.001$) are indicated with **, *** symbols for SCI vs. laminectomy and ##, ### for SCI vs. SCI-T600.

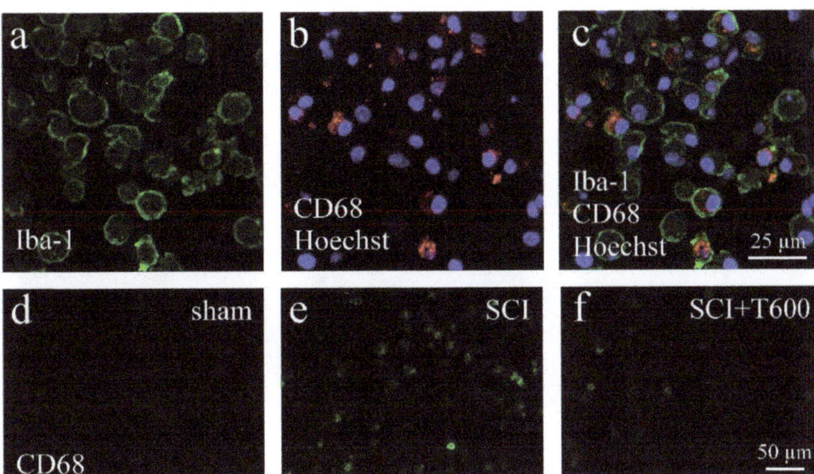

Figure 7. Activation of macrophages after SCI. (**a–c**) Confocal microscopy images show examples of macrophages in the SCI lesion center at 4 dpo. All cells are IR for Iba-1 (**a**, green) and most also for CD68 (**b**, red), shown in combination with nuclear staining (**b,c**, Hoechst 33342, blue). (**d–f**) CD68 IR macrophages in the spinal cord after laminectomy (**d**, sham), SCI (**e**), and SCI 2 × 300 mg/kg TUDCA treatment (**f**). Same magnifications are used in (**a–c**) and in (**d–f**) (scale bars in (**c,f**)).

3.8. Effect of TUDCA Treatment on SCI-Induced Apoptosis

Several groups reported that TUDCA treatment reduced cell death after SCI [22,24–26]. We found previously that the bmSC preparation used here was also cytoprotective [8]. To allow comparison with these publications, we evaluated cellular apoptosis in the lesion center and at 4 mm anterior and posterior to this position (Figure S1). Four days after SCI, we found $11.3 \pm 3.1\%$ (mean $\pm$ SD) of the cell nuclei in the lesion area to be TUNEL positive, whereas sham operated rats had only $0.40 \pm 0.35\%$ apoptosis in the white matter at the respective position ($p < 0.001$). Treatment with two injections of 300 mg/kg TUDCA significantly reduced TUNEL staining in the lesion area to $5.9 \pm 4.2\%$ ($p < 0.05$). In the ventral white matter at positions anterior and posterior of the lesion center, we found a slight, non-significant increase in the number of apoptotic cells at 4 dpo (control: $0.2 \pm 0.3\%$, SCI: $1.1 \pm 0.7\%$), which was not affected by treatment (SCI-T600: $1.1 \pm 0.3\%$). The analysis of TUNEL staining in the chronic phase after SCI revealed continuing high levels of apoptosis without (SCI: $7.6 \pm 18.6\%$) and with TUDCA treatment (SCI-T600: $16.4 \pm 17.1\%$), indicating that bile acid injections had no lasting effect on cell survival ($p > 0.1$). In rats treated with bmSC, the proportion of apoptotic cells at 6 W was lower (SCI-bmSC: $4.1 \pm 7.8\%$, $p < 0.05$), but there was no significant effect due to additional treatment with TUDCA (SCI-T + bmSC: $1.4 \pm 1.0\%$, $p < 0.05$ vs. SCI; $p > 0.1$ vs. SCI-bmSC).

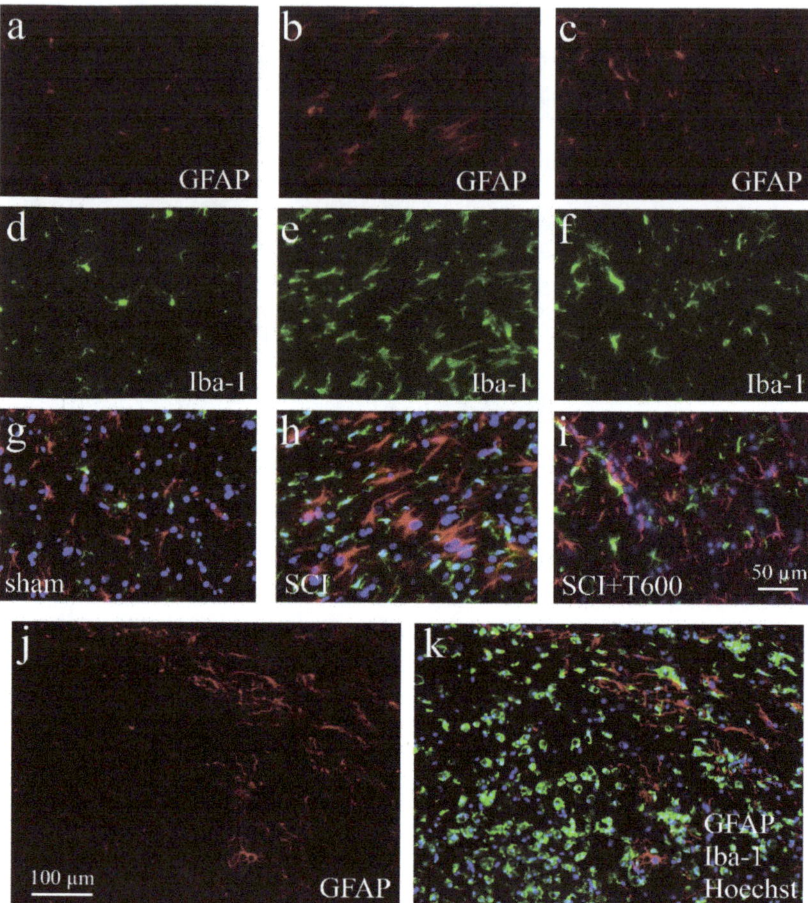

Figure 8. Activation of microglia and astrocytes after SCI. The IF photographs show examples of grey matter at 4 dpo and a distance of ca. 0.4 mm from the lesion center, which were double-stained for astrocytes (GFAP; **a–c**) and microglia (Iba-1; **d–f**). Photographs were superimposed with additional nuclear staining (Hoechst 33342; **g–i**). (**a,d,g**) Section from laminectomy-operated animal. (**b,e,h**) SCI with saline injection; (**c,f,i**) SCI treated with two injections of TUDCA 300 mg/kg. (**j,k**) At 4 dpo under all SCI treatment conditions, astrocytes showed increased GFAP IR but a glial scar had not yet developed. Scale = 50 μm in (**i**) (for (**a–i**), 40× objective) and 100 μm in (**j**) (for (**j,k**), 20× objective). Macrophages fill the lesion center.

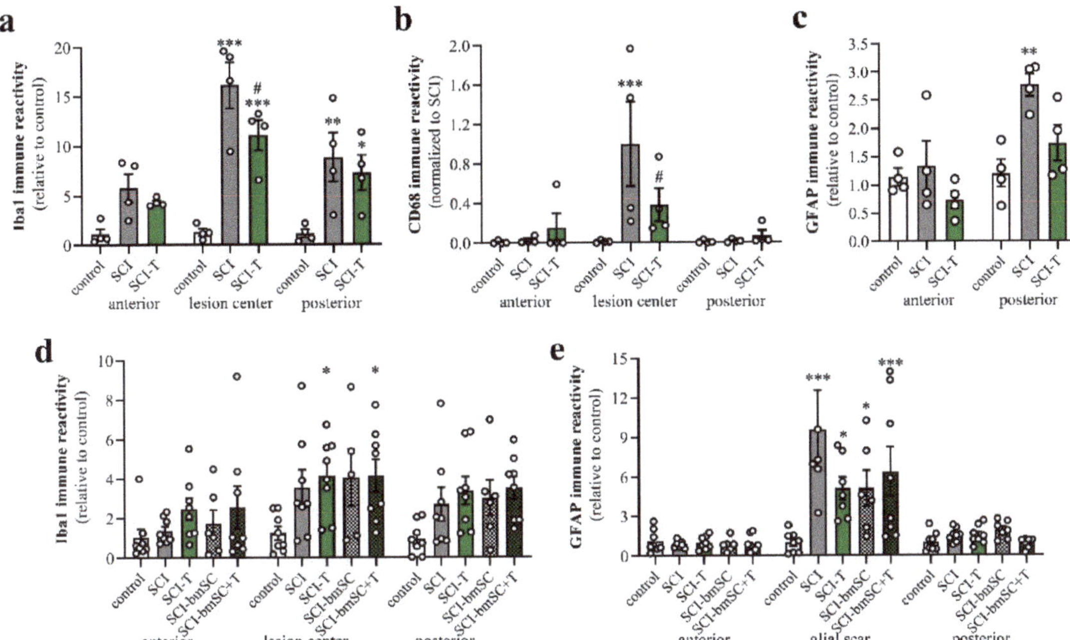

Figure 9. Activation of macrophages, microglia and astrocytes after SCI. Data from spinal cord areas were normalized for IR in the anterior position of sham-operated animals and analyzed with 2-factor ANOVA followed by post hoc Sidak tests. (**a**) Quantification of Iba-1 IR (fluorescence integrated density) at 4 dpo in white matter at 8 mm anterior, 8 mm posterior of the lesion site and in the lesion center [2-factor ANOVA, location effect: $F_{(2, 27)} = 11.7$, $p < 0.01$; treatment effect: $F_{(2, 27)} = 29.9$, $p < 0.001$; interaction: $F_{(4, 27)} = 3.2$, $p < 0.05$]. (**b**) Quantification of CD68 IR at 4 dpo in areas 8 mm anterior and posterior of the lesion site and in the lesion center [2-factor ANOVA, location effect: $F_{(2, 27)} = 6.7$, $p < 0.01$; treatment effect: $F_{(2, 27)} = 3.3$, $p = 0.05$; interaction: $p < 0.05$; data were normalized to SCI, lesion center, as almost no CD68 cells are found outside this area]. (**c**) Quantification of GFAP at 4 dpo in white matter at 8 mm anterior, 8 mm posterior of the lesion site [location effect: $F_{(1, 18)} = 14.2$, $p < 0.01$; treatment effect: $F_{(2, 18)} = 6.6$, $p < 0.01$; interaction: $F_{(2, 18)} = 3.4$, n.s.]. (**d**) Quantification of Iba-1 IR at 6 W in white matter 8 mm anterior, posterior and in the lesion center site [location effect: $F_{(2, 98)} = 5.9$, $p < 0.01$; treatment effect: $F_{(4, 98)} = 5.4$, $p < 0.001$; interaction: $F_{(8, 98)} = 0.3$, n.s.]. (**e**) Quantification of GFAP IR at 6 W in white matter 8 mm anterior and posterior of the lesion center and in the glial scar [location effect: $F_{(2, 97)} = 36.1$, $p < 0.001$; treatment effect: $F_{(4, 97)} = 4.1$, $p < 0.01$; interaction: $F_{(8, 97)} = 2.5$, $p < 0.001$]. Significant differences detected with post hoc comparisons tests ($p < 0.05$, $p < 0.001$, $p < 0.001$) are indicated with *, **, *** for SCI vs. control and # SCI vs. SCI-T600].

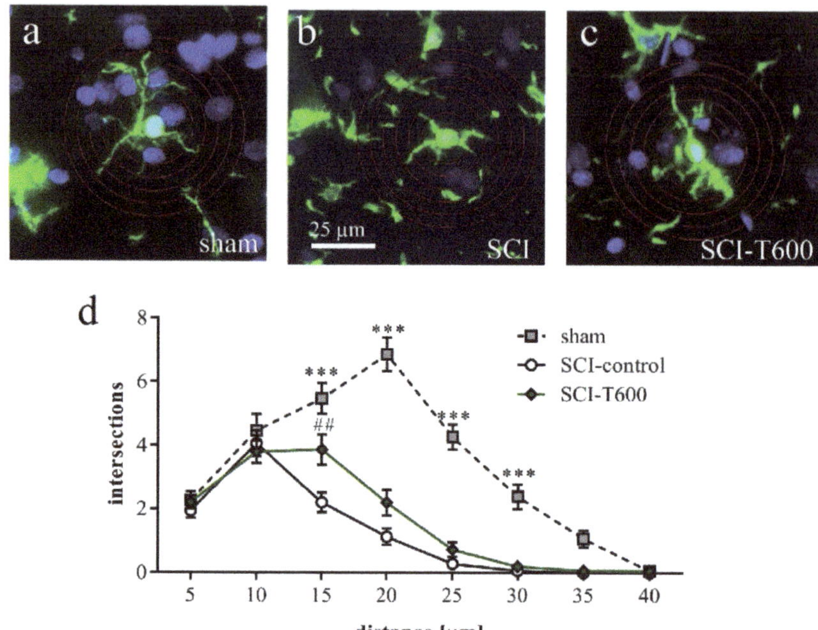

Figure 10. Effect of SCI and TUDCA treatment on microglia morphology. (**a–c**) Examples of Iba-1 IR cells in the spinal cord gray matter at 8 mm distance from the injury site at 4 days after sham operation (**a**), SCI control (**b**) and SCI with TUDCA treatment (**c**) are shown with superimposed rings for Sholl analysis. (**d**) Number of intersections of cellular processes at 5 µm intervals from the cell soma [two-factor ANOVA, distance from soma $F_{(7, 335)} = 78.6$, $p < 0.0001$; treatment effect: $F_{(2, 335)} = 107$, $p < 0.001$; interaction: $F_{(14, 335)} = 12.9$, $p < 0.001$, results of Bonferroni multiple comparison tests are indicated with *** ($p < 0.001$) for sham vs. SCI-control and sham vs. SCI-T600 and with ## ($p < 0.01$) for SCI vs. SCI-T600]. All panels have the same magnification (scale bar in **b**).

4. Discussion

In the present study, we found short term but no lasting benefits of TUDCA treatment on the recovery of bladder control and motor function after SCI in rats. Transcripts of the bile acid receptor TGR5 were detected in the spinal cord at a similar level of expression as in the brain. When combined with bmSC implants, TUDCA applications in the acute phase did not provide additional therapeutic benefit. While we confirmed previously reported anti-inflammatory and cytoprotective effects of TUDCA, our data indicate that the bile acid also reduced expression of genes that are associated with the M2 phenotype. The spinal cord contusion injury caused a transient loss of body weight in the rats, which was not significantly affected by TUDCA- or bmSC-treatment.

4.1. Are Bile Acids Promising for the Treatment of SCI?

There is good evidence that bile acids, specifically TUDCA and the TGR5 ligand INT777, modulate the activity of macrophages and microglia. Activation of TGR5 inhibits NFκB signaling and subsequent expression of inflammatory genes [29,30]. In LPS-treated microglia cells, bile acid treatment reduces nitrite production, expression of pyruvate kinase M2, which can act as a transcription factor, and downstream genes, such as lactate dehydrogenase [18]. Anti-inflammatory treatment is currently the only pharmacological approach for SCI in patients.

So far, five studies have been published to test TUDCA in rat models of SCI. The scientists used a weight drop device that caused a contusion similar to the present experiments [22–24,26] or a compression model, also at vertebral level T9, which was less severe and only damaged the dorsal columns [31]. In these studies, motor recovery was monitored during the subacute phase (5 dpo: [22,23,26]; 7 dpo: [31]; 10 dpo: [24]). To the extent that the data are comparable, i.e., severe T9 contusion injury and evaluation using the BBB scale, our present data confirm the reported outcomes. In addition, we found that TUDCA-treated rats recovered the autonomic control of bladder function earlier than under control conditions or after bmSC injection. However, our main conclusion is that TUDCA treatment does not provide lasting benefits after SCI: Extending the previous studies, we investigated effects in the chronic phase. In all rats, the improvement of motor function reached a limit after 3–4 weeks, which did not significantly differ between treatment groups except for the effect of additional bmSC treatment. High doses of TUDCA (5 × 300 mg/kg) were not more effective than lower ones, an observation that was also made with mice, where one injection of 100 mg/kg TUDCA gave better results than higher doses [25]. Another mouse study has recently been published with even higher doses, using fourteen administrations of 200 mg/kg TUDCA per os [32]. Since the longest treatment regime that we used terminated at 5 dpo, we do not rule out the possibility that continued application would extend the cytoprotective effects into the chronic phase and have therapeutic benefits.

The results with mice are not directly comparable with our data as the histological evaluation at 14 dpo demonstrated only minor effects of the SCI while motor scores were lower (BBB = 5, control SCI) than in our experiments (BBB = 8.2, control SCI; the rat scale was used in the mouse study as well). Behavioral evaluation was performed for up to 14 days, when TUDCA-treated mice reached a motor score of 10. We do not know whether recovery had reached a plateau or would have improved further in the following weeks. It is certainly possible that the continued application of TUDCA extended the beneficial effect into the chronic phase. Hou and colleagues also performed additional experiments regarding the therapeutic mechanism. These showed that TUDCA was not only neuroprotective but also improved axonal growth [32].

Most available data obtained with cell cultures indicates that TUDCA reduces inflammation, production of reactive oxygen species and ER stress by binding TGR5, subsequent cAMP synthesis and PKA activation [13,30,32]. Our rationale for the combinatorial therapy of TUDCA with bmSC was that the moderating influence on macrophages may improve survival of the stem cells. In a recent experiment we found a beneficial effect of bmSC treatment compared to methylprednisolone but were not able to detect the implanted cells later in the tissue [8]. The present experiments revealed no additional benefit of TUDCA with stem cells compared to stem cells alone. While the inhibitory effect of TUDCA on the release of inflammatory mediators was confirmed, there are more potent anti-inflammatory drugs available, which also do not provide a satisfactory therapy of SCI [2,33]. Furthermore, we found that TUDCA treatment rather decreased the alternative monocyte activation in the tissue. It is questionable whether the transient improvement of motor recovery and bladder function observed with TUDCA is meaningful from a clinical point of view.

Since application of 5 × 300 mg/kg TUDCA was associated with more weight gain, but 2 × 100 mg/kg and 2 × 300 mg/kg with less weight gain than observed in SCI-controls (post hoc tests n.s.), this parameter does not seem to be affected by treatment. Animals injected with bmSC showed a similar rate of recovery as the controls. In a previous study using the same type of bmSC, a stronger weight gain was observed with this treatment [8]. In those experiments, the bmSC-treated rats also suffered a more severe weight loss during the first days after SCI, which may have contributed to the relative increase in body weight thereafter. This was not the case in the present study.

4.2. How Does TUDCA Influence the Recovery of Urinary Function?

Autonomic dysfunctions, which have a large impact on the quality of life of SCI patients, include impairments of bladder storage and emptying. Contraction of the bladder is mediated by parasympathetic efferents from the sacral spinal cord via the pelvic nerve to the detrusor muscle of the bladder. Relaxation of the external urethral sphincter is controlled by somatic innervation from sacral segments via the pudendal nerve. These neural circuits are controlled by a coordination center in the reticular formation in the pons and midbrain [34]. Its activity is required for the voiding reflex [35]. The thoracic SCI employed in the present study disconnects the pontine from the sacral micturition centers, and this may cause the rats' inability to urinate spontaneously [36]. Additional sympathetic innervation of the urethra originates in the lumbar segments of the spinal cord. These are also posterior to the SCI and therefore separated from the supraspinal areas. While in human SCI patients, neurogenic bladder dysfunction is generally irreversible [37], all rats recovered their ability to urinate spontaneously. This, in our experience, is generally observed even after the severe contusion injury (2 N) performed with the *Infinite Horizon* impactor. In clinical terms, the transient depression of spinal reflexes caudal to a SCI has been defined as "spinal shock". Why these reflexes return is not completely understood [38].

Severed fibers in the spinal cord that connect with supraspinal centers are unlikely to be restored during the first two weeks after surgery, when the rats recover the ability to urinate spontaneously. Therefore, the most probable explanation for the functional regeneration is that spared fibers in ventral white matter tracts are dysfunctional during the subacute phase after SCI and recover from spinal shock in the subsequent days. Treatment with TUDCA may accelerate tissue remodeling, e.g., by increasing the release of TGFβ [39]. It is also conceivable that plastic changes in the sacral micturition centers occur after the descending innervation is lost. As some neurons express TGR5 [12,40], it is an intriguing possibility that bile acids may affect neuronal plasticity. To our knowledge, this has not been investigated so far. Finally, the observed benefit may have been a non-specific side effect of the systemic anti-inflammatory activity of the TUDCA injections.

4.3. How Does TUDCA Affect the Inflammatory Phenotype after SCI?

Our data on gene expression of CD11b, GFAP, IL-6, and CCL-2 confirm the inhibitory effect of TUDCA on inflammatory pathways. These qRT-PCR measurements at 4 dpo included a spinal cord segment of 2 cm including and surrounding the lesion site. Our histological evaluation of Iba-1 showed significantly lower IR near the lesion center, which is in accordance with the gene expression data. Evaluation of CD68 IR, absent in the non-injured tissue, also demonstrated an effect of bile acid treatment, indicating a lower number of inflammatory macrophages in the lesion site. At a distance of 8 mm, CD68 positive cells were observed in some animals only, without significant differences between groups. These data corroborate previous results where TUDCA reduced Iba-1 IR in the hippocampus of LPS-treated mice [17].

In the context of neuro-inflammation, two phenotypes of macrophages and activated microglia are frequently distinguished, referred to as M1 (pro-inflammatory) and M2 (anti-inflammatory or alternative activation); [41]. The M2 state has been subdivided further to accommodate differential patterns of gene expression. Experimentally, the M1 phenotype may be defined as the outcome of stimulation with TNFα and IFNγ, the M2a phenotype as the response to IL-4, and the M2c phenotype in response to IL-10 [42]. Based on the fact that cAMP induces M2 associated genes of microglia [43] and on histological data with a mouse model of systemic inflammation [17], we expected the TUDCA injections not only to reduce inflammation in the spinal cord [31], but also to stimulate the differentiation of the alternative phenotype of microglia. This was not the case. Rather, we observed at 4 dpo, a significant reduction of two classical M2 markers (arginase-1, IL-4Rα) and no effect on the third (IL-10). In cell culture experiments with microglia, bacterial lipopolysaccharides (LPS) increased IL-4α and IL-10 expression. TUDCA did not change this significantly and had no effect in the absence of LPS [17]. However, in the same study, in vivo injections of TUDCA induced arginase-1 IR and IL-10 mRNA (but not IL-4Rα) in the hippocampus of LPS-treated mice. Thus, the so-called "M2 phenotype" may not be a useful concept [44], and it appears as if TGR5 activation has different effects depending on other prevailing stimuli in the environment. At this point, we have no conclusive concept regarding the role of bile acids or even TGR5 in microglia differentiation.

Bile acids are also cytoprotective. This seems to involve signaling kinases of Akt and PI3K [40,45] and may be independent of the inhibition of NFκB and inflammatory signals. SCI studies with TUDCA showed a reduction of apoptosis [22,25,26,32], and our data with TUNEL staining confirms it. Again, the absence of lasting effects on cell survival and motor recovery suggests that the relevance of this for SCI is limited.

Despite this sobering assessment of our results, TGR5 may still be considered a target in neuropathologies. Prolonged treatment with bile acids [32] is an option to be explored. In rodents and primates, TUDCA is not a physiological, endogenous signal. It is conceivable that other bile acids and synthetic TGR5 ligands [46,47] reduce apoptosis more potently and elicit unknown benefits because cAMP, the second messenger of TGR5, is implicated in multiple pro-regenerative pathways. Inhibitory signals for axonal growth may be overcome by this mechanism [32]. Recently, oleanolic acid was successfully tested in a mouse model of SCI [48]. Bile acid applications are being investigated in other neuropathologies [49]. Apart from TGR5 signaling, bile acids activate a variety of nuclear receptors [15,50], and some of these are also promising targets in the context of SCI [51–54].

5. Conclusions

Treatment of SCI with the bile acid TUDCA reduces inflammation and improves recovery of autonomic and motor functions in the subacute phase. No lasting therapeutic benefits were observed in the chronic phase or in combinatorial treatment with bmSC. Therefore, our data do not support the use of TUDCA as a treatment of SCI, at least when this is done in the acute phase only. The efficacy of alternative agonists of the bile acid receptors TGR5 and FXR should be investigated.

Supplementary Materials: The following are available online at https://www.mdpi.com/article/10.3390/biomedicines10071501/s1, Table S1: Effect of SCI and TUDCA treatment on the expression of genes associated with microglial activation; Figure S1: SCI-induced apoptosis as shown with TUNEL staining.

Author Contributions: Conceptualization, J.M., L.R.-R. and B.W.K.; investigation, S.W., J.M. and C.G.-R.; resources, J.M., L.R.-R., J.P.J.M.d.M. and E.C.W.; writing-original draft preparation, J.M.; writing-review and editing, J.M., L.R.-R. and B.W.K.; visualization, J.M.; supervision, J.M. and L.R.-R.; funding acquisition, J.M. and L.R.-R. All authors have read and agreed to the published version of the manuscript.

Funding: The study was funded by national research grant SAF2017-89366-R (MINECO, AEI/FEDER /UE; Spain) to J.M. and L.R.-R., S.W. received a PhD scholarship from the Chinese Scholarship Council, award 201606300031. Publishing costs were paid by the Hospital Nacional de Parapléjicos/Servicio de Salud de Castilla la Mancha.

Institutional Review Board Statement: The animal study protocol was approved by the Institutional Ethics Committee of the Hospital Nacional de Parapléjicos (163CEEA/2017). All procedures for collection of human bone marrow were approved by the ethics committee of Maastricht University Medical Center (METC 13-2-032).

Informed Consent Statement: Not applicable.

Data Availability Statement: Data supporting the reported results are available from the corresponding author upon reasonable request.

Acknowledgments: The Animal, Cytometry and Microscopy core facilities of the HNP and their staff provided excellent technical service. Natalia Díaz Rubio helped with data evaluation.

Conflicts of Interest: J.P.J.M.d.M. and E.C.W. are co-owners of Neuroplast BV. S.W., C.G.-R., L.R.-R., B.W.K. and J.M. declare that they have no conflict of interest.

Abbreviations

36B4	acidic ribosomal phosphoprotein P0
Akt	serine/threonine protein kinase encoded by the oncogene in retrovirus isolated from stock A strain k mouse thymoma cell line (protein kinase B)
ANOVA	analysis of variance
BBB	Basso/Beattie/Bresnahan locomotor rating scale for the assessment of hind limb motor function in rats
bmSC	bone marrow-derived stromal cells
CCL	chemokine C-C motif ligand
CD	cluster of differentiation
DMSO	dimethyl sulfoxide
dpo	days post operation
EDTA	ehylenediaminetetraacetic acid
GFAP	glial fibrillary acidic protein
GMP	good manufacturing practice
Iba-1	ionized calcium-binding adapter molecule 1
IFN	interferon
IF	immunofluorescence
IgG	immunoglobulin G
IH	Infinite Horizon spinal cord impactor
IL	interleukin
i.p.	intraperitoneal
IR	immunoreactivity
LPS	lipopolysaccharide
NFκB	nuclear factor kappa B
PBS	phosphate buffered saline
PCR	polymerase chain reaction
PFA	paraformaldehyde
PI3K	phosphoinositide 3-kinase
PKA	serine-threonine protein kinase A family
PWT	paw withdrawal threshold
ROI	region of interest
RT	room temperature
s.c.	subcutaneous
SCI	spinal cord injury
SEM	standard error of the mean
T200, T600, T1500	treatment with TUDCA 2 × 100 mg/kg, 2 × 300 mg/k and 5 × 300 mg/kg respectively in the present study
TBS-T	Tris-buffered saline/Tween 20
TGF	transforming growth factor
TGR5	Takeda G protein-coupled receptor-5
TNF	tumor necrosis factor
TRITC	tetramethyl rhodamine iso-thiocyanate
TUDCA	tauroursodeoxycholic acid
TUNEL	terminal deoxynucleotidyl transferase dUTP nick end labeling
W	weeks after SCI

References

1. Donnelly, D.J.; Popovich, P.G. Inflammation and its role in neuroprotection, axonal regeneration and functional recovery after spinal cord injury. *Exp. Neurol.* **2008**, *209*, 378–388. [CrossRef] [PubMed]
2. Falavigna, A.; Quadros, F.W.; Teles, A.R.W.C.C.; Barbagallo, G.; Brodke, D.; Al Mutair, A.; Riew, K.D. Worldwide steroid prescription for acute spinal cord injury. *Glob. Spine J.* **2018**, *8*, 303–310. [CrossRef] [PubMed]
3. Hugenholtz, H.; Gass, D.E.; Dvorak, M.F.; Fewer, D.H.; Fox, R.J.; Izukawa, D.M.; Lexchin, J.; Tuli, S.; Bharatwal, N.; Short, C. High-dose methylprednisolone for acute closed spinal cord injury–only a treatment option. *Can. J. Neurol. Sci.* **2002**, *29*, 227–235. [CrossRef] [PubMed]

4. Tetzlaff, W.; Okon, E.B.; Karimi-Abdolrezaee, S.; Hill, C.E.; Sparling, J.S.; Plemel, J.R.; Plunet, W.T.; Tsai, E.C.; Baptiste, D.; Smithson, L.J.; et al. A systematic review of cellular transplantation therapies for spinal cord injury. *J. Neurotrauma* **2011**, *28*, 1611–1682. [CrossRef] [PubMed]

5. Shende, P.; Subedi, M. Pathophysiology, mechanisms and applications of mesenchymal stem cells for the treatment of spinal cord injury. *Biomed. Pharmacother.* **2017**, *91*, 693–706. [CrossRef]

6. Tsintou, M.; Dalamagkas, K.; Makris, N. Taking central nervous system regenerative therapies to the clinic: Curing rodents versus nonhuman primates versus humans. *Neural Regen. Res.* **2019**, *15*, 425–437.

7. Munter, J.D.; Beugels, J.; Munter, S.; Jansen, L.; Cillero-Pastor, B.; Moskvin, O.; Brook, G.; Pavlov, D.; Strekalova, T.; Kramer, B.W.; et al. Standardized human bone-marrow-derive stem cells infusion improves survival and rocovery in a rat model of spinal cord injury. *J. Neurolog. Sci.* **2019**, *402*, 16–29. [CrossRef]

8. Romero-Ramírez, L.; Wu, S.; de Munter, J.; Wolters, E.C.; Kramer, B.W.; Mey, J. Treatment of rats with spinal cord injury using human bone marrow-derived stromal cells prepared by negative selection. *J. Biomed. Sci.* **2020**, *27*, 35. [CrossRef]

9. Basso, D.M.; Beattie, M.S.; Bresnahan, J.C. A sensitive and reliable locomotor rating scale for open field testing in rats. *J. Neurotrauma* **1995**, *12*, 1–21. [CrossRef]

10. Feng, Y.; Siu, K.; Wang, N.; Ng, K.-M.; Tsao, S.-W.; Nagamatsu, T.; Tong, Y. Bear bile: Dilemma of traditional medicinal use and animal protection. *J. Ethnobiol. Ethnomed.* **2009**, *5*, 2. [CrossRef]

11. Grant, S.; DeMorrow, S. Bile acid signaling in neurodegenerative and neurological disorders. *Int. J. Mol. Sci.* **2020**, *21*, 5982. [CrossRef] [PubMed]

12. Keitel, V.; Gorg, B.; Bidmon, H.J.; Zemtsova, I.; Spomer, L.; Zilles, K.; Haussinger, D. The bile acid receptor TGR5 (Gpbar-1) acts as a neurosteroid receptor in brain. *Glia* **2010**, *58*, 1794–1805. [CrossRef] [PubMed]

13. Eggink, H.M.; Soester, M.R.; Pols, T.W.H. TGR5 ligands as potential therapeutics in inflammatory diseases. *Int. J. Interferon Cytokine Mediat. Res.* **2014**, *6*, 27–38.

14. Pols, T.W.H.; Noriega, L.G.; Nomura, M.; Auwerx, J.; Schoonjans, K. The bile acid membrane receptor TGR5: A valuable metabolic target. *Dig. Dis.* **2011**, *29*, 37–44. [CrossRef]

15. De Marino, S.; Carino, A.; Masullo, D.; Finamore, C.; Marchianò, S.; Cipriani, S.; Di Leva, F.S.; Catalanotti, B.; Novellino, E.; Limongelli, V.; et al. Hyodeoycholic acid derivatives as liver X receptor α and G-protein-coupled bile acid receptor agonists. *Sci. Rep.* **2017**, *7*, 43290. [CrossRef]

16. Chiang, J.Y.L. Bile acid metabolism and signaling. *Compr. Physiol.* **2013**, *3*, 1191–1212.

17. Yanguas-Casás, N.; Barreda-Manso, M.A.; Nieto-Sampedro, M.; Romero-Ramírez, L. TUDCA: An agonist of the bile acid receptor GPBAR1/TGR5 with anti-inflammatory effects in microglial cells. *J. Cell Physiol.* **2017**, *232*, 2231–2245. [CrossRef]

18. Romero-Ramírez, L.; García-Rama, C.; Wu, S.; Mey, J. Bile acids attenuate PKM2 pathway activation in proinflammatory microglia. *Sci. Rep.* **2022**, *12*, 1459. [CrossRef]

19. Rosa, A.I.; Duarte-Silva, S.; Silva-Fernandes, A.; Nunes, M.J.; Carvalho, A.N.; Rodrigues, E.; Gama, M.J.; Rodrigues, C.M.P.; Marciel, P.; Castro-Caldas, M. Tauroursodeoxycholic acid improves motor symptoms in a mouse model of Parkinson's disease. *Mol. Neurobiol.* **2018**, *55*, 9139–9155. [CrossRef]

20. Bhargava, P.; Smith, M.D.; Mische, L.; Harrington, E.; Fitzgerald, K.C.; Martin, K.; Kim, S.; Reyes, A.A.; González-Cardona, J.; Volsko, C.; et al. Bile acid metabolism is altered in multiple sclerosis and supplementation ameliorates neuroinflammation. *J. Clin. Investig.* **2020**, *130*, 3467–3482. [CrossRef]

21. Elia, A.E.; Lalli, S.; Monsurro, M.R.; Sagnelli, A.; Taiello, A.C.; Reggiori, B.; La, B.V.; Tedeschi, G.; Albanese, A. Tauroursodeoxycholic acid in the treatment of patients with amyotrophic lateral sclerosis. *Eur. J. Neurol.* **2016**, *23*, 45–52. [CrossRef] [PubMed]

22. Çolak, A.; Kelten, B.; Sagmanligil, A.; Akdemir, O.; Karaoglan, A.; Sahan, E.; Çelik, Ö.; Barut, S. Tauroursodeoxycholic acid and secondary damage after spinal cord injury in rats. *J. Clin. Neurosci.* **2008**, *15*, 665–671. [CrossRef] [PubMed]

23. Dong, Y.; Miao, L.; Lin, L.; Ding, H. Neuroprotective effects impact on caspase-12 expression of tauroursodeoxycholic acid after acute spinal cord injury in rats. *Int. J. Clin. Exp. Pathol.* **2015**, *8*, 15871–15878. [PubMed]

24. Miao, L.; Dong, L.; Zhou, F.-B.; Chang, Y.-L.; Suo, Z.G.; Ding, H.Q. Protective effect of tauroursodeoxycholic acid on the aotophagy of nerve cells in rats with acute spinal cord injury. *Eur. Rev. Med. Pharmacol. Sci.* **2018**, *22*, 1133–1141.

25. Zhang, Z.; Chen, J.; Chen, F.; Yu, D.; Li, R.; Lv, C.; Wang, H.; Li, H.; Li, J.; Cai, Y. Tauroursodeoxycholic acid alleviates secondary injury in the spinal cord via up-regulation of CIBZ gene. *Cell Stress Chaperones* **2018**, *23*, 560. [CrossRef]

26. Dong, Y.; Yang, S.; Fu, B.; Zhou, S.; Ding, H.; Ma, W. Mechanism of tauroursodeoxycholic acid-mediated neuronal protection after acute spinal cord injury through AKT signaling pathway in rats. *Int. J. Clin. Exp. Pathol.* **2020**, *13*, 2218–2227.

27. Mann, A.; Chesselet, M.-F. Techniques in motor assessment in rodents. In *Movement Disorders—Genetics and Models*; LeDouy, M.S., Ed.; Elsevier: London, UK, 2015.

28. Pikov, V.; Wrathall, J.R. Coordination of the bladder detrusor and the external urethral sphincter in a rat model of spinal cord injury: Effects of injury severity. *J. Neurosci.* **2001**, *21*, 559–569. [CrossRef]

29. Pols, T.W.H.; Nomura, M.; Harach, T.; Lo Sasso, G.; Ooosterveer, M.H.; Thomas, C.; Rizzo, G.; Gioiello, A.; Adorini, L.; Pellicari, R.; et al. TGR5 activation inhibits atherosclerosis by reducing macrophage inflammation and lipid loading. *Cell Metab.* **2011**, *14*, 747–757. [CrossRef]

30. Yanguas-Casás, N.; Barreda-Manso, M.A.; Nieto-Sampedro, M.; Romero-Ramírez, L. Tauroursodeoxycholic acid reduces glial cell activation in an animal model of acute neuroinflammation. *J. Neuroinflamm.* **2014**, *11*, 50. [CrossRef]

31. Kim, S.J.; Ko, W.K.; Jo, M.J.; Arai, Y.; Choi, H.; Kumar, H.; Han, I.B.; Sohn, S. Anti-inflammatory effect of Tauroursodeoxycholic acid in RAW 264.7 macrophages, Bone marrow-derived macrophages, BV2 microglial cells, and spinal cord injury. *Sci. Rep.* **2018**, *8*, 3176. [CrossRef]

32. Hou, Y.; Luan, J.; Huang, T.; Deng, T.; Li, X.; Xiao, Z.; Zhan, J.; Luo, D.; Hou, Y.; Xu, L.; et al. Tauroursodeoxycholic acid alleviates secondary injury in spinal cord injury mice by reducing oxidative stress, apoptosis, and inflammatory response. *J. Neuroinflamm.* **2021**, *18*, 216. [CrossRef] [PubMed]

33. Hayda, E.; Elden, H. Acute spinal cord injury: A review of pathophysiology and potential of non-steroidal ant-inflammatory drugs for pharmacological intervention. *J. Chem. Neuroanat.* **2018**, *87*, 25–31. [CrossRef] [PubMed]

34. Benevento, B.T.; Sipski, M.L. Neurogenic bladder, neurogenic bowel, and sexual dysfunction in people with spinal cord injury. *Phys. Ther.* **2002**, *82*, 601–612. [CrossRef]

35. Fowler, C.J.; Griffiths, D.; de Groat, W.C. The neuronal control of micturition. *Nat. Rev. Neurosci.* **2008**, *9*, 453–466. [CrossRef] [PubMed]

36. Mitsui, T.; Shumsky, J.S.; Lepor, A.C.; Murray, M.; Fisher, I. Transplantation of neuronal and glial restricted precursors into contused spinal cord improves bladder and motor functions, decreases thermal hypersensitivity and modifies intraspinal circuitry. *J. Neurosci.* **2005**, *25*, 9624–9636. [CrossRef] [PubMed]

37. Redshaw, J.D.; Lenjerr, S.M.; Elliott, S.P.; Stoffel, J.T.; Rosenbluth, J.P.; Presson, A.P.; Myers, J.B. Protocol for a randomized clinical trial investigating early sacral nerve stimnulation as an adjunct to standard neurogenic bladder management following acute spinal cord injury. *BMC Urol.* **2018**, *18*, 72. [CrossRef]

38. Ditunno, J.F.; Little, J.W.; Tessler, A.; Burns, A.S. Spinal shock revisited: A four-phase model. *Spinal Cord* **2004**, *42*, 383–395. [CrossRef]

39. Yanguas-Casás, N.; Barreda-Manso, M.A.; Perez-Rial, S.; Nieto-Sampedro, M.; Romero-Ramírez, L. TGFbeta Contributes to the Anti-inflammatory Effects of Tauroursodeoxycholic Acid on an Animal Model of Acute Neuroinflammation. *Mol. Neurobiol.* **2017**, *54*, 6737–6749. [CrossRef]

40. Castro, J.; Harrington, A.M.; Lieu, T.; García-Caraballo, S.; Maddern, J.; Schober, G.; O'Donnell, T.; Grundy, L.; Lumsden, A.L.; Miller, P.; et al. Activation of pruritogenic TGR5, MRGPRA3, and MRGPRC11 on colon-innervating afferents induces visceral hypersensitivity. *JCI Insight* **2019**, *4*, e131712. [CrossRef]

41. Freilich, R.W.; Woodbury, M.E.; Ikezu, T. Integrated expression profiles of mRNA and miRNA in polarized primary murine microglia. *PLoS ONE* **2013**, *8*, e79416. [CrossRef]

42. Siddiqui, T.; Lively, S.; Schlichter, L.C. Complex molecular and functional outcomes of single versus sequential cytokine stimulation of rat microglia. *J. Neuroinflamm.* **2016**, *13*, 66. [CrossRef] [PubMed]

43. Ghosh, M.; Xu, Y.; Pearse, D.D. Cyclic AMP is a key regulator of M1 to M2a phenotypic conversion of microglia in the presence of Th2 cytokines. *J. Neuroinflamm.* **2016**, *13*, 9. [CrossRef] [PubMed]

44. Ransohoff, R.M. A polarizing question: Do M1 and M2 microglia exist? *Nat. Neurosci.* **2016**, *19*, 987–991. [CrossRef] [PubMed]

45. Sun, D.; Gu, G.; Wang, J.; Chai, Y.; Fan, Y.; Yang, M.; Xu, X.; Gao, W.; Li, F.; Yin, D.; et al. Administration of taurourosdoxycholic acid attenuates early brain injury via Akt pathway activation. *Front. Cell. Neurosci.* **2017**, *11*, 193. [CrossRef] [PubMed]

46. Guo, C.; Chen, W.-D.; Wang, Y.-D. TGR5, not only a metabolic regulator. *Front. Physiol.* **2016**, *7*, 646. [CrossRef]

47. Terui, R.; Yanase, Y.; Yokoo, H.; Suhara, Y.; Makishima, M.; Demizu, Y.; Misawa, T. Development of selective TGR5 ligands based on the 5,6,7,7-tetrahydro-5,5,8,8-tetramethylnaphthalene skeleton. *ChemMedChem* **2020**, *16*, 458–462. [CrossRef]

48. Wang, J.-L.; Ren, C.-H.; Feng, J.; Ou, C.-H.; Liu, L. Oleanolic acid inhibits mouse spinal cord injury through suppressing inflammation and apoptosis via the blockage of p38 and JNK MAPKs. *Biomed. Pharmacother.* **2020**, *123*, 109752. [CrossRef]

49. Ackerman, H.D.; Gerhard, G.S. Bile disorders in neurodegenerative disorders. *Front. Aging Neurosci.* **2016**, *8*, 263. [CrossRef]

50. Wang, H.; Chen, J.; Hollister, K.; Sowers, L.C.; Forman, B.M. Endogenous bile acids are ligands for the nuclear receptor FXR/BAR. *Mol. Cell* **1999**, *3*, 543–553. [CrossRef]

51. Mey, J. A new target for CNS injury? The role of retinoic acid signaling after nerve lesions. *J. Neurobiol.* **2006**, *66*, 757–779. [CrossRef]

52. Van Neerven, S.; Kampmann, E.; Mey, J. RAR/RXR and PPAR/RXR signaling in neurological and psychiatric diseases. *Prog. Neurobiol.* **2008**, *85*, 433–451. [CrossRef] [PubMed]

53. Paterniti, I.; Genovese, T.; Mazzon, E.; Crisafulli, C.; Di Paola, R.; Galuppo, M.; Bramanti, P.; Cuzzocrea, S. Liver X receptor agonist treatment regulates inflammatory response after spinal cord trauma. *J. Neurochem.* **2010**, *112*, 611–624. [CrossRef] [PubMed]
54. Fandel, D.; Wasmuht, D.; Ávila-Martín, G.; Taylor, M.D.; Galán-Arriero, I.; Mey, J. Spinal cord injury-induced changes of nuclear receptors PPARalpha and LXRbeta and modulation with oleic acid/albumin treatment. *Brain Res.* **2013**, *1535*, 89–105. [CrossRef] [PubMed]

 *biomedicines*

Article

TIRAP/Mal Positively Regulates TLR8-Mediated Signaling via IRF5 in Human Cells

Kaja Elisabeth Nilsen [1,2,†], Astrid Skjesol [1,2,†], June Frengen Kojen [1,2], Terje Espevik [1,2], Jørgen Stenvik [1,2,3,‡] and Maria Yurchenko [1,2,3,*,‡]

[1] Centre of Molecular Inflammation Research, Norwegian University of Science and Technology, NO-7491 Trondheim, Norway; kaja.e.nilsen@ntnu.no (K.E.N.); astrid.skjesol@gmail.com (A.S.); june.f.kojen@ntnu.no (J.F.K.); terje.espevik@ntnu.no (T.E.); jorgen.stenvik@ntnu.no (J.S.)

[2] Department of Clinical and Molecular Medicine, Norwegian University of Science and Technology, NO-7491 Trondheim, Norway

[3] Department of Infectious Diseases, Clinic of Medicine, St. Olavs Hospital HF, Trondheim University Hospital, NO-7006 Trondheim, Norway

* Correspondence: mariia.yurchenko@ntnu.no

† These authors contributed equally to this work.

‡ These authors contributed equally to this work.

Abstract: Toll-like receptor 8 (TLR8) recognizes single-stranded RNA of viral and bacterial origin as well as mediates the secretion of pro-inflammatory cytokines and type I interferons by human monocytes and macrophages. TLR8, as other endosomal TLRs, utilizes the MyD88 adaptor protein for initiation of signaling from endosomes. Here, we addressed the potential role of the Toll-interleukin 1 receptor domain-containing adaptor protein (TIRAP) in the regulation of TLR8 signaling in human primary monocyte-derived macrophages (MDMs). To accomplish this, we performed *TIRAP* gene silencing, followed by the stimulation of cells with synthetic ligands or live bacteria. Cytokine-gene expression and secretion were analyzed by quantitative PCR or Bioplex assays, respectively, while nuclear translocation of transcription factors was addressed by immunofluorescence and imaging, as well as by cell fractionation and immunoblotting. Immunoprecipitation and Akt inhibitors were also used to dissect the signaling mechanisms. Overall, we show that TIRAP is recruited to the TLR8 Myddosome signaling complex, where TIRAP contributes to Akt-kinase activation and the nuclear translocation of interferon regulatory factor 5 (IRF5). Recruitment of TIRAP to the TLR8 signaling complex promotes the expression and secretion of the IRF5-dependent cytokines IFNβ and IL-12p70 as well as, to a lesser degree, TNF. These findings reveal a new and unconventional role of TIRAP in innate immune defense.

Keywords: TLR8; TIRAP; Akt; human macrophages; IRF5; Akt inhibitors

Citation: Nilsen, K.E.; Skjesol, A.; Frengen Kojen, J.; Espevik, T.; Stenvik, J.; Yurchenko, M. TIRAP/Mal Positively Regulates TLR8-Mediated Signaling via IRF5 in Human Cells. *Biomedicines* **2022**, *10*, 1476. https://doi.org/10.3390/biomedicines10071476

Academic Editor: Krisztina Nikovics

Received: 13 May 2022

Accepted: 17 June 2022

Published: 22 June 2022

Publisher's Note: MDPI stays neutral with regard to jurisdictional claims in published maps and institutional affiliations.

1. Introduction

Toll-like receptors (TLRs) are one of the most studied groups of pattern-recognition receptors (PRRs), which recognize pathogen-associated molecular patterns and damage-associated molecular patterns. TLRs provide protection against both external and internal threats by initiating a pro-inflammatory response as well as activating and guiding the adaptive immune system to mount effector responses that will eliminate and/or ameliorate the problem [1]. Dysfunction or dysregulation of TLR responses can have dire consequences for the host, such as increased susceptibility to infection or excessive life-threatening inflammation. Thus, understanding how these receptors operate holds great potential for guiding preventive measures as well as the development of better treatments and new drugs for infectious and autoimmune diseases, cancer, and cardiovascular disease [2]. Through the years, some TLRs have received more attention than others, often due to the availability of easily accessible models and their convenience for study, but with technological and scientific advances, this gap is closing. TLR8 provides a classic example of a less-studied TLR as it is non-functional in rodents and, thus, application of murine models/KOs is limited [3].

TLR8 is expressed by several immune cells, in particular monocytes, macrophages, myeloid dendritic cells, and neutrophils [4–7]. Localized at the endosomes, TLR8 recognizes ribonuclease T2 degradation products of single-stranded RNA (ssRNA) of various origin: viral, bacterial, protozoan, and (possibly human) endogenous RNA [5,8–12]. With this repertoire of potential ligands, TLR8 might be relevant for the defense against a broad range of infections, as well as for driving autoimmune diseases [3,6,11]. TLR8 ligand-binding in the endosomal lumen induces conformational changes that result in the dimerization of the cytosolic TIR domains. This allows myeloid differentiation primary response gene 88 (MyD88) to bind, followed by the recruitment of interleukin-1 receptor-associated kinase 4 (IRAK4) and interleukin-1 receptor-associated kinase 1 (IRAK1), resulting in the formation of the active Myddosome complex. The signal is transduced via TNF-receptor-associated factor 6 (TRAF6) and transforming growth-factor-β-activated kinase 1 (TAK1), which activates downstream mitogen-activated protein kinase (MAPK) cascades and inhibitor of nuclear-factor kappa B kinase subunit beta (IKKβ), culminating in the activation of transcription factors such as nuclear-factor kappa B (NF-κB) and activator-protein 1 (AP-1) [1]. In human primary macrophages, TLR8-activated TAK1 also signals via IKKβ to induce the nuclear translocation of interferon regulatory factor 5 (IRF5), which is critical for the expression of interferon β (*IFNβ*) and interleukin-12 subunit alpha (*IL-12A*) genes [13].

The phosphoinositide 3-kinase (PI3K)-Akt serine/threonine kinase pathway is well-known to be involved in the regulation of metabolism and survival, and its dysregulation is closely linked to tumor development [14]. It has also been implicated in regulating TLR-mediated responses, although the reports are divergent, with evidence of both pro- and anti-inflammatory effects [15]. Aksoy et al., found that in human monocyte-derived DCs, PI3K negatively regulated the expression of *IFNβ* in TRIF-dependent signaling downstream of TLR3 and TLR4 [16], whereas Guiducci et al., found PI3K–Akt necessary for nuclear translocation of IRF7 and expression of TLR7 and TLR9-induced type I IFNs in human plasmacytoid predendritic cells [17]. These studies highlight the importance of taking cell-type specific differences into account when investigating the role of PI3K–Akt in TLR signaling.

TLR2 and TLR4 require a Toll-internleukin-1-receptor (TIR) domain containing adaptor protein/MyD88 adaptor-like (TIRAP) to attract MyD88 to the signaling complex. For some time, the restricted role of TIRAP for these plasma-membrane-localized TLRs seemed apparent, given the phosphatidylinositol-4,5-bisphosphate (PI(4,5)P2)-binding motif in the N-terminal domain of TIRAP, which attracts this adaptor protein to the plasma membrane where PI(4,5)P2 is abundant [18,19]. However, Bonham et al., have shown that TIRAP is also capable of binding other phosphoinositides, PI(3)P and PI(3,5)P2, on endosomal membranes, and mediates signaling from endosomal TLR7 and TLR9 in murine cells [20]. Recently, some evidence of TIRAP involvement in signaling from TLR7 and TLR9 in human cells was provided by Leszczynska et al., and Zyzak et al. [21,22]. In both studies, TIRAP is suggested to regulate *IFNβ* expression by regulating ERK1/2 (MAPK3/MAPK1) activation. Whether TIRAP could contribute to signaling from TLR8, the third member of the TLR7 subgroup of TLRs, has not yet been explored.

Here we show that TIRAP is regulating the expression and secretion of TLR8-induced IFNβ and IL-12A cytokines in human primary monocytes and monocyte-derived macrophages (MDMs). We propose that TIRAP is recruited to the activated Myddosome, from where it connects to Akt activation, contributing to the nuclear translocation of IRF5 and subsequent expression of *IFNβ* and *IL-12A* genes. In addition, TIRAP can enhance TLR8 signaling via the TAK1-pathway, thus modulating the expression of other cytokines such as TNF.

2. Materials and Methods

2.1. Cells and Reagents

Human buffy coats and serum were from the blood bank at St. Olavs Hospital (Trondheim, Norway), with approval by the Regional Committee for Medical and Health Research Ethics (REC) in Central Norway (no. 2009/2245). Primary human monocytes were isolated from the buffy coat by adherence, as previously described [23]. Monocytes were maintained in RPMI1640 (Sigma, Merck, Darmstadt, Germany), supplemented with 30% of pooled human serum. MDMs (used in the *TIRAP* silencing experiments) were obtained by differentiating cells for 12 days in RPMI1640 with 10% human serum and 20 ng/mL rhM-CSF (#216-MC-025, R&D Systems, Minneapolis, MN, USA). Ultrapure K12 LPS from *E. coli*, thiazoloquinoline compound CL075, and synthetic diacylated lipoprotein FSL-1 (Pam2CGDPKHPKSF) were from InvivoGen (San Diego, CA, USA). For stimulation of the primary cells, LPS and FSL-1 were used at concentration 100 ng/mL, CL075—2 µg/mL. IRAK4 inhibitor PF-06426779 (Merck, Darmstadt, Germany), selective allosteric pan-Akt inhibitors MK-2206 (#1032350-13-2, Axon Medchem, Groningen, Netherlands), Miransertib) and ATP-competitive pan-Akt inhibitor Capivasertib ((#1313881-70-7 and #1143532-39-1, MedChemExpress, Sollentuna, Sweden) were diluted in DMSO at concentration 5 mM and stored at -80 °C; working solutions were prepared in cell-culture media immediately before use. Preparation of THP-1 *TIRAP* KO cells using LentiCRISPRv2 plasmid [24] is described in supplementary materials and methods (available online).

2.2. Antibodies

The following primary antibodies were used: mouse GAPDH (ab9484), rabbit β-tubulin (ab6046) from Abcam (Cambridge, UK); rabbit phospho-Akt Ser473 (D9E XP), phospho-p38 MAPK (T180/Y182), phospho-STAT1 (Tyr701) (58D6), phospho-TAK1 (T184/187) (90C7), phospho-JNK (81E11) (T183/Y185), phospho-p44/42 MAPK (ERK1/2) (Thr202/Tyr204), IκBα (44D4), IRAK1 (D51G7), MyD88 (D80F5), Histone H3 (3H1), and phospho-NF-κB p65 (Ser536) (93H1) from Cell Signaling Technology (Danvers, MA, USA); rabbit PCNA Abs were from Santa Cruz Biotech (Santa Cruz, CA, USA); sheep IRF5 and IRAK4 were from MRC-PPU Reagents (University of Dundee, Dundee, UK); goat TIRAP polyclonal Abs were from Invitrogen (#PA5-18439, Waltham, MA, USA), and mouse STAT1 antibodies were from BD Biosciences (#610185, Wokingham, UK). Secondary antibodies (HRP-linked) were from DAKO Denmark A/S (Glostrup, Denmark).

2.3. siRNA Treatment

Oligos used for silencing were AllStars Negative Control siRNA (SI03650318) and FlexiTube Hs_TIRAP_10 siRNA (SI03075135) (QIAGEN, Germantown, AR, USA). PBMCs were seeded in 24-well plates (NUNC, ThermoFisher Scientific, Waltham, MA, USA), 1.5×10^6 cells per well, and differentiated to MDMs as described above. On day 7 and day 9, cells were transfected by silencing and control oligo (20 nM final concentration) using Lipofectamine 3000 (Invitrogen, ThermoFisher Scientific, Waltham, MA, USA), as suggested by the manufacturer. Cells were stimulated by LPS, CL075 or FSL-1, or used for bacterial infections in 48 h after second transfection.

2.4. RT-qPCR

Total RNA was isolated from the cells using Qiazol reagent (QIAGEN, Germantown, AR, USA), and chloroform extraction was followed by purification on RNeasy Mini columns with DNAse digestion step (QIAGEN). cDNA was prepared with a Maxima First Strand cDNA Synthesis Kit for a quantitative real-time polymerase-chain reaction (RT-qPCR) (ThermoFisher Scientific, Waltham, MA, USA), in accordance with the protocol of the manufacturer, from 400–600 ng of total RNA per sample. Q-PCR was performed using the PerfeCTa qPCR FastMix (Quanta Biosciences, Gaithersburg, MD, USA) in replicates and cycled in a StepOnePlus™ Real-Time PCR cycler (ThermoFisher Scientific, Waltham, MA, USA). The following TaqMan® Gene Expression Assays

(Applied Biosystems®, ThermoFisher Scientific, Waltham, MA, USA) were used: *IFNβ* (Hs01077958_s1), *TNF* (Hs00174128_m1), *TBP* (Hs00427620_m1), *TIRAP* (Hs00364644_m1), *IL-6* (Hs00985639_m1), *IL-1β* (Hs01555410_m1), *IL-12A* (Hs01073447_m1), and *IL-12B* (Hs01011518_m1). The level of *TBP* mRNA was used for normalization and the results presented as a relative expression compared to the control's untreated sample. Relative expression was calculated using Pfaffl's mathematical model [25]. Graphs and statistical analyses were made with GraphPad Prism v9.1.2 (Dotmatics, Bishops Stortford, UK), with additional details provided in the figure legends and statistics paragraph (Section 2.11).

2.5. ELISA and BioPlex Assays

TNF level in supernatants of human macrophages was determined using human TNF-alpha DuoSet ELISA (DY210-05) (R&D Systems, Minneapolis, MN, USA), IFNβ level—using VeriKine-HSTM Human Interferon-Beta Serum ELISA Kit from PBL Assay Science (Piscataway, NJ, USA). Other cytokines (IL-12p70, IL-6, MCP-1, IL-8) were analyzed using BioPlex cytokine assays from Bio-Rad, in accordance with the instructions of the manufacturer, using the Bio-Plex Pro™ Reagent Kit III and Bio-Plex™ 200 System (Bio-Rad, Hercules, CA, USA).

2.6. Cell Fractionation

Stimulated human primary monocytes were detached by accutase treatment and collected by centrifugation. Cell pellets were washed by once by PBS with 2% FCS. For preparation of total lysate, a cell pellet from one of the wells per condition was collected and lysed by RIPA lysis buffer (150 mM NaCl, 50 mM TrisHCl (pH7.5), 1% Triton X100, 5 mM EDTA, protease inhibitors, phosphatase inhibitors). The remaining cells were resuspended in Buffer A (50 mM NaCl, 10 mM HEPES pH = 8, 500 mM sucrose, 1 mM EDTA, 0.2% Triton-X100), and the samples were vortexed and centrifuged (5000 rpm, 5 min, 4 °C). Cytosolic fraction in the supernatant was transferred to clean tubes. Pellets with nuclei were washed with Buffer B (50 mM NaCl, 10 mM HEPES (pH = 8), 25% glycerol, 0.1 mM EDTA), and centrifuged, the supernatants were discarded, and the pellets were resuspended in Buffer C (350 mM NaCl, 10 mM HEPES (pH = 8), 25% glycerol, 0.1 mM EDTA) with added Benzonaze endonuclease (Merck) and incubated on ice for 30 min, followed by centrifugation (15,000 rpm, 15 min, 4 °C) to extract nuclear proteins.

2.7. Western Blotting

Cell lysates for pSTAT1 analysis were prepared by simultaneous extraction of proteins and total RNA using Qiazol reagent (QIAGEN, Germantown, AR, USA), as suggested by the manufacturer. Extracted total RNA was used for RT-qPCR, while protein samples were used for simultaneous analysis of protein expression/post-translational modifications. Protein pellets were dissolved by heating the samples for 10 min at 95 °C in a buffer containing 4 M urea, 1% SDS (Sigma, Merck, Darmstadt, Germany), and NuPAGE® LDS Sample Buffer (4X) (Thermo Fisher Scientific, Waltham, MA, USA), with a final 25 mM DTT in the samples. Otherwise, lysates were made using 1X RIPA lysis buffer. For traditional Western blot analysis, we used pre-cast protein gels NuPAGE™ Novex™. Proteins were transferred to iBlot Transfer Stacks by using the iBlot Gel Transfer Device (ThermoFisher Scientific, Waltham, MA, USA). The blots were developed with the SuperSignal West Femto (ThermoFisher Scientific) and visualized with the LI-COR ODYSSEY Fc Imaging System (LI-COR Biotechnology, Lincoln, NE, USA). For densitometry analysis of the bands, Odyssey Image Studio 5.2 software (LI-COR Biotechnology, Lincoln, NE, USA) was used, and the relative numbers of bands' intensity were normalized to the intensities of the respective loading-control protein (GAPDH, or PCNA, or β-tubulin). Loading-control-protein expression was always performed on the same membrane as the protein of interest.

2.8. Immunoprecipitations

PBMC-derived monocytes for endogenous IPs were lysed using 1 X lysis buffer (150 mM NaCl, 50 mM TrisHCl (pH 8.0), 1 mM EDTA, 1% NP40) and supplemented with EDTA-free Complete Mini protease Inhibitor Cocktail Tablets as well as a PhosSTOP phosphatase-inhibitor cocktail from Roche, with 50 mM NaF and 2 mM Na_3VO_3 (Sigma, Merck, Darmstadt, Germany). Immunoprecipitations (IPs) were carried out on rotator at +4 °C for 4 h by co-incubation of the lysates from the stimulated cells (400–500 μg of protein/IP) with specific anti-TIRAP antibodies covalently coupled to Dynabeads (M-270 Epoxy, Thermo Fisher Scientific, Waltham, MA, USA), as suggested by the manufacturer, followed by extensive washing of the beads in a lysis buffer. Co-precipitated complexes were eluted by heating the samples in a 1× loading buffer (LDS, Invitrogen), without reducing the reagent to minimize the antibodies' leakage to the eluates. Eluates were transferred to clean tubes, followed by the addition of DTT to the 40 mM concentration, heating, and Western blot analysis.

2.9. Immunofluorescence and Scan^R Analysis

Monocytes were isolated from PBMC using CD14 MicroBeads UltraPure (Miltenyi Biotech) and seeded in 96-well glass-bottom plates (P96-1.5H-N, Cellvis, CA, USA) at 50 K cells/well and in 24-well culture plates (250 K cells/well). Monocytes were differentiated into MDMs and transfected two times with a *TIRAP* siRNA and AllStar siRNA control, using the standardized protocol. MDMs were left untreated or stimulated with CL075 (Invivogen) at 2 μg/mL for 60 min, with four technical replicates per condition. Fixation, immunostaining with anti-human IRF5 mAb (Abcam, #10T1) and anti-human p65 XP mAb (Cell Signaling Technology, #8242, Danvers, MA, USA), and Scan^R high-throughput imaging (Olympus Europa SE & Co., Hamburg, Germany) were done, as previously described in detail [13]. Quantification of IRF5 nuclear translocation was done with Scan^R analysis software (v2.8.1) and calculated as a percentage of the positively stained nuclei multiplied by the mean fluorescence-intensity value (MFI) of the positively stained nuclei. Silencing efficiency of the TIRAP gene for each donor ($n = 6$) was examined by RT-qPCR, using the parallel 24-well plates.

2.10. Bacteria and Infection Experiments

Anonymized clinical isolates of GBS, *S. aureus*, and *E. coli* were from a diagnostic collection by the Department of Medical Microbiology, St. Olavs Hospital, Trondheim, Norway. For infection experiments, blood-agar colonies were picked and grown in Todd-Hewitt Broth (GBS) or Tryptic Soy Broth (*E. coli* and *S. aureus*), with vigorous shaking at 37 °C overnight. The bacteria cultures were diluted to the desired density (CFU/mL), based on OD600 measurements and calculations, as previously described in detail [26]. Cultures of MDMs in 24-well plates, with *TIRAP* siRNA or control siRNA pre-treatment, were incubated with bacteria at 37 °C for 60 min, before the killing of all extracellular bacteria with 100 μg/mL gentamicin. The MDM cultures were incubated further for a total challenge time of four hours. Cell lysis and RNA purification was done with the RNeasy 96 Plus kit (QIAGEN), followed by cDNA synthesis with Maxima cDNA synthesis kit (Thermo Fisher Scientific), and qPCR analysis of the cytokine expression.

2.11. Statistical Analysis

Data that were assumed to follow a log-normal distribution was log-transformed prior to statistical analysis. RT-qPCR was log-transformed and analyzed by Repeated Measurements Analysis of Variance (RM-ANOVA), or a mixed model if there was missing data, followed by Holm-Šídák's multiple comparisons post-test. Scan^R data were log-transformed and analyzed with a paired *t*-test (two-sided). ELISA data was analyzed using a Wilcoxon matched-pairs signed-rank test. All graphs and analyses were generated with GraphPad Prism v9.1.2 (Dotmatics, Bishops Stortford, UK).

3. Results

3.1. TIRAP-Silencing Decreases TLR8-Mediated Cytokine Expression and Secretion in Human Primary MDMs

Based on the reported links between endosomal TLRs and TIRAP expression [20–22], we have questioned whether TIRAP could also be involved in the regulation of signaling via TLR8 in human immune cells. To address this, we performed *TIRAP* silencing in human primary MDMs prior to stimulation with the thiazoquinoline compound CL075, which is a synthetic ligand specific for TLR8 in human monocytes and MDMs [27]. Analysis of cytokine mRNA expression following TLR8 stimulation in *TIRAP*-silenced cells showed a significant reduction in *IFNβ* and *IL-12A* mRNA expression, with the most significant effect on *IL-12A* expression (Figure 1). Of the pro-inflammatory cytokines assessed following TLR8 stimulation, *TIRAP* silencing only affected *TNF* mRNA expression, both at two hours and four hours (Figure 1). TLR2 and TLR4 stimulation was conducted in parallel, since TIRAP is an important bridging adaptor for these TLRs [18,28]. Indeed, *TIRAP* silencing led to decreased *TNF*, *IL-6*, *IL-1β*, *IL-12A*, and *IL-12B* mRNA expression following TLR2 and TLR4 stimulation (Figure S1a,b). TLR2 did not induce *IFNβ* mRNA expression or secretion in human MDMs, in agreement with earlier studies [29]. Notably, in the case of TLR2 and TLR4 stimulation, we have not observed a significant effect on the early induction of pro-inflammatory cytokines (Figure S1a,b), which could be due to the still-sufficient amount of residual TIRAP protein for the initiation of signaling, despite a marked decrease in TIRAP-protein expression in silenced MDMs (Figure S1c).

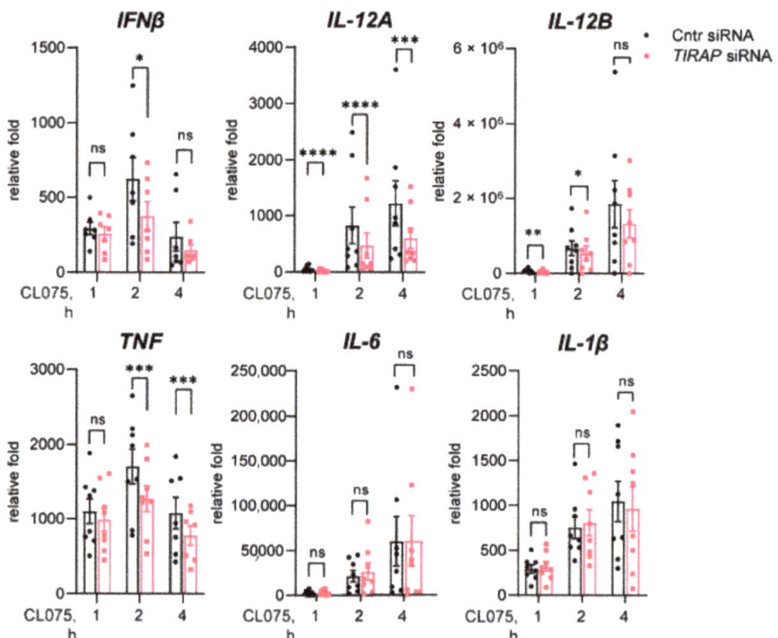

Figure 1. *TIRAP* silencing in primary human MDMs significantly decreases TLR8-mediated *IFNβ* and *IL-12A* expression. Macrophages were transfected with control or *TIRAP*-silencing oligo, followed by stimulation with TLR8 ligand CL075 (2 μg/mL) for the indicated time. RT-qPCR analysis of cytokine-gene expression after stimulation by CL075 in consecutive experiments with cells from different donors (*n* = 6–8). Data for cytokine expression induced by CL075 stimulation were normalized to untreated sample and presented as a mean relative fold change +SEM. Statistical testing was done by 2-way RM-ANOVA including a post-test, as described (* $p < 0.05$, ** $p < 0.01$, *** $p < 0.001$, **** $p < 0.0001$, and ns—non-significant).

To avoid problems with inefficient silencing, knockout-model (KO) systems are widely used. We prepared *TIRAP* KO THP-1 human monocyte/macrophage-like subline by Crispr/Cas9 gene editing, combined with subsequent TLR8 overexpression to achieve more robust type I IFNs induction in these cells. However, there was a noticeable change in TLR4 signaling in *TIRAP* KO THP-1 cells from two weeks (early) to four weeks (late) of cultivation in puromycin-selection medium (Figure S2a,b). TLR4-mediated *IFNβ* expression was inhibited in early *TIRAP* KO cells, while the effect was lost after prolonged cultivation (Figure S2a,b). The response to LPS in *TIRAP*-silenced THP-1 cells was more comparable to early *TIRAP* KO cells (Figure S2a,c). This instability of the KO cells could be due to compensatory mechanisms, and we, thus, considered silencing of the *TIRAP* gene as a better approach for pinpointing the impact of TIRAP in TLR signaling.

TLR2 and TLR4 induce the expression of pro-inflammatory cytokines via formation of the TIRAP/MyD88 complex [30]. Even though TLR4-dependent expression of *IFNβ* relies on TRAM/TRIF signaling from the endosomes, we observed that *TIRAP* silencing significantly reduced LPS-mediated *IFNβ* expression both in THP-1 cells and in primary human MDMs (Figures S1a and S2c) [31].

To follow the protein levels of the secreted cytokines, we performed ELISA or Multiplex assays using supernatants from cells stimulated by TLR ligands for four hours. Indeed, *TIRAP* silencing reduced TLR4- and TLR8-mediated IFNβ and IL-12p70 secretion (Figure 2), and it also decreased the phosphorylation of STAT1 (Figure S3), a transcription factor in IFN-α/β-receptor (IFNAR) signaling that may be used as a surrogate readout for IFNβ secretion [32]. These results are corroborating the mRNA data, suggestive of TIRAP involvement in regulation of IFNβ expression and secretion at later stages of signaling (2–4 h). TLR8-mediated *TNF* mRNA expression was also reduced upon *TIRAP* knockout (Figure 1), yet the level of secreted TNF was not significantly affected (Figure 2). Overall, our results indicate that TIRAP regulates TLR8-mediated signaling in human primary MDMs, with the strongest effect on expression and secretion of IRF5-regulated cytokines IFNβ and IL-12p70, and a less clear effect on the expression and secretion of proinflammatory cytokines.

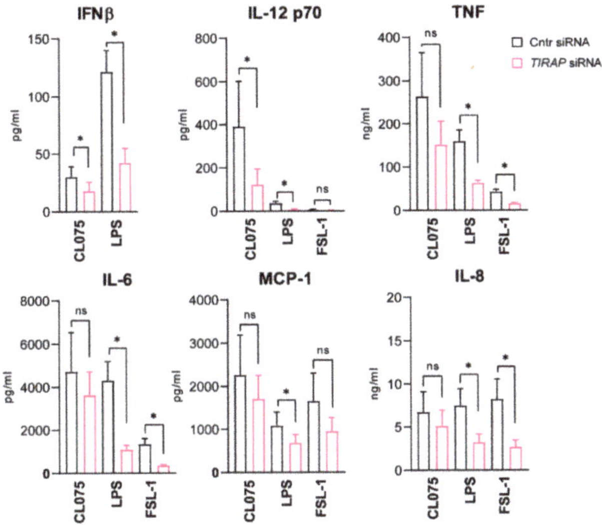

Figure 2. *TIRAP* silencing significantly inhibits TLR8-mediated IFNβ and IL12 p70 secretion by primary human macrophages. IFNβ and TNF secretion in 6–8 consecutive experiments with cells from different donors were analyzed by specific ELISA kits, with other cytokines' secretion addressed by BioPlex assays. Statistical significance evaluated using Wilcoxon matched-pairs signed-rank test, presented as mean with SD, significance levels—* $p < 0.05$, ns—non-significant.

3.2. TIRAP Silencing Inhibits TLR8-Dependent IL-12A Expression in Response to Bacterial Infection

TLR8 can sense ssRNA of bacterial origin [5,9,26]. Given how common bacterial infections are, and the potential serious consequences associated with them, it is essential to understand the mechanistic interactions between the pathogen and immune cells. Group B streptococcus (*S. agalactiae*, GBS) and *S. aureus* are commensal bacteria but hold great invasive potential and can cause serious infections [33–35]. TLR8 was recently shown to be involved in the innate immune responses to these Gram-positive bacteria but had less impact on responses to Gram-negative bacteria [13,26,27]. As such, we wanted to address the importance of TIRAP in the TLR8-mediated responses to GBS and *S. aureus*. *TIRAP*-silenced primary MDMs were stimulated with TLR4 ligand LPS or TLR8 ligand CL075, or infected with clinical isolates of *E. coli*, GBS, or *S. aureus*, for a total of four hours and at two bacterial doses. Expression of *IFNβ*, *TNF*, *IL-12A*, *IL-12B*, and *IL-6* mRNA was analyzed by RT-qPCR (Figure 3). In accordance with our previous findings, successful *TIRAP* silencing reduced the expression of these cytokines following LPS treatment and infection by *E. coli* (Figure 3). *TIRAP* silencing also significantly reduced *IL-12A* mRNA expression following GBS infection, with a similar tendency for *IFNβ* induction, although statistical support was not achieved (Figure 3). These results indicated a clear effect of *TIRAP* silencing on TLR8-mediated *IL-12A* expression, with a similar yet non-significant trend for *IFNβ* mRNA (Figures 1 and 3). In *S. aureus*-infected cells, *TIRAP* silencing significantly reduced *IL-6* and *TNF* mRNA expression, which could reflect the inhibition of TLR2-mediated pro-inflammatory signaling in *TIRAP*-silenced cells (Figure 3). We observed non-significant trends in the reduction in *IFNβ*, *IL-12A*, or *IL-12B* mRNA expression, following an *S. aureus* challenge (Figure 3). *S. aureus* is sensed both by TLR2 and TLR8, and TLR2 activation can suppress TLR8-IRF5-mediated induction of *IFNβ* and *IL-12A* [13], and it is possible that the combined TLR2 and TLR8 activation can diminish the requirement for TIRAP in the TLR8 mediated responses to *S. aureus*. Additional sensing mechanisms of whole live bacteria are also involved, such as complement and Fc receptors. These might compensate for the loss of TIRAP in our experiments, thus explaining the relatively small effects upon a bacterial challenge compared to pure TLR8-agonist stimulation. Thus, a kinetic study would be required to further clarify the impact of TIRAP in the sensing of live bacteria. Moreover, gene silencing has limitations, since it does not completely abrogate the expression or protein level of the silenced gene (Figure S3). Altogether, these results indicate that TIRAP can influence IRF5-mediated TLR8 signaling also during bacterial infection, which is especially clear for the induction of *IL-12A* expression by GBS.

3.3. TLR8-Mediated Nuclear Translocation of IRF5 Is Reduced by TIRAP Silencing

Based on our findings, which show that *TIRAP* silencing had the most prominent inhibitory effect on the expression and secretion of the IRF5-regulated cytokines IFNβ and IL-12p70 (Figures 1–3), we addressed TLR8-mediated nuclear translocation of the transcription factors NF-κB p65/RelA (p65) and IRF5 after the stimulation of cells by the CL075 ligand. NF-κB, a heterodimer with p65 as one of the components of the dimer, mainly induces the expression of pro-inflammatory cytokines, such as *TNF* and *IL-6* [1,36]. In human monocytes and MDMs, TLR8 induces expression of *IFNβ* and *IL-12A* genes in an IRF5-dependent manner, while TNF is less affected by *IRF5* silencing [13,26,27].

MDMs were treated with *TIRAP* siRNA or control siRNA prior to stimulation with the TLR8 ligand CL075, and silencing efficacy was confirmed by RT-qPCR from one of the parallel wells (Figure 4a). NF-κB p65 and IRF5 were stained by specific antibodies in TLR8-stimulated (60 min) cells, and nuclear translocation was assessed by automated high-throughput fluorescence imaging and quantification with ScanˆR (Figure 4). *TIRAP* silencing significantly attenuated the nuclear translocation of IRF5 after TLR8 stimulation, while p65 translocation was unaffected (Figure 4b,c).

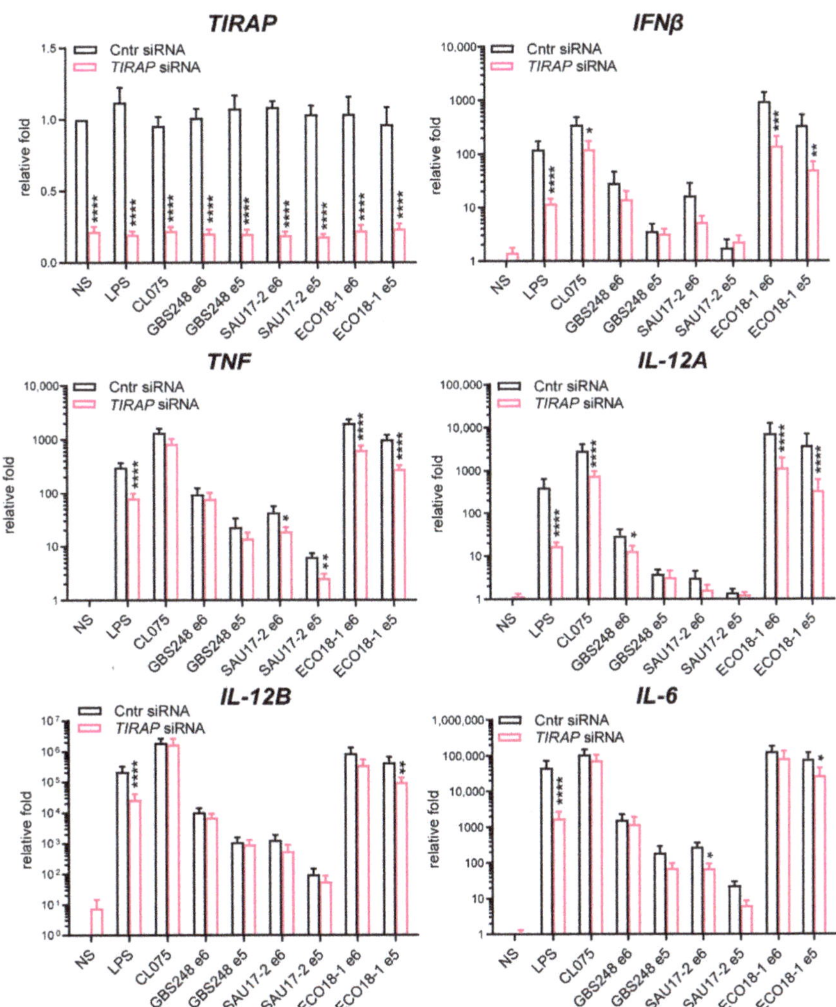

Figure 3. *TIRAP* silencing attenuates cytokine production from MDMs challenged with clinical isolates of *E. coli* (ECO), while affecting mainly *IL-12A* induction by Group B streptococcus (GBS) and pro-inflammatory cytokine induction by *S. aureus* (SAU). MDMs (5–6 donors in consecutive experiments) were pre-treated with *TIRAP* siRNA or control oligo and incubated with LPS (100 ng/mL), CL075 (1 μg/mL), or live bacteria (GBS 248, SAU 17-2, and ECO 18-1) for a total time of four hours. The doses of bacteria were 1×10^5/mL (e5) and 1×10^6/mL (e6). This roughly corresponds to MOI 0.01 and 0.1 for GBS, MOI 0.02 and 0.2 for SAU, and MOI 0.1 and 1.0 for ECO. Gene expression was determined by RT-qPCR, normalized to untreated sample, and presented as a mean relative fold change +SEM. Statistical testing was done with 2-way RM-ANOVA and post-test (* $p < 0.05$, ** $p < 0.01$, *** $p < 0.001$, **** $p < 0.0001$).

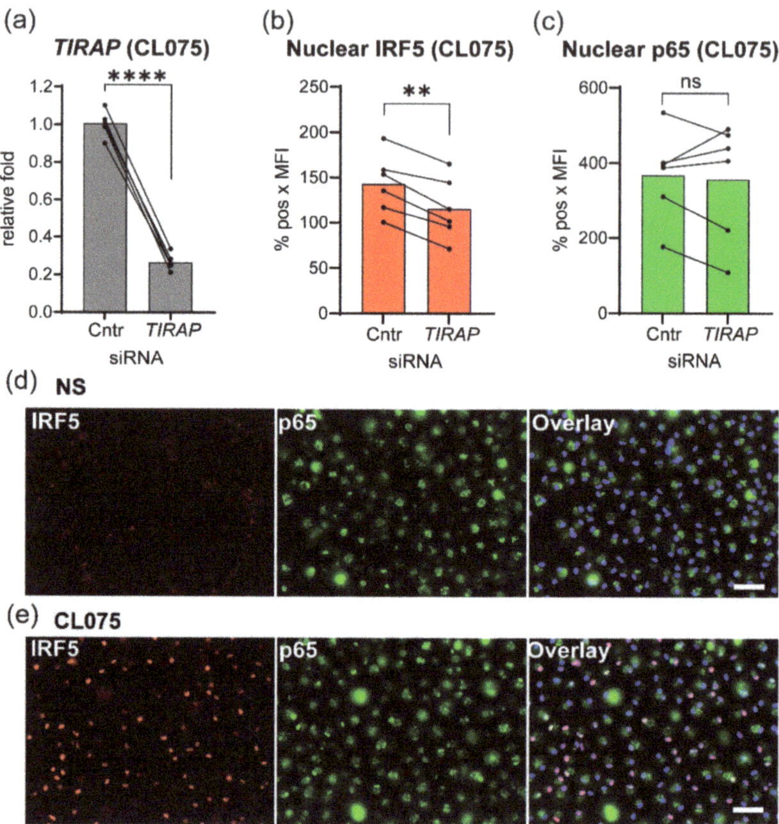

Figure 4. Silencing of *TIRAP* gene inhibits nuclear translocation of IRF5 in 60 min after CL075 stimulation. Experiments were performed on human MDMs (*n* = 6 donors). (**a**) *TIRAP*-silencing efficacy was quantified with RT-qPCR from parallel wells of CL075 stimulated cells using non-stimulated cells for normalization (fold = 1.0). Control or *TIRAP*-silenced cells were stimulated with CL075 (2 µg/mL) for one hour, followed by fixation of cells, double staining of IRF5 (**b**) and NF-kB (p65/RelA) (**c**), DNA staining by Hoechst 3342 for nuclei visualization, and quantitative imaging by high-content screening (Olympus Scan^R system). The level of nuclear IRF5 (**b**) and p65 (**c**) was calculated as the percentage of positively stained nuclei multiplied by the mean fluorescence-intensity value (MFI) of the positively stained nuclei. In non-stimulated cells, the background-staining levels (%pos × MFI) for nuclear IRF5 and p65 were <15 and <73, respectively. (**d**,**e**) Representative immunofluorescent images of non-stimulated (NS) and CL075 stimulated cells used for quantification of IRF5 (red channel) and p65 (green channel) in nuclei (blue channel). Scale bar shown in overlay represents 50 µm. Statistical significance was examined with paired *t*-test (** $p < 0.01$, **** $p < 0.001$).

Phosphorylation of IRFs and other transcription factors typically reflect their active state and is linked to their nuclear translocation and the activation of transcription of target genes [37,38]. To evaluate the potential regulation of IRF5 phosphorylation by TIRAP, we performed electrophoresis using Phos-tag gel [39] (Figure S4), since efficient phospho-IRF5 antibodies are not commercially available. Lysates of LPS-stimulated cells were included as a negative control since TLR4 signaling does not trigger IRF5 translocation in human MDMs, but rather it activates the IRF3-transcriptional factor in a TRAM/TRIF-dependent manner [13,26,40]. No clear alterations in the phosphorylation pattern of IRF5 or p65 in *TIRAP*-silenced cells after CL075 stimulation were revealed (Figure S4). Thus,

TIRAP expression positively regulated the nuclear translocation of IRF5 downstream to TLR8 (Figure 4), while having no effect on p65 phosphorylation or nuclear translocation (Figure S4).

3.4. TIRAP Co-Precipitates with the Myddosome Complex Induced by TLR8 Dimerisation

TIRAP interacts with activated TLR2 and TLR4 dimers and recruits the MyD88 signaling adaptor, allowing formation of the Myddosome complex [18]. The role of TIRAP in the TLR8-MyD88 complex has not previously been addressed, and there are no published data on possible TIRAP recruitment to the TLR8-activated Myddosome. To address the recruitment of TIRAP to the TLR8-Myddosome complex, we performed immunoprecipitations with TIRAP-specific antibody-coated beads and lysates from human primary MDMs, stimulated with LPS (positive control), or CL075 (Figure 5). Indeed, TIRAP co-precipitated with MyD88, IRAK4, and IRAK1, the core signaling proteins of the Myddosome complex [1], not only in LPS-stimulated cells, but also upon stimulation via TLR8 (Figure 5).

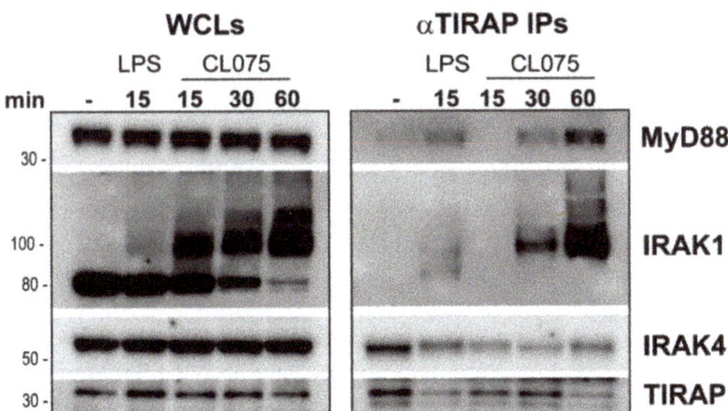

Figure 5. TIRAP is recruited to TLR8-initiated MyD88 and IRAK1/4-signaling complex. Endogenous TIRAP was immunoprecipitated for four hours from lysates (whole cell lysates—WCLs) of human MDMs: untreated or stimulated by LPS (100 ng/mL) or CL075 (2 μg/mL) for indicated time. LPS stimulation was applied as a positive control for TIRAP recruitment to the activated Myddosome. Cellular lysates were analyzed in parallel to control for input, with WB for MyD88, IRAK1, IRAK4, and TIRAP. A representative experiment is shown from a total of four consecutive experiments with different donors.

To validate the IRAK4 band in the TIRAP precipitates (due to the IRAK4 size of 50–52 kDa, which is close to the size of IgG heavy chains), we examined TIRAP co-precipitations with IRAKs and MyD88 from LPS-stimulated THP-1 cells, using an IRAK4 inhibitor (PF-06426779) that induces a band size shift of IRAK4 due to the inhibition of IRAK4 autophosphorylation (Figure S5). Pre-treatment with the IRAK4 inhibitor decreased LPS-mediated IRAK1 posttranslational modifications and TAK1 phosphorylation in the lysates, resulting in the expected IRAK4 band-size shift in the precipitates (Figure S5). This shows that IRAK4 staining is specific for the chosen IPs conditions (Figure 5).

The shared time point of 15 min for the LPS and CL075-stimulated cells demonstrates that TIRAP recruitment to IRAKs and MyD88 was delayed for CL075-stimulated cells when compared with LPS-stimulated cells (Figure 5). Overall, the extent of IRAK1 modification in 15 min of TLR stimulation in MDMs showed great donor variation (not shown), as expected. IP results with lysates from a donor with a fast and strong response to CL075 were selected to demonstrate that even with fast IRAK1 activation (already within 15 min), TIRAP was not co-precipitating with the Myddosome-complex molecules (Figure 5). In contrast, TIRAP was recruited to TLR4-activated Myddosome within 15 min, even though this experiment revealed only weak IRAK1 modification following LPS stimulation (Figure 5). These observations are in line with the well-established role of TIRAP as adaptor that connects MyD88 to TLR4 [28,41].

Thus, TIRAP recruitment to the Myddosome complex was delayed and more prominent at 30–60 min after CL075 stimulation, when compared to LPS stimulation (Figure 5). Overall, we suggest that recruitment of TIRAP to TLR8 occurs after the Myddosome formation is initiated. This is in line with the concept of the direct interaction of endosomal TLRs with the signaling adaptor MyD88 [28,41]. TIRAP may, thus, not be required for connecting MyD88 to TLR8 to initiate the signaling but is rather recruited to the activated TLR8-signaling complex, and, thus, subsequently regulate downstream signaling. Overall, our data show that TIRAP is attracted to the TLR8 Myddosome and further support the hypothesis that TIRAP is involved in the regulation of TLR8 signaling.

3.5. TLR8-Mediated Akt Phosphorylation Is Negatively Affected by TIRAP Silencing

To gain further insight into the role of TIRAP downstream of the TLR4 and TLR8 Myddosomes, we analyzed the phosphorylation/activation state of signaling intermediates in *TIRAP*-silenced human MDMs (Figure 6). *TIRAP* silencing was expected to have an inhibitory effect on the activation/phosphorylation of TLR4-mediated MyD88-dependent signaling molecules. Upon TLR ligation, MyD88 is recruited to TLR4 in a TIRAP-dependent manner [28], and IRAK4 kinase is subsequently attracted to MyD88 via death domain (DD) interactions, followed by IRAK1 recruitment and activation. As a result, IRAK1 is phosphorylated and poly-ubiquitinated, which induces the shift of IRAK1 band size from 80 kDa to 100 kDa or a significant reduction in the 80 kDa band [42–44]. Active IRAK1 forms the complex as well as promotes the phosphorylation and activation of TAK1-mitogen-activated kinase kinase kinase (MAPKKK) that acts upstream and induces ERK1/2, JNK1/2, and p38 MAPK phosphorylation and activation, while TAK1 also activates the canonical IKK complex (reviewed in [45]). IKKβ is crucial for IκBα (nuclear factor of kappa light-polypeptide gene enhancer in the B-cells inhibitor, α) phosphorylation, which leads to its degradation that is required for the activation of NF-κB (reviewed in [46]). IKKβ is also critical for the activation of IRF5 in TLR8 signaling [13,37]. Both p38MAPK and JNK1/2 positively regulate the transcriptional activity of the AP-1 (ATF-2-c-jun) transcriptional complex [47], which together with IRFs (IRF3 for TLR4 and IRF5 for TLR8) and NF-kB translocate to the nucleus and activate type I IFN promoters [13,40].

As could be seen from Figure 6a,c, in cells stimulated with LPS for 15–30 min, *TIRAP* silencing reduced the phosphorylation of TAK1, ERK1/2, JNK1/2, and p38 MAPK as well as the post-translational modification of IRAK1 (100 kDa band), and resulted in less effective degradation of IκBα. Silencing of *TIRAP* had some inhibitory effect on TLR8-mediated TAK1 and p38 MAPK phosphorylation, while not affecting IRAK1 posttranslational modifications or the degradation of IκBα (Figure 6a,b). Indeed, de-phosphorylation of TAK1 in *TIRAP*-silenced cells was faster when compared to control cells upon CL075 stimulation (Figure 6a,b), which may indicate a possible role of TIRAP in the stabilization of IRAK1/TABs/TAK1 signaling complex and increased TAK1-mediated cytokine production.

PI3Ks and its downstream target, serine/threonine-kinase Akt (PKB), is activated by many receptors, including TLRs, and is known to regulate macrophage survival and migration as well as the response to different metabolic and inflammatory signals in macrophages [48–50]. Phosphorylation of Akt (serine 473, S473) reflects a fully activated Akt kinase [51]. The most consistent effect of *TIRAP* silencing upon the ligation of TLR8 across PBMCs from several donors was the decreased phosphorylation of Akt, particularly 45–60 min after TLR8 activation (Figure 6a,b), which correlates with the timeframe when TIRAP is recruited to the TLR8-induced Myddosome (Figure 5). These results indicate that TLR8 signaling may be coupled to the PI3K/Akt pathway, and that TIRAP positively regulates TLR8-mediated activation of Akt (Figure 6a,b). In comparison, in LPS-stimulated cells, *TIRAP* silencing had not much effect on the phosphorylation of Akt (Figure 6a,c).

Previously, Guiducci et al., reported that TLR7 stimulation induces phosphorylation of Akt, and inhibition of Akt reduces nuclear translocation of IRF7 and type I IFNs' induction [17]. Thus, positive regulation of TLR8-mediated IRF5 nuclear translocation by TIRAP (Figure 4) might be mechanistically linked to the regulation of Akt (Figure 6a,b).

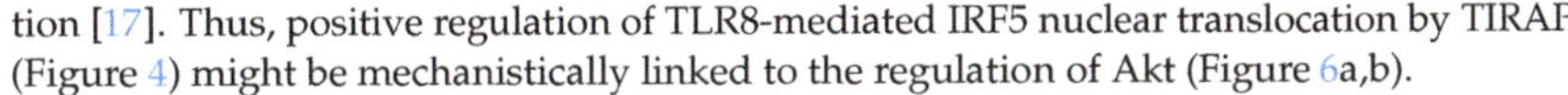

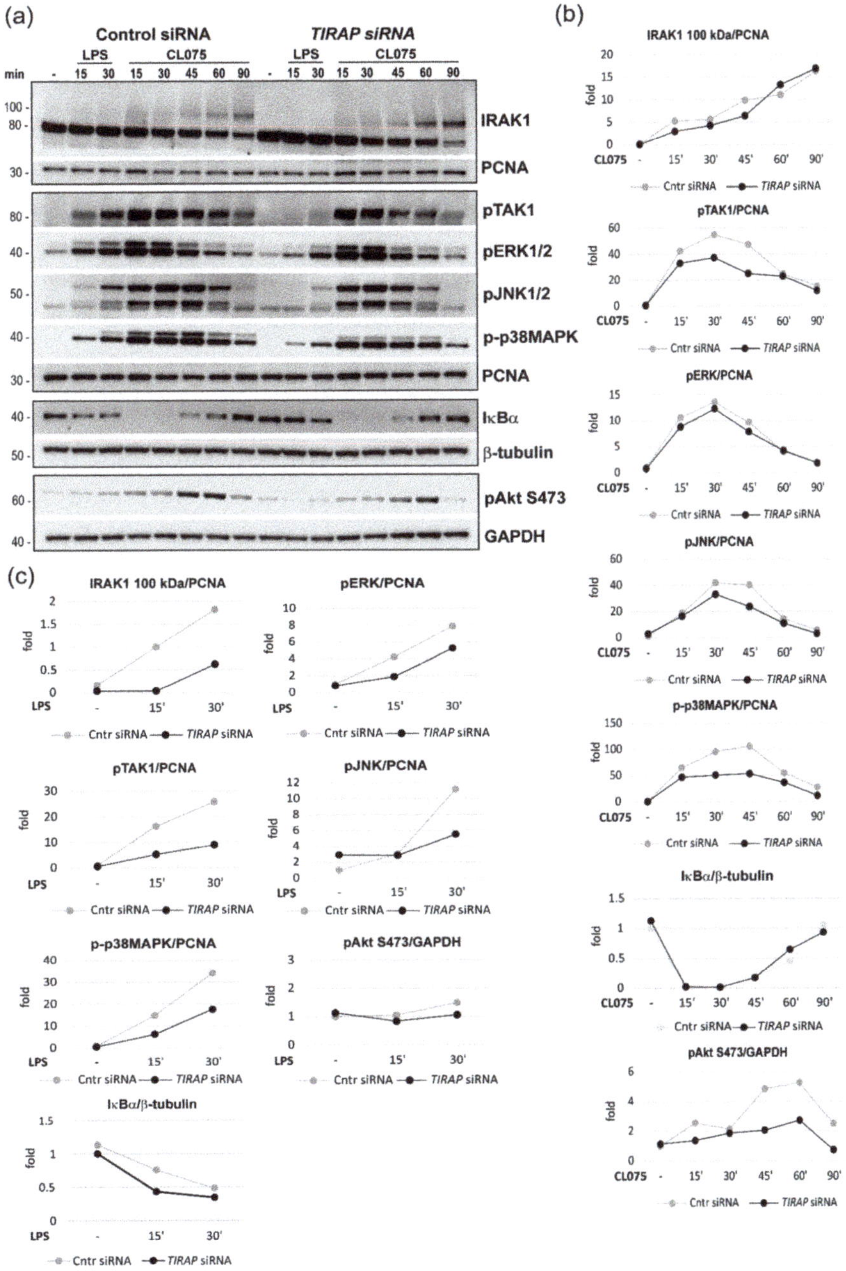

Figure 6. Silencing of *TIRAP* consistently inhibits TLR8-mediated phosphorylation of Akt S473. (**a**) Western blotting of lysates from MDMs treated with a control oligo or *TIRAP*-specific siRNA

oligo and stimulated with 100 ng/mL LPS or 2 µg/mL CL075. The antibodies used are indicated on the figure, and GAPDH or PCNA are equal-loading controls. (**b**) Graphs show quantifications of protein levels relative to GAPDH or PCNA for CL075-stimulated cells and (**c**) LPS-stimulated cells. Representative image and graphs for one of four donors. Densitometry analysis and normalization to loading control was done using LiCor Odyssey software.

3.6. p38 MAPK Inhibition Is Not Affecting TLR8-Mediated IRF5 and p65 Nuclear Translocation

TIRAP silencing resulted in reduced p38 MAPK phosphorylation, upon stimulation by TLR8 ligand (Figure 6). To investigate the role of p38 MAPK in TLR8 signaling, we pre-treated monocytes with a selective p38 MAPK inhibitor BIRB 796 and several other control inhibitors, followed by the stimulation of cells with CL075 (1 µg/mL) for one and two hours Overall, inhibition of p38 MAPK had no effect on nuclear translocation of IRF5 or p65 (Figure S6a). In contrast, TAK1 kinase inhibitor (5z-7-oxozeaenol) blocked IRF5 translocation, but not p65 translocation, while inhibiting IKKβ with a IKKII–VIII inhibitor that blocked both p65 and IRF5 translocation, as shown in our earlier study [13]. These data suggest that decreased phosphorylation of p38 MAPK does not explain the reduction in IRF5 nuclear translocation upon *TIRAP* silencing. Still, inhibition of p38 MAPK strongly inhibited the expression of *IFNβ* and *TNF* mRNA in 2 h of CL075 stimulation, similar to TAK1 inhibition (Figure S6b). Since p65 nuclear translocation was not affected by the p38 MAPK inhibitor, this could be explained by attenuation of the AP-1 transcriptional complex activity in BIRB796-treated cells, which would result in reduced cytokine induction according to the established role for AP-1 in cytokines' promoter activity [47]. Overall, we conclude that TIRAP may regulate TLR8 signaling via two distinct pathways: an Akt pathway and the TAK1 pathway that enhances p38 MAPK activation. Both pathways contribute to cytokine induction. The regulation of *IFNβ* and *IL-12A* expression by modulation of nuclear IRF5 levels is the most marked effect of *TIRAP* silencing, which could not be explained by the decreased p38 MAPK activation in silenced cells. Thus, we proceeded with testing the effect of Akt inhibition on TLR8-mediated *IFNβ* and *IL-12A* expression as well as IRF5 nuclear translocation.

3.7. Akt Inhibition Decreases TLR8-Mediated Expression of IFNβ and IL-12A Genes

To further examine the role of Akt in TLR8 signaling, and particularly in the regulation of *IFNβ* and *IL-12A* expression, we used specific Akt inhibitors. Two allosteric inhibitors (MK-2206 and Miransertib) and one ATP-competitive Akt inhibitor (Capivasertib) had a similar inhibitory effect on the TLR8-mediated expression of *IFNβ* and *IL-12A* (Figure S7). We proceeded with the inhibitor MK-2206 and pre-treated primary human monocytes prior to stimulation with CL075 (Figure 7). Indeed, Akt inhibition resulted in a significant decrease in *IFNβ* and *IL-12A* mRNA in two hours of TLR8 stimulation, with no effect on *TNF* mRNA expression (Figure 7). Of the TLR4-mediated responses, Akt inhibition resulted in the reduced *TNF* expression, without effect on *IFNβ* and *IL-12A* expression, suggesting a different mechanism for the regulation of TLR4-mediated *IFNβ* expression by TIRAP.

Due to the previously detected link between activation of TAK1 and IRF5, we addressed the possible impact of Akt inhibition on the phosphorylation of TAK1 S172 and the downstream phosphorylation of p38 MAPK (Figure S6a,b). Akt inhibition was efficient, since the allosteric Akt inhibitor blocked S473 phosphorylation, as previously revealed [14]. Interestingly, Akt inhibition rather increased the phosphorylation of TAK1 and downstream p38 MAPK after both TLR4 and TLR8 stimulation (Figure S8a,b). At the same time, Akt inhibition by MK-2206 reduced the TLR8-mediated phosphorylation of STAT1 (Figure S8c), which correlates with *IFNβ* gene expression (Figure 7). Overall, these results suggest that Akt is involved in the positive regulation of TLR8 signaling, leading to the expression of *IFNβ* and *IL-12A*, although mechanistically it appears not to be mediated by increased TAK-1 activation. In contrast to TLR8, inhibition of Akt had no significant effect on TLR4-mediated *IFNβ* and *IL-12A* induction.

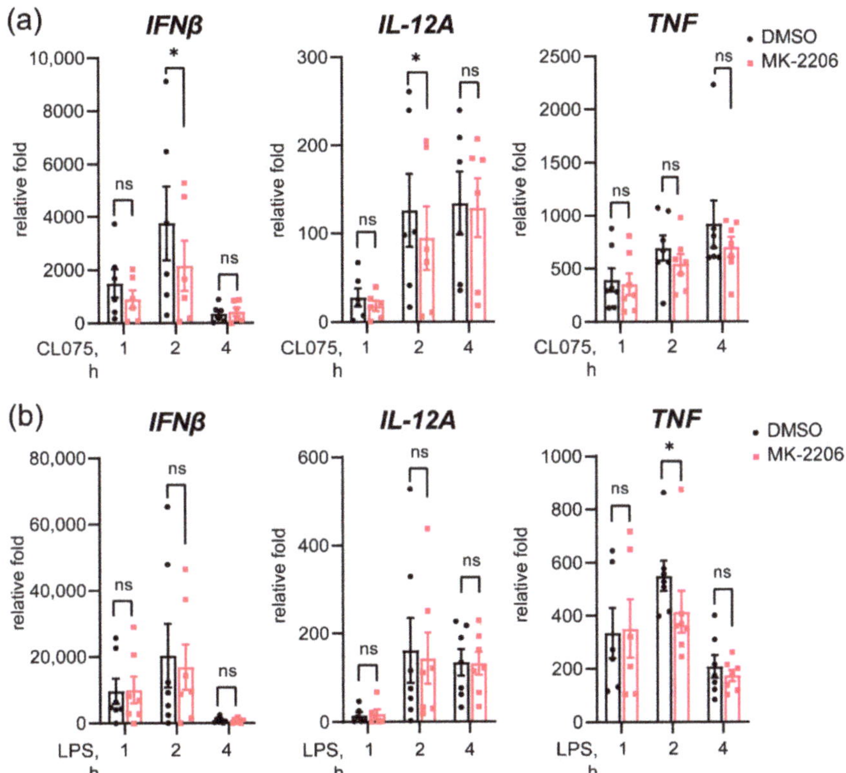

Figure 7. Akt inhibition in MDMs significantly reduces TLR8-mediated expression of *IFNβ* and *IL-12A* genes, while having no effect on TLR4-mediated cytokine expression. RT-qPCR analysis of cytokine expression after pre-treatment with Akt inhibitor MK-2206 (2 μM) and stimulation by (**a**) CL075 (2 μg/mL) or (**b**) LPS (100 ng/mL). Gene expression normalized to unstimulated sample and presented as a mean relative fold change +SEM. Statistical testing was done by 2-way RM-ANOVA including a post-test, as described (* $p < 0.05$, and ns—non-significant).

3.8. Akt Inhibition Decreases TLR8-Mediated Nuclear Translocation of IRF5

As we have shown, nuclear translocation of IRF5 in TLR8-stimulated *TIRAP* silenced cells was significantly reduced without a clear effect on IRF5 phosphorylation (Figure 4). We, thus, investigated the effect of Akt activity on nuclear translocation and phosphorylation of IRF5 in MDMs using subcellular fractionation (Figure 8). LPS-stimulated cells were included as a negative control for IRF5 nuclear translocation, and anti-phospho-Akt (S473) was used to demonstrate an efficient Akt blockade. Histone 3 levels were analyzed for the normalization of nuclear extracts, while GAPDH used to control for the possible contamination of nuclear extracts with cytosolic content, which was not the case (Figure 8). Overall, inhibition of Akt markedly reduced IRF5 nuclear translocation in 60 min of TLR8 stimulation (Figure 8). As with *TIRAP* silencing, the total phosphorylation pattern of IRF5 in total lysates of monocytes was not altered upon Akt inhibition (Figure S9). Together, these data suggest TIRAP is involved in a crosstalk between TLR8 and Akt, which contributes to IRF5 nuclear translocation and the expression of *IFNβ* and *IL-12A* genes (Figure 9).

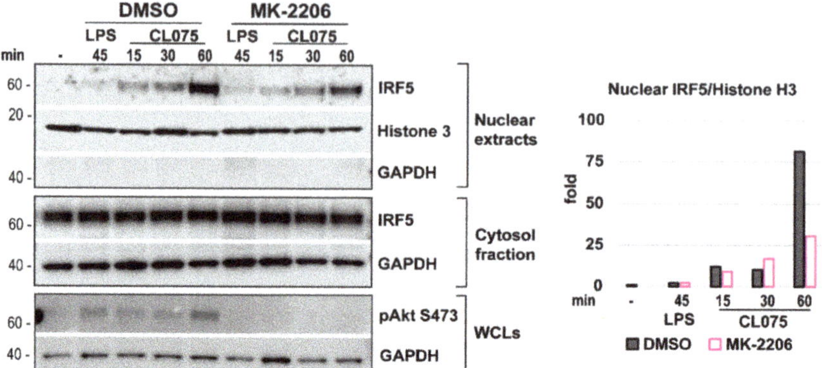

Figure 8. Inhibition of Akt reduced IRF5 nuclear translocation in human monocytes. Western blot analysis of cytosolic fraction and nuclear extracts from cells pre-treated with MK-2206 (2 µM) was followed by stimulation with CL075 (2 µg/mL) or LPS (100 ng/mL). LPS stimulation was applied for negative control. IRF5 levels in nuclear extracts were normalized based on Histone 3 bands' intensity (graph), while GAPDH Western blot was performed to control for potential contamination of nuclear extracts with cytosol content. To control for Akt inhibition efficacy, Akt (S473) phosphorylation level was addressed in WCLs (whole cell lysates) from parallel wells. Representative of three consecutive experiments.

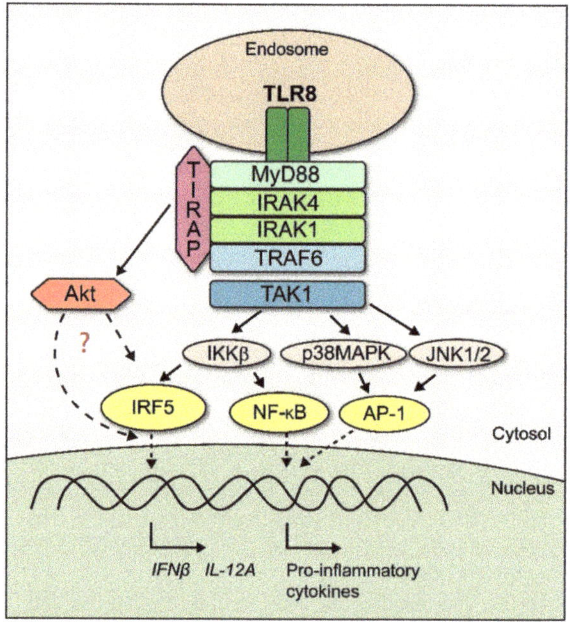

Figure 9. Model showing TIRAP involvement in regulation of TLR8 signaling. Recruitment of TIRAP to Myddosome promotes Akt activation and facilitates nuclear translocation of IRF5 as well as expression and secretion of IRF5-dependent cytokines IFNβ and IL-12p70.

4. Discussion

TIRAP/Mal is a critical bridging adaptor that connects MyD88 to TLR2 and TLR4 at the plasma membrane. However, it is now clear that the role of TIRAP in TLR signaling is much more complex (reviewed in [18]), with even some TLR-independent functions discovered for TIRAP (reviewed in [52]).

In its N-terminal part, TIRAP contains a phosphoinositide (PI)-binding domain (PBD), which interacts with phosphatidylinositol 4,5-bisphosphate (PtdIns(4,5)P2)-enriched membranes [19,20,53]. Bonham et al., demonstrated that TIRAP PBD is also capable of binding PtdIns(3)P on endosomal membranes, and when murine-bone-marrow-derived macrophages (BMDMs) are challenged with natural ligands (influenzas virus and Herpes simplex virus), TIRAP regulates signaling via TLR7 and TLR9, with a particular impact on IFNα expression [20].

Here, we show that TLR8 also utilizes TIRAP in its IRF5-dependent signaling pathway in human primary monocytes and MDMs, which has not previously been reported. Moreover, our findings suggest that TIRAP plays an unconventional role in TLR8 signaling and most likely is recruited after the formation of the proximal TLR8-Myddosome complex, subsequently enhancing Akt/PKB activation (Figure 9).

To investigate the potential involvement of TIRAP in TLR8 signaling, we based our study on *TIRAP* silencing in human primary phagocytes. Despite the quite high variability in kinetics and magnitude of TLR signaling in the human primary cells from healthy human subjects, which can be of genetic as well as non-genetic causes, our experiments provide some important advantages over studies with cell lines. The response of primary cells more accurately reflects the natural human-cell biology and host–pathogen interactions, so it is, therefore, of higher relevance.

Our data on the mRNA expression and cytokine secretion of IFNβ and IL-12A in *TIRAP*-silenced cells showed a significant reduction in these responses upon TLR8 stimulation, while the effect on pro-inflammatory cytokines was not as clear. We have recently demonstrated that TLR8 is a dominant TLR in the response to the Gram-positive bacteria, such as GBS and *S. aureus*, in human primary monocytes and MDMs [13,26,27]. We, therefore, addressed the contribution of TIRAP in the response to Gram-positive and Gram-negative bacterial infections in MDMs and revealed a partial dependency of TIRAP in the regulation of cytokine production with all the examined bacteria. The impact of TIRAP in *E. coli*-induced cytokine production could be mainly attributed to LPS-activated TLR4-signaling, while a partial reduction in TNF and IL-6 production induced by *S. aureus* may involve TLR2 signaling. Genetic *TIRAP* deficiency in humans impairs both TLR4 and TLR2 signaling, though the TLR2-response in macrophages and susceptibility to *S. aureus* infections can be rescued in vivo by lipoteichoic acid (LTA)-specific IgG antibodies, likely due to a compensatory mechanism via Fc-receptor (CD32) engagement [54]. We have shown earlier that the GBS-induced cytokine production in myeloid cell cultures is almost entirely TLR8-mediated [27], and the significant reduction in *IL-12A* expression, thus, demonstrates a role of TLR8–TIRAP signaling during a challenge with a viable Gram-positive bacterium. Even though induction of *IL-12A* and *IFNβ* by the Gram-positive bacteria is mainly TLR8 dependent [27], our data only revealed a tendency of attenuated *IFNβ* expression after *TIRAP* silencing, which did not reach statistical support, possibly due to underpowered statistics. It is also possible that the selected time point for gene-expression analysis was sub-optimal, or that the triggering of several signaling mechanism by the whole live bacteria (e.g., TLR2, TLR8, Fc-receptors, complement receptors, etc.) could compensate for the reduced TIRAP levels. Moreover, considerable levels of TIRAP protein remain in the cells after *TIRAP* silencing. The effect of *TIRAP* silencing was more prominent for TLR2 and TLR4 signaling with prolonged stimulation, which may imply that the levels of TIRAP in silenced cells are sufficient to initiate proximal signaling but not to sustain the cytokine production. Generation of *TIRAP* KO THP-1 cells was done as an alternative strategy, but the phenotype appeared unstable, possibly due to compensatory signaling mechanisms. Thus, the real contribution of TIRAP to TLR8-mediated cytokine induction might be more prominent, both for purified agonists and whole bacteria, and this issue, thus, warrants further studies.

As we have already noted, TIRAP appears to be more important for the expression of IRF5-dependent genes *IFNβ* and *IL-12A* than for TLR8-regulated pro-inflammatory cytokines. A similar differential regulation of *IFNβ* by TIRAP was also observed by Zyzak et al., in TLR9 signaling in human PBMCs and the microglia cell line [22], and by Lesczcynska et al., in TLR7 signaling in the human dendritic cell line [21]. Both reports implicate ERK1/2 in the TIRAP-dependent effects observed. Similar to our findings, Leszczynska et al., conclude that the TIRAP-dependent effects on *IFNβ* expression are mediated by IRF7, whereas Zyzak et al. link the non-canonical NF-kB pathway to IFNs-expression regulation [21,22]. However, especially in the latter report [22], most of the research into the mechanisms shown are done in murine cells.

TLR8-mediated *IFNβ* and *IL-12A* gene expression in human primary monocytes and macrophages is dependent on activation of IRF5 [13,27]. Our findings provide evidence that TIRAP is involved in the TLR8-IRF5 signaling mechanism. In contrast to the findings by Zyzak et al. [22] and Leszczynska et al. [21] regarding TLR9 and TLR7, *TIRAP* silencing in human MDMs did not alter ERK1/2 activation but, consistently, reduced the phosphorylation of Akt kinase. The PI3K–Akt pathway can regulate cellular metabolism and survival, and its dysregulation is firmly linked to tumor development [14]. The PI3K–Akt pathway is also implicated in the regulation of TLR signaling, with evidence of both pro- and anti-inflammatory effects [15,50,55,56]. Guiducci et al., showed that nuclear translocation of IRF7 and type I IFNs expression is enhanced by PI3K–Akt signaling, following TLR7 and TLR9 activation in human pDCs [17]. Lima et al., reported crosstalk between TLR9 and PI3Kγ in human PBMCs [57], while Sarkar et al., demonstrated that TLR3-dependent activation of the PI3K–Akt axis induces IRF3 phosphorylation and nuclear translocation in HEK293 cells [58]. Similar to these findings, here we show that Akt is involved in the regulation of *IFNβ* and *IL-12A* induction upon TLR8-IRF5 signaling. However, Akt could not be linked to TLR4-induced *IFNβ* and *IL-12A* expression in human MDMs.

Lopez-Pelaez et al., previously identified that IKKβ phosphorylates Ser462 in IRF5 to induce nuclear translocation and subsequent expression of *IFNβ* in a TLR7-stimulated human pDC cell line [37]. We find that Akt inhibition and *TIRAP*-silencing reduced IRF5 nuclear translocation upon TLR8 stimulation, though IRF5 phosphorylation appeared unaffected. However, it is possible that changes in the phosphorylation of the specific sites in IRF5 may not be detected by the analysis of the total phosphorylation pattern of IRF5.

Overall, we found that *TIRAP* silencing most consistently decreased the phosphorylation of Akt in human MDMs, and that inhibition of Akt had a comparable effect with *TIRAP* silencing. It might be possible that recruitment of TIRAP followed by Akt activation is adding another layer to the regulation of IRF5 activation and IRF5-dependent gene expression, either by direct phosphorylation of IRF5, or by regulating the activity of transport proteins involved in IRF5 nuclear translocation. Recently, a TLR adaptor interacting with SLC15A4 on the lysosome (TASL) was identified as a critical endosomal adapter for IRF5 activation in TLR7-9 signaling in human cells [59], and further studies are necessary to deduce the precise signaling events upstream of IRF5 activation as well as the specific roles of TIRAP and Akt in these pathways.

As an endosomal ssRNA-sensing receptor, TLR8 is relevant both for viral and bacterial infections [5,9,13,26,27,60–62]. IFNs are particularly important during viral infections, as they induce the expression of gene-encoding proteins with anti-viral effects, such as inhibiting the viral replication, assembly, and release of the virus particle [1]. Thus, further exploration of the role of TIRAP in virus-induced TLR8 signaling (such as Influenza A Virus, HIV, West Nile Virus [61]) is warranted. Cell-type specific differences in the expression level and utilization of TIRAP in TLR8 signaling should also be addressed. Furthermore, SNPs in TIRAP are associated to the incidence and severity of several diseases, such as tuberculosis, HIV, and systemic lupus erythematosus (SLE) [18] as well as infections/diseases in which TLR8 is likely to play a role [63–65]. Thus, understanding the contribution of TIRAP to the fine-tuning of TLR8 signaling might be of significant clinical relevance.

Biomedicines **2022**, *10*, 1476

Supplementary Materials: The following supporting information can be downloaded at https://www.mdpi.com/article/10.3390/biomedicines10071476/s1. Supplementary Materials and Methods. Figure S1: TIRAP silencing in primary human MDMs significantly decreases TLR2- and TLR4-mediated cytokines mRNA expression. Figure S2: *TIRAP* silencing could be a more relevant approach than knockout to evaluate the fine-tuning of TLRs-mediated signaling by TIRAP. Figure S3: *TIRAP* silencing reduces STAT1 phosphorylation in LPS- or CL075-stimulated human MDMs. Figure S4: Total phosphorylation pattern of IRF5 in CL075-stimulated cells is not affected by *TIRAP* silencing. Figure S5: Band shift for IRAK4 protein in TIRAP IPs, induced by the pre-treatment of cells by the PF-06426779 IRAK4 inhibitor, supports the specificity of IRAK4 staining for the selected IPs and WB conditions. Figure S6. Pharmacological inhibition of p38 MAPK does not block the TLR8-induced nuclear translocation of IRF5 or p65, but it still suppresses TLR8-induced cytokine transcription. Figure S7: Tested Akt inhibitors have a similar inhibitory effect on TLR8-mediated *IFNβ* and *IL-12A* expression. Figure S8: Inhibition of Akt strongly decreased TLR8-mediated phosphorylation of Y701 in STAT1, with no inhibitory effect on TLR4- or TLR8-mediated MAPKs phosphorylation. Figure S9: Akt inhibition had no effect on the total phosphorylation pattern of the IRF5-transcription factor.

Author Contributions: Conceptualization, M.Y., T.E. and J.S.; methodology, A.S., J.F.K., K.E.N., M.Y. and J.S.; software, J.S. and M.Y.; formal analysis, A.S., J.S., J.F.K., K.E.N. and M.Y.; investigation, A.S., J.F.K., K.E.N., M.Y. and J.S.; resources, M.Y., T.E. and J.S.; writing—original draft preparation, K.E.N.; writing—review and editing, M.Y., T.E., A.S. and J.S.; visualization, M.Y., J.S. and K.E.N.; supervision, J.S. and M.Y.; funding acquisition, M.Y., T.E. and J.S. All authors have read and agreed to the published version of the manuscript.

Funding: This research was funded by the Research Council of Norway through its Centers of Excellence funding scheme, grant 223255/F50 (to T.E.), by the Liaison Committee for Education, Research and Innovation in Central Norway, grant 90794301 (to M.Y.), and by the Liaison Committee for Education, Research, and Innovation in Central Norway, grant 90162400 (to J.S.).

Institutional Review Board Statement: The use of human buffy coats and serum from the blood bank at St. Olavs Hospital (Trondheim, Norway) was approved by the Regional Committee for Medical and Health Research Ethics (REC) in Central Norway (no. 2009/2245).

Informed Consent Statement: Not applicable.

Data Availability Statement: All the data for manuscript is provided in main text or supplementary materials.

Acknowledgments: The Scan^R imaging and analysis was carried out at the Cellular and Molecular Imaging Core Facility (CMIC), Norwegian University of Science and Technology.

Conflicts of Interest: The authors declare no conflict of interest.

References

1. Fitzgerald, K.A.; Kagan, J.C. Toll-like Receptors and the Control of Immunity. *Cell* **2020**, *180*, 1044–1066. [CrossRef] [PubMed]
2. Li, D.; Wu, M. Pattern recognition receptors in health and diseases. *Signal Transduct. Target. Ther.* **2021**, *6*, 291. [CrossRef] [PubMed]
3. Sarvestani, S.T.; Williams, B.R.; Gantier, M.P. Human Toll-like receptor 8 can be cool too: Implications for foreign RNA sensing. *J. Interferon Cytokine Res.* **2012**, *32*, 350–361. [CrossRef]
4. Marques, J.T.; Williams, B.R. Activation of the mammalian immune system by siRNAs. *Nat. Biotechnol.* **2005**, *23*, 1399–1405. [CrossRef]
5. Cervantes, J.L.; La Vake, C.J.; Weinerman, B.; Luu, S.; O'Connell, C.; Verardi, P.H.; Salazar, J.C. Human TLR8 is activated upon recognition of Borrelia burgdorferi RNA in the phagosome of human monocytes. *J. Leukoc. Biol.* **2013**, *94*, 1231–1241. [CrossRef] [PubMed]
6. Cervantes, J.L.; Weinerman, B.; Basole, C.; Salazar, J.C. TLR8: The forgotten relative revindicated. *Cell. Mol. Immunol.* **2012**, *9*, 434–438. [CrossRef] [PubMed]
7. Hornung, V.; Rothenfusser, S.; Britsch, S.; Krug, A.; Jahrsdorfer, B.; Giese, T.; Endres, S.; Hartmann, G. Quantitative expression of toll-like receptor 1–10 mRNA in cellular subsets of human peripheral blood mononuclear cells and sensitivity to CpG oligodeoxynucleotides. *J. Immunol.* **2002**, *168*, 4531–4537. [CrossRef] [PubMed]
8. Tanji, H.; Ohto, U.; Shibata, T.; Miyake, K.; Shimizu, T. Structural reorganization of the Toll-like receptor 8 dimer induced by agonistic ligands. *Science* **2013**, *339*, 1426–1429. [CrossRef]
9. Eigenbrod, T.; Pelka, K.; Latz, E.; Kreikemeyer, B.; Dalpke, A.H. TLR8 Senses Bacterial RNA in Human Monocytes and Plays a Nonredundant Role for Recognition of Streptococcus pyogenes. *J. Immunol.* **2015**, *195*, 1092–1099. [CrossRef]

10. Coch, C.; Hommertgen, B.; Zillinger, T.; Dassler-Plenker, J.; Putschli, B.; Nastaly, M.; Kummerer, B.M.; Scheunemann, J.F.; Schumak, B.; Specht, S.; et al. Human TLR8 Senses RNA From Plasmodium falciparum-Infected Red Blood Cells Which Is Uniquely Required for the IFN-gamma Response in NK Cells. *Front. Immunol.* **2019**, *10*, 371. [CrossRef]

11. Guiducci, C.; Gong, M.; Cepika, A.M.; Xu, Z.; Tripodo, C.; Bennett, L.; Crain, C.; Quartier, P.; Cush, J.J.; Pascual, V.; et al. RNA recognition by human TLR8 can lead to autoimmune inflammation. *J. Exp. Med.* **2013**, *210*, 2903–2919. [CrossRef]

12. Greulich, W.; Wagner, M.; Gaidt, M.M.; Stafford, C.; Cheng, Y.; Linder, A.; Carell, T.; Hornung, V. TLR8 Is a Sensor of RNase T2 Degradation Products. *Cell* **2019**, *179*, 1264–1275.e1213. [CrossRef] [PubMed]

13. Bergstrom, B.; Aune, M.H.; Awuh, J.A.; Kojen, J.F.; Blix, K.J.; Ryan, L.; Flo, T.H.; Mollnes, T.E.; Espevik, T.; Stenvik, J. TLR8 Senses Staphylococcus aureus RNA in Human Primary Monocytes and Macrophages and Induces IFN-beta Production via a TAK1-IKKbeta-IRF5 Signaling Pathway. *J. Immunol.* **2015**, *195*, 1100–1111. [CrossRef] [PubMed]

14. Lazaro, G.; Kostaras, E.; Vivanco, I. Inhibitors in AKTion: ATP-competitive vs allosteric. *Biochem. Soc. Trans.* **2020**, *48*, 933–943. [CrossRef] [PubMed]

15. Ruse, M.; Knaus, U.G. New players in TLR-mediated innate immunity: PI3K and small Rho GTPases. *Immunol. Res.* **2006**, *34*, 33–48. [CrossRef]

16. Aksoy, E.; Vanden Berghe, W.; Detienne, S.; Amraoui, Z.; Fitzgerald, K.A.; Haegeman, G.; Goldman, M.; Willems, F. Inhibition of phosphoinositide 3-kinase enhances TRIF-dependent NF-kappa B activation and IFN-beta synthesis downstream of Toll-like receptor 3 and 4. *Eur. J. Immunol.* **2005**, *35*, 2200–2209. [CrossRef]

17. Guiducci, C.; Ghirelli, C.; Marloie-Provost, M.A.; Matray, T.; Coffman, R.L.; Liu, Y.J.; Barrat, F.J.; Soumelis, V. PI3K is critical for the nuclear translocation of IRF-7 and type I IFN production by human plasmacytoid predendritic cells in response to TLR activation. *J. Exp. Med.* **2008**, *205*, 315–322. [CrossRef] [PubMed]

18. Bernard, N.J.; O'Neill, L.A. Mal, more than a bridge to MyD88. *IUBMB Life* **2013**, *65*, 777–786. [CrossRef] [PubMed]

19. Barnett, K.C.; Kagan, J.C. Lipids that directly regulate innate immune signal transduction. *Innate Immun.* **2020**, *26*, 4–14. [CrossRef]

20. Bonham, K.S.; Orzalli, M.H.; Hayashi, K.; Wolf, A.I.; Glanemann, C.; Weninger, W.; Iwasaki, A.; Knipe, D.M.; Kagan, J.C. A promiscuous lipid-binding protein diversifies the subcellular sites of toll-like receptor signal transduction. *Cell* **2014**, *156*, 705–716. [CrossRef]

21. Leszczynska, E.; Makuch, E.; Mitkiewicz, M.; Jasyk, I.; Narita, M.; Gorska, S.; Lipinski, T.; Siednienko, J. Absence of Mal/TIRAP Results in Abrogated Imidazoquinolinones-Dependent Activation of IRF7 and Suppressed IFNbeta and IFN-I Activated Gene Production. *Int. J. Mol. Sci.* **2020**, *21*, 8925. [CrossRef]

22. Zyzak, J.; Mitkiewicz, M.; Leszczynska, E.; Reniewicz, P.; Moynagh, P.N.; Siednienko, J. HSV-1/TLR9-Mediated IFNbeta and TNFalpha Induction Is Mal-Dependent in Macrophages. *J. Innate Immun.* **2020**, *12*, 387–398. [CrossRef] [PubMed]

23. Husebye, H.; Aune, M.H.; Stenvik, J.; Samstad, E.; Skjeldal, F.; Halaas, O.; Nilsen, N.J.; Stenmark, H.; Latz, E.; Lien, E.; et al. The Rab11a GTPase controls Toll-like receptor 4-induced activation of interferon regulatory factor-3 on phagosomes. *Immunity* **2010**, *33*, 583–596. [CrossRef] [PubMed]

24. Sanjana, N.E.; Shalem, O.; Zhang, F. Improved vectors and genome-wide libraries for CRISPR screening. *Nat. Methods* **2014**, *11*, 783–784. [CrossRef] [PubMed]

25. Pfaffl, M.W. A new mathematical model for relative quantification in real-time RT-PCR. *Nucleic Acids Res.* **2001**, *29*, e45. [CrossRef] [PubMed]

26. Ehrnstrom, B.; Beckwith, K.S.; Yurchenko, M.; Moen, S.H.; Kojen, J.F.; Lentini, G.; Teti, G.; Damas, J.K.; Espevik, T.; Stenvik, J. Toll-Like Receptor 8 Is a Major Sensor of Group B Streptococcus But Not Escherichia coli in Human Primary Monocytes and Macrophages. *Front. Immunol.* **2017**, *8*, 1243. [CrossRef] [PubMed]

27. Moen, S.H.; Ehrnstrom, B.; Kojen, J.F.; Yurchenko, M.; Beckwith, K.S.; Afset, J.E.; Damas, J.K.; Hu, Z.; Yin, H.; Espevik, T.; et al. Human Toll-like Receptor 8 (TLR8) Is an Important Sensor of Pyogenic Bacteria, and Is Attenuated by Cell Surface TLR Signaling. *Front. Immunol.* **2019**, *10*, 1209. [CrossRef]

28. Horng, T.; Barton, G.M.; Medzhitov, R. TIRAP: An adapter molecule in the Toll signaling pathway. *Nat. Immunol.* **2001**, *2*, 835–841. [CrossRef]

29. Oosenbrug, T.; van de Graaff, M.J.; Haks, M.C.; van Kasteren, S.; Ressing, M.E. An alternative model for type I interferon induction downstream of human TLR2. *J. Biol. Chem.* **2020**, *295*, 14325–14342. [CrossRef]

30. Kawasaki, T.; Kawai, T. Toll-like receptor signaling pathways. *Front. Immunol.* **2014**, *5*, 461. [CrossRef]

31. Kagan, J.C.; Su, T.; Horng, T.; Chow, A.; Akira, S.; Medzhitov, R. TRAM couples endocytosis of Toll-like receptor 4 to the induction of interferon-beta. *Nat. Immunol.* **2008**, *9*, 361–368. [CrossRef] [PubMed]

32. Yurchenko, M.; Skjesol, A.; Ryan, L.; Richard, G.M.; Kandasamy, R.K.; Wang, N.; Terhorst, C.; Husebye, H.; Espevik, T. SLAMF1 is required for TLR4-mediated TRAM-TRIF-dependent signaling in human macrophages. *J. Cell Biol.* **2018**, *217*, 1411–1429. [CrossRef]

33. Anwar, S.; Prince, L.R.; Foster, S.J.; Whyte, M.K.; Sabroe, I. The rise and rise of Staphylococcus aureus: Laughing in the face of granulocytes. *Clin. Exp. Immunol.* **2009**, *157*, 216–224. [CrossRef] [PubMed]

34. Grice, E.A.; Kong, H.H.; Conlan, S.; Deming, C.B.; Davis, J.; Young, A.C.; Program, N.C.S.; Bouffard, G.G.; Blakesley, R.W.; Murray, P.R.; et al. Topographical and temporal diversity of the human skin microbiome. *Science* **2009**, *324*, 1190–1192. [CrossRef] [PubMed]

35. Barcaite, E.; Bartusevicius, A.; Tameliene, R.; Kliucinskas, M.; Maleckiene, L.; Nadisauskiene, R. Prevalence of maternal group B streptococcal colonisation in European countries. *Acta Obstet. Gynecol. Scand.* **2008**, *87*, 260–271. [CrossRef] [PubMed]
36. Duan, T.; Du, Y.; Xing, C.; Wang, H.Y.; Wang, R.F. Toll-Like Receptor Signaling and Its Role in Cell-Mediated Immunity. *Front. Immunol.* **2022**, *13*, 812774. [CrossRef] [PubMed]
37. Lopez-Pelaez, M.; Lamont, D.J.; Peggie, M.; Shpiro, N.; Gray, N.S.; Cohen, P. Protein kinase IKKbeta-catalyzed phosphorylation of IRF5 at Ser462 induces its dimerization and nuclear translocation in myeloid cells. *Proc. Natl. Acad. Sci. USA* **2014**, *111*, 17432–17437. [CrossRef]
38. Mogensen, T.H. IRF and STAT Transcription Factors—From Basic Biology to Roles in Infection, Protective Immunity, and Primary Immunodeficiencies. *Front. Immunol.* **2018**, *9*, 3047. [CrossRef]
39. Nagy, Z.; Comer, S.; Smolenski, A. Analysis of Protein Phosphorylation Using Phos-Tag Gels. *Curr. Protoc. Protein Sci.* **2018**, *93*, e64. [CrossRef]
40. Fitzgerald, K.A.; Rowe, D.C.; Barnes, B.J.; Caffrey, D.R.; Visintin, A.; Latz, E.; Monks, B.; Pitha, P.M.; Golenbock, D.T. LPS-TLR4 signaling to IRF-3/7 and NF-kappaB involves the toll adapters TRAM and TRIF. *J. Exp. Med.* **2003**, *198*, 1043–1055. [CrossRef]
41. Horng, T.; Barton, G.M.; Flavell, R.A.; Medzhitov, R. The adaptor molecule TIRAP provides signalling specificity for Toll-like receptors. *Nature* **2002**, *420*, 329–333. [CrossRef]
42. Motshwene, P.G.; Moncrieffe, M.C.; Grossmann, J.G.; Kao, C.; Ayaluru, M.; Sandercock, A.M.; Robinson, C.V.; Latz, E.; Gay, N.J. An oligomeric signaling platform formed by the Toll-like receptor signal transducers MyD88 and IRAK-4. *J. Biol. Chem.* **2009**, *284*, 25404–25411. [CrossRef]
43. Vollmer, S.; Strickson, S.; Zhang, T.; Gray, N.; Lee, K.L.; Rao, V.R.; Cohen, P. The mechanism of activation of IRAK1 and IRAK4 by interleukin-1 and Toll-like receptor agonists. *Biochem. J.* **2017**, *474*, 2027–2038. [CrossRef]
44. Emmerich, C.H.; Ordureau, A.; Strickson, S.; Arthur, J.S.; Pedrioli, P.G.; Komander, D.; Cohen, P. Activation of the canonical IKK complex by K63/M1-linked hybrid ubiquitin chains. *Proc. Natl. Acad. Sci. USA* **2013**, *110*, 15247–15252. [CrossRef]
45. Sato, S.; Sanjo, H.; Takeda, K.; Ninomiya-Tsuji, J.; Yamamoto, M.; Kawai, T.; Matsumoto, K.; Takeuchi, O.; Akira, S. Essential function for the kinase TAK1 in innate and adaptive immune responses. *Nat. Immunol.* **2005**, *6*, 1087–1095. [CrossRef] [PubMed]
46. Kawai, T.; Akira, S. Signaling to NF-kappaB by Toll-like receptors. *Trends Mol. Med.* **2007**, *13*, 460–469. [CrossRef] [PubMed]
47. Chang, L.; Karin, M. Mammalian MAP kinase signalling cascades. *Nature* **2001**, *410*, 37–40. [CrossRef]
48. Covarrubias, A.J.; Aksoylar, H.I.; Horng, T. Control of macrophage metabolism and activation by mTOR and Akt signaling. *Semin. Immunol.* **2015**, *27*, 286–296. [CrossRef] [PubMed]
49. Song, G.; Ouyang, G.; Bao, S. The activation of Akt/PKB signaling pathway and cell survival. *J. Cell Mol. Med.* **2005**, *9*, 59–71. [CrossRef] [PubMed]
50. Troutman, T.D.; Bazan, J.F.; Pasare, C. Toll-like receptors, signaling adapters and regulation of the pro-inflammatory response by PI3K. *Cell Cycle* **2012**, *11*, 3559–3567. [CrossRef]
51. Alessi, D.R.; Andjelkovic, M.; Caudwell, B.; Cron, P.; Morrice, N.; Cohen, P.; Hemmings, B.A. Mechanism of activation of protein kinase B by insulin and IGF-1. *EMBO J.* **1996**, *15*, 6541–6551. [CrossRef] [PubMed]
52. Belhaouane, I.; Hoffmann, E.; Chamaillard, M.; Brodin, P.; Machelart, A. Paradoxical Roles of the MAL/Tirap Adaptor in Pathologies. *Front. Immunol.* **2020**, *11*, 569127. [CrossRef] [PubMed]
53. Kagan, J.C. Defining the subcellular sites of innate immune signal transduction. *Trends Immunol.* **2012**, *33*, 442–448. [CrossRef] [PubMed]
54. Israel, L.; Wang, Y.; Bulek, K.; Della Mina, E.; Zhang, Z.; Pedergnana, V.; Chrabieh, M.; Lemmens, N.A.; Sancho-Shimizu, V.; Descatoire, M.; et al. Human Adaptive Immunity Rescues an Inborn Error of Innate Immunity. *Cell* **2017**, *168*, 789–800.e710. [CrossRef]
55. Troutman, T.D.; Hu, W.; Fulenchek, S.; Yamazaki, T.; Kurosaki, T.; Bazan, J.F.; Pasare, C. Role for B-cell adapter for PI3K (BCAP) as a signaling adapter linking Toll-like receptors (TLRs) to serine/threonine kinases PI3K/Akt. *Proc. Natl. Acad. Sci. USA* **2012**, *109*, 273–278. [CrossRef]
56. Hamerman, J.A.; Pottle, J.; Ni, M.; He, Y.; Zhang, Z.Y.; Buckner, J.H. Negative regulation of TLR signaling in myeloid cells—implications for autoimmune diseases. *Immunol. Rev.* **2016**, *269*, 212–227. [CrossRef]
57. Lima, B.H.F.; Marques, P.E.; Gomides, L.F.; Mattos, M.S.; Kraemer, L.; Queiroz-Junior, C.M.; Lennon, M.; Hirsch, E.; Russo, R.C.; Menezes, G.B.; et al. Converging TLR9 and PI3Kgamma signaling induces sterile inflammation and organ damage. *Sci. Rep.* **2019**, *9*, 19085. [CrossRef]
58. Sarkar, S.N.; Peters, K.L.; Elco, C.P.; Sakamoto, S.; Pal, S.; Sen, G.C. Novel roles of TLR3 tyrosine phosphorylation and PI3 kinase in double-stranded RNA signaling. *Nat. Struct. Mol. Biol.* **2004**, *11*, 1060–1067. [CrossRef]
59. Heinz, L.X.; Lee, J.; Kapoor, U.; Kartnig, F.; Sedlyarov, V.; Papakostas, K.; Cesar-Razquin, A.; Essletzbichler, P.; Goldmann, U.; Stefanovic, A.; et al. TASL is the SLC15A4-associated adaptor for IRF5 activation by TLR7-9. *Nature* **2020**, *581*, 316–322. [CrossRef]
60. Sartorius, R.; Trovato, M.; Manco, R.; D'Apice, L.; De Berardinis, P. Exploiting viral sensing mediated by Toll-like receptors to design innovative vaccines. *NPJ Vaccines* **2021**, *6*, 127. [CrossRef]
61. Martinez-Espinoza, I.; Guerrero-Plata, A. The Relevance of TLR8 in Viral Infections. *Pathogens* **2022**, *11*, 134. [CrossRef] [PubMed]
62. de Marcken, M.; Dhaliwal, K.; Danielsen, A.C.; Gautron, A.S.; Dominguez-Villar, M. TLR7 and TLR8 activate distinct pathways in monocytes during RNA virus infection. *Sci. Signal* **2019**, *12*, eaaw1347. [CrossRef] [PubMed]

63. Thada, S.; Horvath, G.L.; Muller, M.M.; Dittrich, N.; Conrad, M.L.; Sur, S.; Hussain, A.; Pelka, K.; Gaddam, S.L.; Latz, E.; et al. Interaction of TLR4 and TLR8 in the Innate Immune Response against *Mycobacterium Tuberculosis*. *Int. J. Mol. Sci.* **2021**, *22*, 1560. [CrossRef] [PubMed]
64. Sakaniwa, K.; Shimizu, T. Targeting the innate immune receptor TLR8 using small-molecule agents. *Acta Crystallogr. D Struct. Biol.* **2020**, *76*, 621–629. [CrossRef]
65. Meas, H.Z.; Haug, M.; Beckwith, M.S.; Louet, C.; Ryan, L.; Hu, Z.; Landskron, J.; Nordbo, S.A.; Tasken, K.; Yin, H.; et al. Sensing of HIV-1 by TLR8 activates human T cells and reverses latency. *Nat. Commun.* **2020**, *11*, 147. [CrossRef]

Review

Critical Review on Toxicological Mechanisms Triggered by Inhalation of Alumina Nanoparticles on to the Lungs

Samir Dekali [1,*], **Alexandra Bourgois** [2] **and Sabine François** [2]

[1] French Armed Forces Biomedical Research Institute (IRBA), Department of Biological Radiation Effects/Unit of Emerging Technological Risks, 1 Place du Général Valérie André, BP 73, CEDEX, 91223 Brétigny-sur-Orge, France

[2] French Armed Forces Biomedical Research Institute (IRBA), Department of Biological Radiation Effects/Radiobiology Unit, 1 Place du Général Valérie André, BP 73, CEDEX, 91223 Brétigny-sur-Orge, France

* Correspondence: samir.dekali@gmail.com

Abstract: Alumina nanoparticles (Al_2O_3 NPs) can be released in occupational environments in different contexts such as industry, defense, and aerospace. Workers can be exposed by inhalation to these NPs, for instance, through welding fumes or aerosolized propellant combustion residues. Several clinical and epidemiological studies have reported that inhalation of Al_2O_3 NPs could trigger aluminosis, inflammation in the lung parenchyma, respiratory symptoms such as cough or shortness of breath, and probably long-term pulmonary fibrosis. The present review is a critical update of the current knowledge on underlying toxicological, molecular, and cellular mechanisms induced by exposure to Al_2O_3 NPs in the lungs. A major part of animal studies also points out inflammatory cells and secreted biomarkers in broncho-alveolar lavage fluid (BALF) and blood serum, while in vitro studies on lung cells indicate contradictory results regarding the toxicity of these NPs.

Keywords: alumina nanoparticles; inflammation; fibrosis; toxicity; lung; broncho-alveolar fluid; aluminosis

Citation: Dekali, S.; Bourgois, A.; François, S. Critical Review on Toxicological Mechanisms Triggered by Inhalation of Alumina Nanoparticles on to the Lungs. *Biomedicines* **2022**, *10*, 2664. https://doi.org/10.3390/biomedicines10102664

Academic Editor: Alberto Ricci

Received: 30 September 2022
Accepted: 19 October 2022
Published: 21 October 2022

Publisher's Note: MDPI stays neutral with regard to jurisdictional claims in published maps and institutional affiliations.

1. Introduction

Alumina nanoparticles (Al_2O_3 NPs) are among the most widely used and produced particles in the world, with a wide range of interesting applications such as biosensors, desalination, high-risk pollutants detection, capacitors, solar cell devices, and photonic crystals [1]. These NPs can also be retrieved as pollutants in the environment.

Al_2O_3 NPs are produced in huge quantities as by-products of water treatment (water treatment residuals), bauxite processing (red mud), and hard and brown coal burning in power plants (fly ash) [2]. Welding fumes and propellant combustion residues also seem to be the main sources of Al_2O_3 NPs occupational exposures [3,4]. Nanoparticle forms are particularly studied in toxicological research because of their increasing use and the concerns they raise. Indeed, unique physico-chemical properties of NPs such as small size (<100 nm) and high specific surface area, confer high surface reactivity and uncertainty toward their potential toxicity [5]. Workers' exposure to Al_2O_3 NPs can mainly occur by inhalation. Occupational exposures to these dusts are described in the literature to have deleterious health effects on the respiratory and nervous systems [3,6–9]. Therefore, it is necessary to improve understanding of NPs toxicity in order to redesign strategies to mitigate/reduce environmental and/or health impact [10]. Due to the lack of studies, research efforts are needed to better explore and thus protect predisposed populations such as workers.

This review focuses on biological effects described on the pulmonary system and associated described pathologies. Clinical, animal, and in vitro studies are successively presented in this paper. Briefly, workers exposed to Al_2O_3 NPs developed pneumoconiosis, also named "aluminosis". They also presented local respiratory symptoms such as cough or shortness of breath, inflammation in the lung parenchyma, probably long-term pulmonary fibrosis, and increased risk of developing lung cancer [3,8,9,11]. Pulmonary aluminosis is a rare form of pneumoconiosis caused by aluminum or alumina powders [12,13]. Consequently, the majority of animal studies have focused on pro-inflammatory mechanisms triggered by the inhalation of Al_2O_3 NPs [14,15]. Results showed increases in total protein, neutrophils, lymphocytes, and lactate dehydrogenase (LDH) concentration in broncho-alveolar lavages (BALF), corroborating pro-inflammatory effects of Al_2O_3 NPs and suggesting potential permeabilization of the alveolo-capillary barrier. Moreover, various pro-inflammatory cytokines were secreted in animal BALFs: TNF-α, IL-6, IL-1β, IL-8, MIP-2, and IL-33 [14–16]. However, to the best of our knowledge, inflammatory mechanisms are poorly studied in vitro. This review also critically presents contradictory results on the cytotoxic and genotoxic effects of Al_2O_3 NPs on lung cells.

2. Clinical Studies on Health Effects after Inhalation of Alumina Particles

A recent study was conducted on fifteen male students aged (mean, range) 24, 19–31 with normal lung function in an inhalation chamber for human exposure to atmospheric particulate matter (PM) [9]. The aim was to examine the effects of short-term exposure to Al_2O_3 particles (3.2 μm, crystallinity unknown) on inflammatory markers in induced sputum in healthy volunteers. Controlled inhalations with exposure levels commonly seen in primary aluminum production were used (below toxicological reference values, i.e., 2 h at 3.9 mg/m^3). Authors showed 24 h after exposure increases in polymorphonuclear neutrophils (PNNs), total proteins, and IL-8 concentrations in the sputum, suggesting a marked pulmonary inflammation. Moreover, microarray analyses of mRNA abundance collected from sputum macrophages and pathway analysis showed changes in the expression of 46 genes identified in three major biological process groups: cell–cell signaling, gene expression, RNA damage and repair, and regulation of connective tissue assembly. Only localized respiratory effects were reported in this study, and no systemic effects were observed.

A prevalence of pulmonary fibrosis 300 times higher than that observed in the general population was evaluated in workers exposed to alumina for 25 years [17]. However, these results are questionable because of the exposure of subjects to other particulate compounds that may have contributed to the onset of these respiratory pathologies. Indeed, significant quantities of asbestos fibers have also been found in the lungs of some subjects who have developed pulmonary fibrosis. Workers' exposures to fumes from aluminum welding, which may contain alumina, has also led to the appearance of pneumoconiosis and pulmonary fibrosis in some cases [3,11]. A higher frequency of respiratory diseases, such as chronic obstructive pulmonary disease (COPD), has been observed in aluminum welders. In addition, decreases in lung function (forced expiratory volume in a second (FEV1)) have been observed in workers in the aluminum industry exposed to dust containing mainly alumina [18]. Another study evaluated the lung tissue of fourteen workers exposed to hard metals and aluminum oxide [19]. Among them, five workers underwent transbronchial biopsy showing diffuse interstitial inflammatory changes: two of them were asymptomatic, one had clinically evident disease with severe giant cell inflammation, and two other workers showed local inflammation.

Although some studies on alumina exposure have concluded the induction of respiratory pathologies and a decrease in lung function in humans, other studies have revealed contradictory results. Despite respiratory symptoms (cough, shortness of breath) observed in workers (preparers of powders for propulsion systems using duralium) exposed to aluminum and alumina NPs, no link between these compounds and the appearance of diseases respiratory problems could be demonstrated [8]. Similarly, the follow-up of a cohort of 521 men working in the production of abrasives (exposure to alumina particles, silicon carbides, and formaldehyde) 72 over 25 years did not show a significant increase in total or cancer mortality or the incidence of non-malignant respiratory disease [20]. Finally, monitoring of the pulmonary function of welders using aluminum did not show any deleterious effect, although respiratory symptoms were observed in these workers [21].

A summary of the results of clinical studies performed on health effects after inhalation exposure to alumina particles is available in Table 1.

Table 1. Summary of clinical studies on health effects after inhalation of alumina particles.

References	Studied Atmospheres	Population (Size)	Health/Biological Effects	Particle Size	Particle Concentration	Exposure Time
Sikkeland et al., 2016 [9]	Al_2O_3 particles	Healthy volunteers (15)	Increases in neutrophils, total proteins, and IL-8 concentrations in the sputum. Localized respiratory effects, no systemic effect.	3.2 μm	3.9 mg/m^3	2 h
Jederlinic et al., 1990 [17]	Complex aerosols containing Al_2O_3 particles	Workers (9)	Prevalence of pulmonary fibrosis 300 times higher than that observed in the general population	Unknown	Unknown	25 years
Hull et al., 2002 [3]	Aluminum welding fumes	Workers (2)	Pneumoconiosis cases	10 nm–1 μm (Al, aggregates)	Unknown	22–24 years
Vallyathan et al., 1982 [11]	Aluminum welding fumes	Worker (1)	Pulmonary fibrosis case	Unknown	Unknown	Unknown
Townsend et al., 1985 [18]	Aluminum welding fumes (containing mainly Al_2O_3 particles)	Workers (1142)	Higher frequency of respiratory diseases, such as chronic obstructive pulmonary disease (COPD). Decrease in lung function.	Unknown	Unknown	Unknown
Schwarz et al., 1998 [19]	Complex aerosols containing Al_2O_3 particles and hard metals	Workers (14)	Diffuse pulmonary interstitial inflammatory changes in five workers (2 asymptomatic, 1 symptomatic with giant cell inflammation, and 2 with local inflammation).	Unknown	Unknown	Unknown
Hunter et al., 1944 [8]	Complex aerosols containing aluminum and alumina NPs	Workers (50)	Respiratory symptoms (cough, shortness of breath). No proven correlation with appearance of respiratory diseases.	0.23–0.5 μm (Total particles)	400–2430 /cm^3 (Total particles)	6–39 years
Edling et al., 1987 [20]	Complex aerosols containing alumina particles, silicon carbides, and formaldehyde	Workers (521)	No increase in total or cancer mortality nor incidence of non-malignant respiratory disease.	Unknown	1 mg/m^3 (Total particles)	25 years
Sjorgen et al., 1985 [21]	Aluminum welding fumes	Workers (259)	Respiratory symptoms with no alteration of lung function.	<1 μm (Total particles)	0–42 mg/m^3 (Total particles)	1–41 years

3. Animal Studies of Pulmonary Biological Effects Triggered by Alumina Nanoparticles

In order to study the pulmonary toxicity of particles, different animal exposure methods can be implemented. Inhalation exposure is the most physiological method. However, this is expensive, and it requires extensive technical skills to set up the exposure system, the reproducibility of aerosol generation, and to characterize the physico-chemistry of generated aerosols. Another commonly used method to study in vivo lung toxicity of particles is intratracheal instillation (IT). This technique allows administrating of precise and known doses of particle suspensions [22]. IT is less constrained than inhalation but not representative of environmental exposure conditions. It is also possible to use other methods of exposure, such as nasal instillation, IT aspiration, oropharyngeal aspiration, or oral exposure. These techniques are less physiological and rarely used in literature.

Although possessing important anatomical and physiological similarities with humans, large mammals such as monkeys and pigs are very rarely used for lung toxicity studies due to ethical aspects and their cost. The rat is the alternative recommended by Organisation for Economic Co-operation and Development (OECD) because it is a more qualified model, easier to implement, and remains representative of the human respiratory system [23–25]. This animal model is preferred to the mouse model because it has more anatomical similarities with humans than with mice [26]. This animal model is the most commonly used to study the pulmonary toxicity of alumina particles. These studies are presented considering the kind of animal exposure, i.e., inhalation or IT/intranasal exposures.

3.1. Nose-Only and Whole-Body Inhalation Exposures

Kim and colleagues reported pro-inflammatory effects of alumina nanoparticles after repeated nose-only inhalation on Sprague-Dawley rats [15]. After 28 days of exposure (5 days/week) to Al_2O_3 NPs concentrations ranged between 0.2 and 5 mg/m^3 (size 11.94 nm; unknown crystallinity), they showed increases in the total number of cells, neutrophils, lymphocytes, LDH, TNF-α, IL-6 in BALF. Moreover, they reported histopathological lesions with alveolar macrophage accumulation in four and eight cases of the 5 mg/m^3 group during exposure and recovery, respectively. Pro-inflammatory effects significantly decreased after 28 days of exposure, but neutrophils and LDH concentrations remained significantly elevated compared to control groups. Results obtained in this study demonstrated a strong inflammatory potential of these NPs when inhaled, and authors suggested a no-observed-adverse-effect level of 1 mg/m^3 concentration. Recently, Wistar rats were also nose-only exposed to a high concentration of Al_2O_3 NPs (size 13 nm; γ/δ crystallinity; 20.0–22.1 mg/m^3 aerosol) following two exposure scenarios: single (4 h) or repeated exposures (4 h/day for 4 days) [14]. After repeated exposures, total proteins and LDH concentrations in BALF were significantly increased, suggesting that the alveolo-capillary barrier was damaged. Additionally, a marked pro-inflammatory reaction was observed with increased concentrations of neutrophils, macrophages, IL-1β, TNF-α, GRO/KC, and MIP-2. Moreover, another study showed that after 7 days of inhalation exposure of mice to Al_2O_3 NPs (size 40 nm; unknown crystallinity; dose of 0.4 mg/m^3) in whole-body chamber emphysema and small airway remodeling in lungs can occur, accompanied by enhanced inflammation and apoptosis [16]. Authors demonstrated that protein tyrosine phosphatase, non-receptor type 6 (PTPN6), was down-regulated and Signal Transducer and Activator of Transcription 3 (STAT3) phosphorylated in response to Al_2O_3 NPs exposure, culminating in increased expression of the apoptotic marker Programmed cell death protein 4 (PDCD4). Moreover, IL-6 and IL-33 concentrations were significantly increased in BALF. Therefore, a decrease in PTPN6 may have deleterious effects at the molecular, cellular, and tissue levels, leading to the initiation of inflammation and apoptosis, ultimately resulting in the development of COPD-like lesions.

3.2. Intratracheal or Intranasal Exposures

IT or nasal instillation are the best alternatives to inhalation exposures for pulmonary toxicity studies in animals. These techniques notably offer the possibility of exposing animals in a less costly way, with good control of the dose administered in bolus and technically less complex than exposure by inhalation. In Sprague-Dawley rats exposed by IT to 40 mg of Al_2O_3 particles (size 4.37 µm, γ/α crystallinity), an increase in the number of cells, mainly macrophages but also neutrophils, and fibronectin concentrations were shown in BALF [27]. These concentrations remain increased twelve months after instillation, suggesting a persistence of the inflammatory phenomenon induced by Al_2O_3 NPs. These results are consistent with those obtained by inhalation but provide information on the persistence over time of the observed acute effects. BALF analysis of Wistar rats exposed to Al_2O_3 NPs (6.3 nm, crystallinity unknown, 0.5 mL at 300 cm^2/mL) by IT also induced acute pulmonary inflammation [7]. Increases in polymorphonuclear cells were measured in BALF 24 h after exposure. In addition, Al_2O_3 NPs showed hemolytic potential. The effects observed in this study seem to be correlated with the surface properties of Al_2O_3 NPs (surface charge represented by the zeta potential in particular). Indeed, a correlation between the zeta potential and the influx of granulocytes or the hemolytic power has been observed for NPs with a high zeta potential [7].

Exposure by nasal instillation of Sprague-Dawley rats demonstrated other deleterious effects of alumina NPs (size and crystallinity unknown, 1–40 mg/kg) [28]. A slight dose-dependent inflammation was observed, but impairment of alveolo-capillary barrier permeability was also revealed by the increase in total protein concentration in BALF. The fibrotic potential of Al_2O_3 particles was assessed in Sprague-Dawley rats by IT and in NMRI mice by intraperitoneal injection [29]. The effects of different particles (different microparticle sizes; variable crystal polymorphs α, γ, δ, and χ) were evaluated. The study concluded that the particles usually used for the manufacture of aluminum (α and γ) had no fibrotic effect, whereas other particles tested could induce this type of lesion. Moreover, this study did not show any link between cytotoxic and fibrotic effects. Potential cardiac effects of Al_2O_3 NPs (size 11 nm; α crystallinity; 30 mg/kg/day; over 14 days) were studied on Sprague-Dawley rats after IT [30]. This study revealed adverse effects resulting in electrocardiogram (ECG) disorders and an increase in myocardial (LDH, triglycerides, creatine phosphokinase, cholesterol, nitric oxide) and inflammatory (TNF-α) damage markers. A decrease in antioxidants was also measured (reduced glutathione and superoxide dismutase) in animal serum.

A summary of the results of in vivo studies after inhalation/instillation exposure to alumina particles is available in Table 2.

Table 2. Summary of animal studies.

References	Exposure Method	Animal Model	Biological Effects	Primary Particle Size	Particle Concentration	Particle Crystallinity	Exposure Time
Kim et al., 2018 [15]	Nose-only inhalation	Rats	Pro-inflammatory effects. Increases in neutrophils, lymphocytes, LDH, TNF-α, and IL-6 in BALF. Histopathological lesions with alveolar macrophage accumulation.	11.94 nm	0.2–5 mg/m^3	Unknown	28 days (5 days/week)
Bourgois et al., 2021 [14]	Nose-only inhalation	Rats	Pro-inflammatory effects. Increases in neutrophils, macrophages, IL-1 β, TNF-α, GRO/KC, and MIP-2 in BALF. Histopathological lesions with neutrophil accumulation at the interstitial and alveolar level, as well as by a thickening of alveolar partitions.	13 nm	20–22.1 mg/m^3	γ/δ	4 h; 4 h/day during 4 days
Li et al., 2017 [16]	Whole-body inhalation	Mice	Emphysema, small airway remodeling, enhanced inflammation, and apoptosis. PTN6 down-regulation, STAT3 phosphorylation, and PDCD4 apoptotic marker increased expression. Increases in IL-6 and IL-33 concentrations in BALF. COPD-like lesions.	40 nm	0.4 mg/m^3	Unknown	7 days
Tornling et al., 1993 [27]	Intra-tracheal instillation	Rats	Increases in neutrophils, macrophages, and fibronectin concentrations in BALF. Persistence of this phenomenon 12 months after intra-tracheal instillation.	4.37 μm	40 mg	γ/α	–
Cho et al., 2012 [7]	Intra-tracheal instillation	Rats	Increase in neutrophil concentration. Hemolytic potential of alumina NPs.	6.3 nm	150 cm^2 (0.5 mL at 300 cm^2/mL)	Unknown	–
Kwon et al., 2013 [28]	Nasal instillation	Rats	Increase in total proteins, LDH, IL-6, and TNF-α concentrations in BALF.	Unknown	1–40 mg/kg	Unknown	–
Ess et al., 1993 [29]	Intra-tracheal instillation	Rats	Particles usually used for the manufacture of aluminum (α and γ) had no fibrotic effect, whereas other particles tested could induce this type of lesion. No link between cytotoxicity and fibrotic effect.	<11 μm	50 mg/0.5 mL 1% suspension	α, γ, δ, and χ	–
El-Hussainy et al., 2016 [30]	Intra-tracheal instillation	Rats	ECG disorders and an increase in myocardial (LDH, triglycerides, creatine phosphokinase, cholesterol, nitric oxide) and inflammatory (TNF-α) damage markers in blood serums. Decrease in antioxidants (reduced glutathione and superoxide dismutase) in animal serum.	11 nm	30 mg/kg/day	α	14 days

4. In Vitro Studies of Cytotoxic Mechanisms Induced by Alumina Nanoparticles Exposure on Lung Cells

A study conducted on murine fibroblasts (L929 cell line) and normal human skin fibroblasts (BJ cells) incubated with 10 to 400 µg/mL of γ-Al_2O_3 particles (NPs fraction of 56–91 nm and agglomerates fraction of 106–220 nm) for 24 h did not show a decrease in cell viability nor apoptosis induction. However, the authors demonstrated the same penetration dynamics of Al_2O_3 particles in both cell types [31]. These results were confirmed by more recent experiments on human alveolar A549 epithelial cells and skin keratinocytes HaCaT exposed for 24 h to three different Al_2O_3 particles (primary sizes 14 nm, 111 nm, and 750 nm; α- and α/δ crystallinities; concentrations ranging from 10 to 50 mg/L) [32]. Particles were internalized by cells in the cytoplasm but not detected in nuclei and did not exert toxicity. A comparison of the cytotoxicity induced by different metal oxide NPs (Al_2O_3, CeO_2, TiO_2, and ZnO) on human lung cell lines (A549 carcinoma cells and L-132 normal cells) concluded lower cytotoxicity of NPs of alumina compared to the other NPs tested [33]. After 72 h incubation with Al_2O_3 NPs, no modification of the proliferation and cell viability were observed (size 20 nm, crystallinity unknown, NPs concentrations ranging from 1 to 1000 µg/mL). Moreover, the authors did not report a significant increase in LDH nor reactive oxygen species (ROS) production. Park and colleagues studied the toxicity of three types of synthesized aluminum oxide nanoparticles (AlONPs): γ-aluminum oxide hydroxide nanoparticles (γ-AlOHNPs), γ- and α-AlONPs (diameter 180–200 nm, exposure concentrations 5 and 20 µg/mL) [34]. They exposed for 24 h six human cell lines to NPs, including bronchial epithelial BEAS-2B cells, and showed that γ-AlOHNPs induced the greatest toxicity by decreasing ATP production and normalized cell index (ICN: parameter taking into account cell number, morphology, and cell adhesion), and increasing LDH release. They postulated that low stability in biological and hydroxyl groups of γ-AlOHNPs plays an important role in their cytotoxicity and bioaccumulation. Conflicting results are available in the literature regarding the effects of alumina particle exposure on cell proliferation. A decrease in ICN of bronchial epithelial cells (cell line 169HBE14o-) was shown after exposure to Al_2O_3 NPs (size less than 50 nm, crystallinity unknown) for 48 h [35]. Conversely, the exposure of pleural cells (NCI-H460 cell line) at similar concentrations of alumina NPs (14 nm, unknown crystallinity) did not show any significant modification of the ICN [36].

These studies seem to show that Al_2O_3 particles' physico-chemical parameters (chemistry, size, etc.) and the cellular model used to determine the observed cytotoxicity. Particle size also seems to play the main role in cytotoxicity mechanisms. Indeed, human A549 alveolar epithelial cells were respectively exposed to Al_2O_3 NPs (sizes 10 nm and 50 nm; γ and γ/δ crystallinities) and titanium dioxide particles (TiO_2, 5 nm, and 200 nm) for two and five days. Cell metabolism and cell proliferation were then studied using Alamarblue® and clonogenic assays. Contrary to Kim et al., they showed that Al_2O_3 NPs were more cytotoxic than TiO_2 [33,37]. Smaller NPs (according to their primary size) exhibiting higher relative surface area than larger particles also induced more toxic effects, but this was not correlated with measured hydrodynamic particle sizes (diameter of the NPs and/or agglomerates in a biological medium). This suggests that the more important the specific surface area of NPs is, the more Al_2O_3 NPs can be cytotoxic. However, the cytotoxicity of Al_2O_3 NPs may not only depend on particle size. Indeed, a study investigated genotoxic effects on human fibroblasts of different particles containing alumina (NPs, microparticles, and fibers) and did not put evidence of differences in the induction of micronuclei [38]. Authors hypothesized that biological effects after exposure to particles would appear to depend on chemical composition as well as size, shape, and cell type. Recently, Bourgois et al. also exposed 24 h A549 alveolar epithelial cells to Al_2O_3 particles [39]. They did not show any effects of different particle sizes and crystallinities on normalized cell index, cell viability, reduced glutathione, and double DNA strand breaks.

Although results of studies on Al_2O_3 NPs cytotoxicity do not seem to show systematically significant decreases in cell viability, these particles can induce different biological effects. It was shown that Al_2O_3 NPs (sizes 10–20 nm, crystallinity γ/α, concentrations ranging from 1 to 250 μg/mL) exposure for 24 h led to increases in mRNA and protein expression of VCAM-1, ICAM-1, and ELAM-1 in endothelial cells and increased adhesion of activated monocytes [40]. This study suggests the pro-inflammatory effects of alumina in nanoparticle form. Recently, experiments were performed on human bronchial epithelial (HBE) cells in order to characterize microRNA expression using microarrays after Al_2O_3 NPs exposure for 24 h (size distribution between 5 and 100 nm, unknown crystallinity, concentrations of 50 and 250 μg/mL) [41]. A homologous miRNA in Homo sapiens and Mus musculus, miR-297, was significantly up-regulated following exposures to Al_2O_3 NPs compared to control cells. Moreover, a few studies have reported a genotoxic potential of alumina NPs in vitro. An increased frequency of micronuclei and chromosomal aberrations has been observed on primary cultures of human fibroblasts exposed to alumina NPs (size 0.2 μm, crystallinity unknown, concentration ranging from 0.1 to 10 mg/culture flask). Nevertheless, the genotoxic effects induced were less important than those obtained in parallel with cobalt-chromium (CoCr) NPs, and no increase in DNA double-strand breaks was demonstrated in the presence of alumina [38]. Different results were obtained in another study, which showed that at low concentrations (10 μM to 1 mM), the alumina NPs (size and crystallinity unknown) were at the origin of genotoxic effects without decreasing cell viability (except at the highest concentrations) on human peripheral blood lymphocytes [42]. Indeed, these NPs increased single-strand breaks and oxidative DNA damage (2,6-diamino-4-hydroxy-5-N-methylformamidopyrimidine and 7,8-dihydro-8-oxo-2′deoxyguanosine). A third study has demonstrated the induction of DNA strand breaks by Al_2O_3 NPs (13 and 50 nm, crystallinities unknown) after incubation with Chinese hamster lung fibroblasts [43]. Moreover, significant oxidative stress (decrease in glutathione, activity of superoxide dismutase, malondialdehyde, and total antioxidant capacity) was demonstrated after exposure to concentrations ranging from 15 to 60 μg/mL. Comparison of cytotoxic and genotoxic effects of four different NPs (oxides of cobalt, iron, silicon, and aluminum) on human lymphocytes in a recent study has, however, demonstrated that Al_2O_3 NPs cause less damage to DNA than the other NPs studied [44]. Nevertheless, they do significantly increase the production of reactive oxygen species and lead to a significant decrease in reduced glutathione at a concentration of 100 μg/mL. Negative results were obtained during reverse mutation tests on bacteria in the presence of alumina NPs (sizes less than 50 nm, crystallinity unknown), leading to the conclusion of an absence of mutagenic potential of these NPs [45].

Other specific effects of alumina NPs have been demonstrated in vitro. Thereby, alumina NPs (8–12 nm, crystallinity unknown, concentrations ranging from 1 μM to 10 mM) can induce a decrease in the expression of tight junction proteins [46]. A pre-incubation of HBMEC cells (Human Brain Microvascular Endothelial Cells, brain cell line) with glutathione blocks this effect, meaning that it could be a consequence of a phenomenon related to oxidative stress induced by exposure to NPs. Furthermore, a study carried out on erythrocytes of different species (human, rat, and rabbit) highlighted evidence of a hemolytic power of alumina NPs (13 nm, less than 50 nm, and nanofibers 2–6 nm by 200–400 nm, crystallinities unknown) [47]. Although alumina NPs are metal oxides known for their antimicrobial properties, they have only limited antimicrobial properties [48]. Only high concentrations (1000 μg/mL) have a moderate effect on bacterial proliferation (*Escherichia coli*).

A summary of the results of in vitro studies exploring the cytotoxic effects of alumina particles is available in Table 3.

Table 3. Summary of in vitro studies.

References	Cell Model	Biological Effects	Primary Particle Size	Particle Concentration	Particle Crystallinity	Exposure Time
Radziun et al., 2011 [31]	Murine fibroblasts (L929 cell line) and normal human skin fibroblasts (BJ cells)	No decrease in cell viability or apoptosis induction. NPs internalization in both cell types.	50–80 nm	10–400 µg/mL	γ	24 h
Bohme et al., 2014 [32]	Human alveolar epithelial cells (A549 cell line) and human skin keratinocytes (HaCaT cell line)	Internalization in cell cytoplasm, no detection in cell nuclei, no cytotoxicity.	14 nm 111 nm 750 nm	10–50 mg/L	α and α/δ	24 h
Kim et al., 2010 [33]	Human lung cell lines (A549 carcinoma cells and L-132 normal cells)	Lower cytotoxicity of NPs of alumina compared to the other metal oxide NPs tested (CeO_2, TiO_2, and ZnO).	20 nm	0.5–1000 µg/mL	Unknown	24 h, 48 h, and 72 h
Park et al., 2016 [34]	Six human cell lines, including bronchial epithelial BEAS-2B cells	γ-aluminum oxide hydroxide nanoparticles induced greatest toxicity compared to γ- and α-Al_2O_3 NPs.	180–200 nm	5 and 20 µg/mL	α and γ	24 h
Otero-Gonzalez et al., 2012 [35]	Human bronchial epithelial cells (169HBE14o- cell line)	Decrease in normalized cell index and cell viability at the highest concentrations.	<50 nm	100–1000 mg/mL	Unknown	48 h
Simon-Vazquez et al., 2016 [36]	Human pleural cells (NCI-H460 cell line)	No modification of normalized cell index.	14 nm	15, 63, and 500 µg/mL	Unknown	48 h
Wei et al., 2014 [37]	Human alveolar epithelial cells (A549 cell line)	Smallest NPs more cytotoxic (inhibition of cell proliferation). Hydrodynamic diameter does not influence cytotoxicity. Al_2O_3 NPs more cytotoxic compared to TiO_2 NPs.	10 and 50 nm	0.1–10 mg/mL	γ and γ/δ	2 and 5 days
Tsaousi et al., 2010 [38]	Primary human fibroblasts	No induction of micronuclei and no increase in DNA double-strand breaks. Size and shape of Al_2O_3 nano-objects do not influence genotoxicity (micronuclei and chromosomal aberration).	0.2 nm and 2 µm and alumina fibers (0.9 µm diameter, 12.03 µm length)	1.33–133.33 µg/cm^2	Unknown	24 h
Oesterling et al., 2008 [40]	Primary pulmonary artery endothelial cells, human umbilical vein endothelial cells, and monocytes	Increases in mRNA and protein expression of VCAM-1, ICAM-1, and ELAM-1 increased adhesion of activated monocytes.	10–20 nm	1–250 µg/mL	α/γ	24 h
Yun et al., 2020 [41]	Human bronchial epithelial cells	Up-regulation of homologous miRNA in *Homo sapiens* and *Mus musculus* miR-297.	5–100 nm (scanning electron microscopy)	50 and 250 µg/mL	Unknown	24 h
Sliwinska et al., 2015 [42]	Human peripheral blood lymphocytes	Increased single-strand breaks and oxidative DNA damage (2,6-diamino-4-hydroxy-5-N-methyl formamidopyrimidine and 7,8-dihydro-8-oxo-2′deoxyguanosine).	Unknown	From 10 µM to 1 mM	Unknown	24 h
Zhang et al., 2017 [43]	Chinese hamster lung fibroblasts, *Salmonella typhimurium*	Genotoxicity of Al_2O_3 NPs (Ames test, Comet test, Micronucleus assay, Sperm deformity test). Antioxidant decreases.	13 nm; 50 nm	0.5–5000 µg/mL	Unknown	12 h, 24 h, and 48 h
Rajiv et al., 2016 [44]	Human lymphocytes	Al_2O_3 NPs cause less damage to DNA than the other NPs studied (Co_3O_4, Fe_2O_3, and SiO_2 NPs). Al_2O_3 NPs exposures induced significant increases in reactive oxygen species production.	<50 nm	10–100 µg/mL	Unknown	24 h

Table 3. *Cont.*

References	Cell Model	Biological Effects	Primary Particle Size	Particle Concentration	Particle Crystallinity	Exposure Time
Pan et al., 2010 [45]	*Salmonella typhimurium*	Negative reverse mutation assay: absence of mutagenic potential.	<50 nm	10–1000 µg/plate	Unknown	72 h
Chen et al., 2008 [46]	Human Brain Microvascular Endothelial Cells (HBMEC cell line)	Decrease in the expression of tight junction proteins related to oxidative stress induced.	8–12 nm	From 1 µM to 10 mM	Unknown	24 h
Vinardell et al., 2015 [47]	Erythrocytes (Human, Rat, Rabbit)	Hemolytic power of Al_2O_3 NPs.	13 nm; <50 nm; Nanofibers (2–6 nm diameter, 200–400 nm length)	2.5–40 mg/mL	Unknown	1 h, 3 h, and 24 h
Sadiq et al., 2009 [48]	*Escherichia coli*	Weak antimicrobial power at high concentration.	<50 nm	10–1000 µg/mL	γ	24 h
Bourgois et al., 2019 [39]	Human alveolar epithelial cells (A549 cell line)	No effect on cell index, cell viability, reduced glutathione, and double DNA strand breaks. Internalization of NPs in cytoplasm.	10 nm; 13 nm; 500 nm	1.56–200 µg/cm^2	γ and γ/δ	24 h

5. Discussion

Biological mechanisms of lung toxicity and physiopathology triggered by exposure to alumina particles are still unclear and not sufficiently studied. The small number of cohort studies complicates the identification of clear exposure–response relationships for respiratory diseases [49]. In human studies, the time of population exposure and associated comorbidities and medical background (asthma, smoker/non-smoker, etc.) are often unknown. These studies mainly addressed worker populations. However, a recent work was carried out by Sikkeland et al. on the sputum of healthy volunteers never-smokers, with no allergy and respiratory diseases, free from respiratory infections 4 weeks prior and with a standardized FVC (Forced Vital Capacity)/FEV 1 ratio of 80 ± 1.9 [9]. To the best of our knowledge, it is to date the only existing study addressing specifically inflammatory effects of Al_2O_3 particles (neutrophils and IL-8 increased concentrations) and localized respiratory effects on humans. However, the duration of exposure is only 2 h, while occupational exposure may be for longer durations. Interestingly, they performed sputum collection as increase in neutrophil concentration was characterized among other workers exposed to other pollutants in several different industries, such as paper mills, popcorn factories, cement industry, pig farming, fish feed production, and waste handling. Sputum collection was realized until 24 h after exposure because collecting induced sputum several times within 48 h would be problematic since the sputum induction process may also lead to inflammation [50]. Consequently, sputum collection may be considered to analyze early lung pro-inflammatory effects on humans after exposure, but no further. To analyze the chronic inflammatory response, BALF collection on healthy human volunteers is not considered ethically acceptable as it is invasive and painful. However, such analysis remains essential on anesthetized animals to correctly describe pro-inflammatory mechanisms potentially involved in lung diseases.

Human studies are often not specific to alumina particle toxicity, as workers may inhale a mix of pollutants in the occupational environment. Interestingly, Mazzoli-Rocha et al. exposed by whole-body inhalation BALB/c mice to dust (mainly Al_2O_3 particles) collected in an aluminum-producing facility, and they showed impaired lung mechanics associated with inflammation (influx of polymorphonuclear cells) [51]. In order to improve the knowledge of pro-inflammatory effects caused specifically by alumina particle exposure, two studies were performed recently by nose-only inhalation. On the one hand, Kim et al. exposed rats for one month to different concentrations (ranging from 0.2 to 5 mg/m^3) of Al_2O_3 NPs, showing strong inflammatory cytokine secretion in BALF [15]. However, particle crystallinity was not characterized by the authors. Several physico-chemical properties of Al_2O_3 NPs are often missing and/or not sufficiently characterized in scientific studies. Particle concentration, size distribution, surface chemistry, and NPs crystallinity seem to have a great impact on biological effects [52]. Alumina has several crystalline phases, and transitions between them occur as follows: γ-$Al_2O_3 \rightarrow \delta$-$Al_2O_3 \rightarrow \theta$-$Al_2O_3 \rightarrow \alpha$-$Al_2O_3$ [53]. Three other crystal forms also exist but are in the minority: η, χ, and κ [54]. Crystalline phase α-Al_2O_3 is the thermodynamically stable one, whereas γ, δ, and θ phases correspond to transition metastable alumina particles [53,55]. Therefore Al_2O_3 NPs crystallinity is an important physico-chemical parameter to characterize, as it was shown that these NPs could induce or not fibrotic effects depending on their crystallinities [29]. This was also demonstrated in several human cell lines, including bronchial or alveolar epithelial cells, that cytotoxicity could be modulated depending on the crystallinities of Al_2O_3 particles [34,39]. On the other hand, Bourgois et al. exposed rats by nose-only inhalation to a strong concentration of γ/δ-Al_2O_3 NPs, also showing increased inflammatory response after five days [14]. However, only one elevated concentration (20 mg/m^3) of these well-characterized NPs was administrated to animals for only early analysis of pro-inflammatory effects. Different concentrations of Al_2O_3 particles could also be administered to rats in order to better establish dose-effects curves and consequently to build regulatory toxicological values. To date, in France, the average exposure limit value is 10 mg/m^3 for total alumina dusts, which corresponds to the regulatory limit for the metal aluminum. Therefore, animal studies will play a major role in establishing new occupational exposure limit values for alumina particles and nanoparticles.

Pro-inflammatory mechanisms triggered by nose-only inhalation exposure of Al_2O_3 NPs seem to involve an increase in neutrophils, lymphocytes, and macrophages afflux in BALF in association with pro-inflammatory cytokines secretion and LDH release [14,15]. This result was also observed after whole-body inhalation of Al_2O_3 NPs [16]. Some authors hypothesized that Al_2O_3 NPs could stimulate the NFκB pathway [36]. NFκB can contribute to inflammasome regulation, which is involved in IL-1β synthesis [56,57]. This pathway is also known to be activated in the lungs of patients with COPD [58]. The release of IL-1β in BALF may be linked to TNF-α secretion observed in several studies [14,15,28,30]. It has also been demonstrated that in the context of acute inflammation, IL-1β contributes to TNF-α-mediated chemokine release and neutrophil recruitment to the lung [59]. However, IL-6 secretion was not systematically increased in BALF after inhalation of Al_2O_3 NPs. We hypothesize that it may be attributed to Al_2O_3 NPs concentration administrated to animals, as Li et al. showed increased IL-6 concentration after seven days of exposure to 0.4 mg/m^3, while no IL-6 increase was found by Bourgois et al. after four days of exposure to roughly 20 mg/m^3 [14,16]. However, the crystallinity of Al_2O_3 NPs used by Li et al. is unknown, and exposure durations are different between both studies. Consequently, the conclusion about the mechanism triggering IL-6 secretion is hard to explain because studies cannot rigorously be compared. Another study was realized on C57Bl/6 J male mice exposed to aluminum oxide-based nanowhiskers (3.3 ± 0.6 mg/m^3) using a dynamic whole-body exposure chamber for 2 or 4 weeks [60]. These sub-chronic exposures induced an increase in lung macrophage concentration but did not induce an increase in pro-inflammatory cytokines release (i.e., IL-6, IFN-γ, MIP-1α, TNF-α, and MIP-2). This result is contradictory with previous other studies performed on spherical Al_2O_3 NPs where pro-inflammatory cytokines (i.e., IL-6, IL-1β, TNF-α, and MIP-2) were released in BALF after one or four weeks of exposure [14,15]. Therefore, the nano-objects shape could also play an important role in Al_2O_3 NPs toxicity and associated pro-inflammatory effects on the lungs. Some other cytokines, such as MIP-2 and GRO/KC, may play a role in the early pulmonary inflammation contributing to polymorphonuclear cell recruitment within 24 h after exposure. Several studies demonstrated down-regulation of their secretion in BALF or nasal fluid lavage after several days [14,61]. Overall, in order to study chronic toxic and pro-inflammatory effects of Al_2O_3 NPs, it would be interesting to perform longer studies or to house animals longer after inhalation exposure. These studies may allow bettering determining if pro-inflammatory effects are reversible or if diseases such as COPD, emphysema, or pulmonary fibrosis may occur. As alumina is classified in "aluminum production" as carcinogenic to humans (Group 1) by the International Agency for Research on Cancer, long-term studies are essential. To the best of our knowledge, only one long-term study showed that up to one year after intra-tracheal exposure of rats to Al_2O_3 particles [29]. None of the five aluminas (α- and γ- crystalline phases) used for primary aluminum production showed any fibrogenic potential, while chemical grade Al_2O_3 particles or laboratory-produced samples induced fibrogenic lesions in the lung parenchyma. It would also be interesting to explore after pro-inflammatory and pro-fibrogenic effects of Al_2O_3 particles after inhalation that might modify Al_2O_3 particles' lung burden and, consequently, biological effects compared to intra-tracheal instillation exposure.

Lung pro-inflammatory mechanisms triggered specifically by Al_2O_3 particles are not sufficiently explored in in vitro studies. To the best of our knowledge, only Osterling et al. have investigated mRNA and protein expression of adhesion molecules of monocytes on endothelial cells (VCAM-1, ICAM-1, and ELAM-1) [40]. In order to reduce animal experiments and to better understand pro-inflammatory mechanisms and chronic effects, new 3D in vitro models have been recently developed. A recent literature review highlights the benefits of using 3D co-culture models to investigate the complexity of cellular interactions during pulmonary inflammation [62]. A specific in vitro mini-lung fibrosis model equipped with non-invasive real-time monitoring of cell mechanics was developed [63]. This in vitro model combined a co-culture of three cell types: epithelial and endothelial cell lines incubated with primary fibroblasts from idiopathic pulmonary fibrosis patients. Cells are cultivated on a biomimetic ultrathin basement (biphasic elastic thin for air–liquid culture conditions, BETA) membrane (<1 μm) developed with unique properties, including biocompatibility, permeability, and high elasticity (<10 kPa) for cell culturing under air–liquid interface (ALI). This cellular model may allow us to study more precisely pro-inflammatory or pro-fibrogenic mechanisms following exposures to Al_2O_3 NPs, taking into account the elasticity of the alveolo-capillary barrier in ALI and real-time measurements. Other studies also suggest using cell co-culture, including fibroblasts, to investigate the inflammatory and pro-fibrogenic effects of inhaled components. For instance, Barosova et al. recently published the development of a three-dimensional alveolar model consisting of human primary alveolar epithelial cells, fibroblasts, and endothelial cells, with or without macrophages [64]. Cell co-cultures are cultivated on bicameral chambers in ALI and mimic the alveolo-capillary barrier. Pulmonary cells can be exposed with the help of specific commercialized devices to particle mist. This type of cellular model could be interesting in exploring long-term cytotoxic and pro-inflammatory mechanisms in vitro.

6. Conclusions

This review is a critical update of the current knowledge on underlying toxicological, molecular, and cellular mechanisms induced by exposure to Al_2O_3 NPs on the lungs. Human and animal studies point out that inhalation of Al_2O_3 particles can induce aluminosis, local respiratory symptoms (cough, shortness of breath), and pro-inflammatory response and may trigger long-term pulmonary fibrosis. Not enough cohort studies and clinical studies on healthy volunteers are performed to better understand these mechanisms and to establish clear exposure–response relationships. In studies with animals or cells, physico-chemistry of Al_2O_3 particles has to be extensively analyzed and published in order to improve the understanding of related biological effects. Inhalation exposures are closer to realistic environmental exposures, and long-term animal studies are necessary to determine whether pro-inflammatory reactions may reverse or turn into fatal diseases such as pulmonary fibrosis. Three-dimensional co-culture models may also allow studying these underlying pro-inflammatory and cytotoxic mechanisms for several weeks of exposure, as it was recently performed [64,65]. Consequently, in order to improve the analysis of pro-inflammatory and pro-fibrogenic effects, a combination of long-term animal studies by inhalation exposures and the use of dedicated 3D co-culture models is needed.

Author Contributions: Writing—original draft preparation, S.D. and A.B.; writing—review and editing, S.D. and S.F. All authors have read and agreed to the published version of the manuscript.

Funding: This research received no external funding.

Informed Consent Statement: Not applicable.

Acknowledgments: Authors kindly acknowledge Marco Valente for proofreading the manuscript.

Conflicts of Interest: The authors declare no conflict of interest.

References

1. Liu, S.; Tian, J.; Zhang, W. Fabrication and application of nanoporous anodic aluminum oxide: A review. *Nanotechnology* **2021**, *32*, 222001. [CrossRef] [PubMed]
2. Jacukowicz-Sobala, I.; Ociński, D.; Kociołek-Balawejder, E. Iron and aluminium oxides containing industrial wastes as adsorbents of heavy metals: Application possibilities and limitations. *Waste Manag. Res.* **2015**, *33*, 612–629. [CrossRef]
3. Hull, M.J.; Abraham, J.L. Aluminum welding fume-induced pneumoconiosis. *Hum. Pathol.* **2002**, *33*, 819–825. [CrossRef] [PubMed]
4. Brown, D.M.; Brown, A.M.; Willitsford, A.H.; Dinello-Fass, R.; Airola, M.B.; Siegrist, K.M.; Thomas, M.E.; Chang, Y. Lidar measurements of solid rocket propellant fire particle plumes. *Appl. Opt.* **2016**, *55*, 4657–4669. [CrossRef] [PubMed]
5. Riediker, M.; Zink, D.; Kreyling, W.; Oberdörster, G.; Elder, A.; Graham, U.; Lynch, I.; Duschl, A.; Ichihara, G.; Ichihara, S.; et al. Particle toxicology and health—Where are we? *Part. Fibre Toxicol.* **2019**, *16*, 19. [CrossRef] [PubMed]

6. Xing, M.; Zou, H.; Gao, X.; Chang, B.; Tang, S.; Zhang, M. Workplace exposure to airborne alumina nanoparticles associated with separation and packaging processes in a pilot factory. *Environ. Sci. Process. Impacts* **2015**, *17*, 656–666. [CrossRef]

7. Cho, W.S.; Duffin, R.; Thielbeer, F.; Bradley, M.; Megson, I.L.; Macnee, W.; Poland, C.A.; Tran, C.L.; Donaldson, K. Zeta potential and solubility to toxic ions as mechanisms of lung inflammation caused by metal/metal oxide nanoparticles. *Toxicol. Sci.* **2012**, *126*, 469–477. [CrossRef] [PubMed]

8. Hunter, D.; Milton, R.; Perry, K.M.A.; Thompson, D.R. Effect of Aluminium and Alumina on the Lung in Grinders of Duralumin Aeroplane Propellers. *Occup. Environ. Med.* **1944**, *1*, 159–164. [CrossRef]

9. Sikkeland, L.; Alexis, N.E.; Fry, R.C.; Martin, E.; Danielsen, T.E.; Søstrand, P.; Kongerud, J. Inflammation in induced sputum after aluminium oxide exposure: An experimental chamber study. *Occup. Environ. Med.* **2016**, *73*, 199–205. [CrossRef] [PubMed]

10. Buchman, J.T.; Hudson-Smith, N.V.; Landy, K.M.; Haynes, C.L. Understanding nanoparticle toxicity mechanisms to inform redesign strategies to reduce environmental impact. *Accounts Chem. Res.* **2019**, *52*, 1632–1642. [CrossRef] [PubMed]

11. Vallyathan, V.; Bergeron, W.N.; Robichaux, P.A.; Craighead, J.E. Pulmonary fibrosis in an aluminum arc welder. *Chest* **1982**, *81*, 372–374. [CrossRef] [PubMed]

12. Kuman Oyman, E.; Hatman, E.A.; Karagül, D.A.; Kılıçaslan, Z. A current example of historical cases: Occupational pulmonary aluminosis. *Turk. Thorac. J.* **2021**, *22*, 83–85. [CrossRef]

13. Smolkova, P.; Nakladalova, M. The etiology of occupational pulmonary aluminosis—The past and the present. *Biomed. Pap. Med. Fac. Univ. Palacky Olomouc Czech Repub.* **2014**, *158*, 535–538. [CrossRef]

14. Bourgois, A.; Saurat, D.; de Araujo, S.; Boyard, A.; Guitard, N.; Renault, S.; Fargeau, F.; Frederic, C.; Peyret, E.; Flahaut, E.; et al. Nose-only inhalations of high-dose alumina nanoparticles/hydrogen chloride gas mixtures induce strong pulmonary pro-inflammatory response: A pilot study. *Inhal. Toxicol.* **2021**, *33*, 308–324. [CrossRef] [PubMed]

15. Kim, Y.S.; Chung, Y.H.; Seo, D.S.; Choi, H.S.; Lim, C.H. Twenty-eight-day repeated inhalation toxicity study of aluminum oxide nanoparticles in male sprague-dawley rats. *Toxicol. Res.* **2018**, *34*, 343–354. [CrossRef] [PubMed]

16. Li, X.; Yang, H.; Wu, S.; Meng, Q.; Sun, H.; Lu, R.; Cui, J.; Zheng, Y.; Chen, W.; Zhang, R.; et al. Suppression of PTPN6 exacerbates aluminum oxide nanoparticle-induced COPD-like lesions in mice through activation of STAT pathway. *Part. Fibre Toxicol.* **2017**, *14*, 53. [CrossRef] [PubMed]

17. Jederlinic, P.J.; Abraham, J.L.; Churg, A.; Himmelstein, J.S.; Epler, G.R.; Gaensler, E.A. Pulmonary fibrosis in aluminum oxide workers. Investigation of nine workers, with pathologic examination and microanalysis in three of them. *Am. Rev. Respir. Dis.* **1990**, *142*, 1179–1184. [CrossRef]

18. Townsend, M.C.; Enterline, P.E.; Sussman, N.B.; Bonney, T.B.; Rippey, L.L. Pulmonary function in relation to total dust exposure at a bauxite refinery and alumina-based chemical products plant. *Am. Rev. Respir. Dis.* **1985**, *132*, 1174–1180.

19. Schwarz, Y.; Kivity, S.; Fischbein, A.; Abraham, J.L.; Fireman, E.; Moshe, S.; Dannon, Y.; Topilsky, M.; Greif, J. Evaluation of workers exposed to dust containing hard metals and aluminum oxide. *Am. J. Ind. Med.* **1998**, *34*, 177–182. [CrossRef]

20. Edling, C.; Jarvholm, B.; Andersson, L.; Axelson, O. Mortality and cancer incidence among workers in an abrasive manufacturing industry. *Br. J. Ind. Med.* **1987**, *44*, 57–59. [CrossRef]

21. Sjogren, B.; Ulfvarson, U. Respiratory symptoms and pulmonary function among welders working with aluminum, stainless steel and railroad tracks. *Scand. J. Work Environ. Health* **1985**, *11*, 27–32. [CrossRef] [PubMed]

22. Antonini, J.M.; Roberts, J.R.; Schwegler-Berry, D.; Mercer, R.R. Comparative microscopic study of human and rat lungs after overexposure to welding fume. *Ann. Occup. Hyg.* **2013**, *57*, 1167–1179. [PubMed]

23. OECD/OCDE. *Guidance Document on Acute Inhalation Toxicity Testing*; OECD Publishing: Paris, France, 2009.

24. OECD/OECD. *Ligne Directrice de L'ocde pour les Essais de Produits Chimiques—Toxicité Aiguë par Inhalation*; OECD Publishing: Paris, France, 2009.

25. OCDE/OECD. *Ligne Directrice de L'ocde pour les Essais de Produits Chimiques—Toxicité Subaiguë par Inhalation: Étude Sur 28 Jours*; OECD Publishing: Paris, France, 2009.

26. Frohlich, E.; Salar-Behzadi, S. Toxicological assessment of inhaled nanoparticles: Role of in vivo, ex vivo, in vitro, and in silico studies. *Int. J. Mol. Sci.* **2014**, *15*, 4795–4822. [CrossRef] [PubMed]

27. Tornling, G.; Blaschke, E.; Eklund, A. Long term effects of alumina on components of bronchoalveolar lavage fluid from rats. *Br. J. Ind. Med.* **1993**, *50*, 172–175. [CrossRef]

28. Kwon, J.-T.; Seo, G.-B.; Lee, M.; Kim, H.-M.; Shim, I.; Jo, E.; Kim, P.; Choi, K. Pulmonary toxicity assessment of aluminum oxide nanoparticles via nasal instillation exposure. *Korean J. Environ. Health Sci.* **2013**, *39*, 48–55. [CrossRef]

29. Ess, S.M.; Steinegger, A.F.; Ess, H.J.; Schlatter, C. Experimental study on the fibrogenic properties of different types of alumina. *Am. Ind. Hyg. Assoc. J.* **1993**, *54*, 360–370. [CrossRef]

30. El-Hussainy, E.H.M.; Hussein, A.M.; Abdel-Aziz, A.; El-Mehasseb, I. Effects of aluminum oxide (Al_2O_3) nanoparticles on ECG, myocardial inflammatory cytokines, redox state, and connexin 43 and lipid profile in rats: Possible cardioprotective effect of gallic acid. *Can. J. Physiol. Pharmacol.* **2016**, *94*, 868–878. [CrossRef]

31. Radziun, E.; Wilczynska, J.D.; Ksiazek, I.; Nowak, K.; Anuszewska, E.L.; Kunicki, A.; Olszyna, A.; Zabkowski, T. Assessment of the cytotoxicity of aluminium oxide nanoparticles on selected mammalian cells. *Toxicol. Vitr.* **2011**, *25*, 1694–1700. [CrossRef]

32. Bohme, S.; Stark, H.J.; Meissner, T.; Springer, A.; Reemtsma, T.; Kuhnel, D.; Busch, W. Quantification of Al2O3 nanoparticles in human cell lines applying inductively coupled plasma mass spectrometry (neb-ICP-MS, LA-ICP-MS) and flow cytometry-based methods. *J. Nanopart. Res.* **2014**, *16*, 2592. [CrossRef]

33. Kim, I.S.; Baek, M.; Choi, S.J. Comparative cytotoxicity of Al_2O_3, CeO_2, TiO_2 and ZnO nanoparticles to human lung cells. *J. Nanosci. Nanotechnol.* **2010**, *10*, 3453–3458. [CrossRef]

34. Park, E.J.; Lee, G.H.; Yoon, C.; Jeong, U.; Kim, Y.; Cho, M.H.; Kim, D.W. Biodistribution and toxicity of spherical aluminum oxide nanoparticles. *J. Appl. Toxicol.* **2016**, *36*, 424–433. [CrossRef] [PubMed]

35. Otero-Gonzalez, L.; Sierra-Alvarez, R.; Boitano, S.; Field, J.A. Application and validation of an impedance-based real time cell analyzer to measure the toxicity of nanoparticles impacting human bronchial epithelial cells. *Environ. Sci. Technol.* **2012**, *46*, 10271–10278. [CrossRef] [PubMed]

36. Simon-Vazquez, R.; Lozano-Fernandez, T.; Davila-Grana, A.; Gonzalez-Fernandez, A. Analysis of the activation routes induced by different metal oxide nanoparticles on human lung epithelial cells. *Future Sci. OA* **2016**, *2*, FSO118. [CrossRef]

37. Wei, Z.; Chen, L.; Thompson, D.M.; Montoya, L.D. Effect of particle size on in vitro cytotoxicity of titania and alumina nanoparticles. *J. Exp. Nanosci.* **2014**, *9*, 625–638. [CrossRef]

38. Tsaousi, A.; Jones, E.; Case, C.P. The in vitro genotoxicity of orthopaedic ceramic (Al_2O_3) and metal (CoCr alloy) particles. *Mutat. Res.* **2010**, *697*, 1–9. [CrossRef]

39. Bourgois, A.; Crouzier, D.; Legrand, F.X.; Raffin, F.; Boyard, A.; Girleanu, M.; Favier, A.L.; Francois, S.; Dekali, S. Alumina nanoparticles size and crystalline phase impact on cytotoxic effect on alveolar epithelial cells after simple or HCl combined exposures. *Toxicol. Vitr.* **2019**, *59*, 135–149. [CrossRef]

40. Oesterling, E.; Chopra, N.; Gavalas, V.; Arzuaga, X.; Lim, E.J.; Sultana, R.; Butterfield, D.A.; Bachas, L.; Hennig, B. Alumina nanoparticles induce expression of endothelial cell adhesion molecules. *Toxicol. Lett.* **2008**, *178*, 160–166. [CrossRef]

41. Yun, J.; Yang, H.; Li, X.; Sun, H.; Xu, J.; Meng, Q.; Wu, S.; Zhang, X.; Yang, X.; Li, B.; et al. Up-regulation of miR-297 mediates aluminum oxide nanoparticle-induced lung inflammation through activation of Notch pathway. *Environ. Pollut.* **2020**, *259*, 113839. [CrossRef]

42. Sliwinska, A.; Kwiatkowski, D.; Czarny, P.; Milczarek, J.; Toma, M.; Korycinska, A.; Szemraj, J.; Sliwinski, T. Genotoxicity and cytotoxicity of ZnO and Al_2O_3 nanoparticles. *Toxicol. Mech. Methods* **2015**, *25*, 176–183. [CrossRef]

43. Zhang, Q.; Wang, H.; Ge, C.; Duncan, J.; He, K.; Adeosun, S.O.; Xi, H.; Peng, H.; Niu, Q. Alumina at 50 and 13 nm nanoparticle sizes have potential genotoxicity. *J. Appl. Toxicol.* **2017**, *37*, 1053–1064. [CrossRef]

44. Rajiv, S.; Jerobin, J.; Saranya, V.; Nainawat, M.; Sharma, A.; Makwana, P.; Gayathri, C.; Bharath, L.; Singh, M.; Kumar, M.; et al. Comparative cytotoxicity and genotoxicity of cobalt (II, III) oxide, iron (III) oxide, silicon dioxide, and aluminum oxide nanoparticles on human lymphocytes in vitro. *Hum. Exp. Toxicol.* **2016**, *35*, 170–183. [CrossRef] [PubMed]

45. Pan, X.; Redding, J.E.; Wiley, P.A.; Wen, L.; McConnell, J.S.; Zhang, B. Mutagenicity evaluation of metal oxide nanoparticles by the bacterial reverse mutation assay. *Chemosphere* **2010**, *79*, 113–116. [CrossRef] [PubMed]

46. Chen, L.; Yokel, R.A.; Hennig, B.; Toborek, M. Manufactured aluminum oxide nanoparticles decrease expression of tight junction proteins in brain vasculature. *J. Neuroimmune Pharmacol.* **2008**, *3*, 286–295. [CrossRef] [PubMed]

47. Vinardell, M.P.; Sordé, A.; Díaz, J.; Baccarin, T.; Mitjans, M. Comparative effects of macro-sized aluminum oxide and aluminum oxide nanoparticles on erythrocyte hemolysis: Influence of cell source, temperature, and size. *J. Nanoparticle Res.* **2015**, *17*, 1–10. [CrossRef]

48. Sadiq, I.M.; Chowdhury, B.; Chandrasekaran, N.; Mukherjee, A. Antimicrobial sensitivity of Escherichia coli to alumina nanoparticles. *Nanomedicine* **2009**, *5*, 282–286. [CrossRef]

49. Benke, G.; Abramson, M.; Sim, M. Exposures in the alumina and primary aluminium industry: An historical review. *Ann. Occup. Hyg.* **1998**, *42*, 173–189. [CrossRef]

50. Holz, O.; Richter, K.; Jörres, R.A.; Speckin, P.; Mücke, M.; Magnussen, H. Changes in sputum composition between two inductions performed on consecutive days. *Thorax* **1998**, *53*, 83–86. [CrossRef]

51. Mazzoli-Rocha, F.; Santos, A.N.D.; Fernandes, S.; Normando, V.M.F.; Malm, O.; Saldiva, P.H.N.; Picanço-Diniz, D.L.W.; Faffe, D.S.; Zin, W.A. Pulmonary function and histological impairment in mice after acute exposure to aluminum dust. *Inhal. Toxicol.* **2010**, *22*, 861–867. [CrossRef]

52. Rivera Gil, P.; Oberdörster, G.; Elder, A.; Puntes, V.; Parak, W.J. Correlating physico-chemical with toxicological properties of nanoparticles: The present and the future. *ACS Nano* **2010**, *4*, 5527–5531. [CrossRef]

53. Piriyawong, V.; Thongpool, V.; Asanithi, P.; Limsuwan, P. Preparation and characterization of alumina nanoparticles in deionized water using laser ablation technique. *J. Nanomater.* **2012**, *2012*, 2. [CrossRef]

54. Matori, K.A.; Wah, L.C.; Hashim, M.; Ismail, I.; Zaid, M.H. Phase transformations of α-alumina made from waste aluminum via a precipitation technique. *Int. J. Mol. Sci.* **2012**, *13*, 16812–16821. [CrossRef]

55. Meda, L.; Marra, G.; Galfetti, L.; Inchingalo, S.; Severini, F.; de Luca, L. Nano-composites for rocket solid propellants. *Compos. Sci. Technol.* **2005**, *65*, 769–773. [CrossRef]

56. Liu, T.; Zhang, L.; Joo, D.; Sun, S.C. NF-κB signaling in inflammation. *Signal. Transduct. Target Ther.* **2017**, *2*, 17023. [CrossRef] [PubMed]

57. Lamkanfi, M.; Dixit, V.M. Mechanisms and functions of inflammasomes. *Cell* **2014**, *157*, 1013–1022. [CrossRef]

58. Barnes, P.J. Inflammatory mechanisms in patients with chronic obstructive pulmonary disease. *J. Allergy Clin. Immunol.* **2016**, *138*, 16–27. [CrossRef] [PubMed]

59. Saperstein, S.; Huyck, H.; Kimball, E.; Johnston, C.; Finkelstein, J.; Pryhuber, G. The effects of interleukin-1beta in tumor necrosis factor-alpha-induced acute pulmonary inflammation in mice. *Mediators Inflamm.* **2009**, *2009*, 958658. [CrossRef] [PubMed]

60. Adamcakova-Dodd, A.; Stebounova, L.V.; O'Shaughnessy, P.T.; Kim, J.S.; Grassian, V.H.; Thorne, P.S. Murine pulmonary responses after sub-chronic exposure to aluminum oxide-based nanowhiskers. *Part. Fibre Toxicol.* **2012**, *9*, 22. [CrossRef] [PubMed]

61. Pirela, S.V.; Miousse, I.R.; Lu, X.; Castranova, V.; Thomas, T.; Qian, Y.; Bello, D.; Kobzik, L.; Koturbash, I.; Demokritou, P. Effects of laser printer-emitted engineered nanoparticles on cytotoxicity, chemokine expression, reactive oxygen species, dna methylation, and dna damage: A comprehensive in vitro analysis in human small airway epithelial cells, macrophages, and lymphoblasts. *Environ. Health Perspect.* **2016**, *124*, 210–219. [PubMed]

62. Osei, E.T.; Booth, S.; Hackett, T.L. What have in vitro co-culture models taught us about the contribution of epithelial-mesenchymal interactions to airway inflammation and remodeling in asthma? *Cells* **2020**, *9*, 1694. [CrossRef] [PubMed]

63. Doryab, A.; Taskin, M.B.; Stahlhut, P.; Groll, J.; Schmid, O. Real-time measurement of cell mechanics as a clinically relevant readout of an in vitro lung fibrosis model established on a bioinspired basement membrane. *Adv. Mater.* **2022**, *34*, 2205083. [CrossRef]

64. Barosova, H.; Maione, A.G.; Septiadi, D.; Sharma, M.; Haeni, L.; Balog, S.; O'Connell, O.; Jackson, G.R.; Brown, D.; Clippinger, A.J.; et al. Use of epialveolar lung model to predict fibrotic potential of multiwalled carbon nanotubes. *ACS Nano* **2020**, *14*, 3941–3956. [CrossRef] [PubMed]

65. Kasper, J.Y.; Hermanns, M.I.; Unger, R.E.; Kirkpatrick, C.J. A responsive human triple-culture model of the air-blood barrier: Incorporation of different macrophage phenotypes. *J. Tissue Eng. Regen. Med.* **2017**, *11*, 1285–1297. [CrossRef] [PubMed]

 *biomedicines*

Article

Long-Term Anxiety-like Behavior and Microbiota Changes Induced in Mice by Sublethal Doses of Acute Sarin Surrogate Exposure

Sabine François [1,†], Stanislas Mondot [2,†], Quentin Gerard [3,4], Rosalie Bel [3], Julie Knoertzer [3], Asma Berriche [3,5], Sophie Cavallero [1], Rachid Baati [6], Cyrille Orset [4], Gregory Dal Bo [3,*,†] and Karine Thibault [3,*,†]

1. Department of Radiation Biological Effects, Armed Forces Biomedical Research Institute, 91220 Bretigny sur Orge, France; sfm.francois@gmail.com (S.F.); sophie.cavallero@def.gouv.fr (S.C.)
2. Micalis Institute, AgroParisTech, Université Paris-Saclay, INRAE, 78350 Jouy-en-Josas, France; stanislas.mondot@inrae.fr
3. Department of Toxicology and Chemical Risks, Armed Forces Biomedical Research Institute, 91220 Bretigny sur Orge, France; gerard@cyceron.fr (Q.G.); rosalie.bel@outlook.com (R.B.); julie.knoertzer@def.gouv.fr (J.K.); asmaberriche@gmail.com (A.B.)
4. Institut Blood and Brain@caen-normandie Cyceron, Caen-Normandie University, UNICAEN, INSERM, UMR-S U1237, Physiopathology and Imaging of Neurological Disorders (PhIND), 14000 Caen, France; orset@cyceron.fr
5. CEA, 92260 Fontenay aux Roses, France
6. ICPEES UMR CNRS 7515, Institut de Chimie des Procédés, pour l'Energie, l'Environnement, et la Santé, 67000 Strasbourg, France; rachid.baati@unistra.fr
* Correspondence: greg.dal-bo@chemdef.fr (G.D.B.); karine.thibault@def.gouv.fr (K.T.)
† These authors contributed equally to this work.

Citation: François, S.; Mondot, S.; Gerard, Q.; Bel, R.; Knoertzer, J.; Berriche, A.; Cavallero, S.; Baati, R.; Orset, C.; Dal Bo, G.; et al. Long-Term Anxiety-like Behavior and Microbiota Changes Induced in Mice by Sublethal Doses of Acute Sarin Surrogate Exposure. *Biomedicines* **2022**, *10*, 1167. https://doi.org/10.3390/biomedicines10051167

Academic Editor: María Morell Hita

Received: 25 February 2022
Accepted: 16 May 2022
Published: 18 May 2022

Publisher's Note: MDPI stays neutral with regard to jurisdictional claims in published maps and institutional affiliations.

Abstract: Anxiety disorder is one of the most reported complications following organophosphorus (OP) nerve agent (NA) exposure. The goal of this study was to characterize the long-term behavioral impact of a single low dose exposure to 4-nitrophenyl isopropyl methylphosphonate (NIMP), a sarin surrogate. We chose two different sublethal doses of NIMP, each corresponding to a fraction of the median lethal dose (one mild and one convulsive), and evaluated behavioral changes over a 6-month period following exposure. Mice exposed to both doses showed anxious behavior which persisted for six-months post-exposure. A longitudinal magnetic resonance imaging examination did not reveal any anatomical changes in the amygdala throughout the 6-month period. While no cholinesterase activity change or neuroinflammation could be observed at the latest timepoint in the amygdala of NIMP-exposed mice, important modifications in white blood cell counts were noted, reflecting a perturbation of the systemic immune system. Furthermore, intestinal inflammation and microbiota changes were observed at 6-months in NIMP-exposed animals regardless of the dose received. This is the first study to identify long-term behavioral impairment, systemic homeostasis disorganization and gut microbiota alterations following OP sublethal exposure. Our findings highlight the importance of long-term care for victims of NA exposure, even in asymptomatic cases.

Keywords: 4-nitrophenyl isopropyl methylphosphonate; mood disorder; anxiety; sarin surrogate; sublethal exposure; dysbiosis; microbiota

1. Introduction

Several recent high-profile uses of chemical warfare agents (CWA) derived from organophosphorus (OP) compounds have brought them back to attention in the past few years. These include their use in the Syrian conflict (2013–2017) and three separate targeted attacks between 2017 and 2020 resulting in Kim Jong-Nam's death and the poisonings of Sergei Skripal and Alexei Navalny [1]. These events all took place more than two decades after the first documented uses of OP nerve agents (NA) during the Iran–Iraq armed conflict and in two terrorist attacks in Japan, for which long-term neurological sequelae are still emerging [2,3]. However, the NA dose exposure could not be quantified in either of these reports, and while the described NA exposure was combined with sulfur mustard in Talabani et al. [3], similar symptoms were observed particularly in visual (lacrimation, impaired ability to focus, ocular pain, etc.) and neuropathophysiological areas (headache, numbness, confusion, agitation, etc.) as well as in less-suspected organs (abdominal pain, nausea, and a higher prevalence of respiratory infection). Post-traumatic stress responses were also highly enhanced in sarin-exposed Tokyo victims, correlating with observations made in Gulf War (GW) veterans exposed to high doses of OP (dichlorvos) and carbamate (lindane, bendiocard) pesticides and pyridostigmine bromide (PB) pills and/or NA [4]. These reports therefore highlight the need to meticulously investigate the long-term effects of NA exposure.

Pathophysiologically, NA, OP and carbamate pesticides as well as PB inhibit cholinesterase (ChE), leading to toxic hypercholinergy throughout the body. At the cellular level, acetylcholine accumulation appears at neuromuscular junctions and in synapses, inducing any number of symptoms including fasciculation, hypersecretion, muscle contractions, tetany, tremors and convulsion [5]. Acute exposure to large doses of OP results in muscle paralysis, respiratory distress and ultimately death in only a few minutes post-exposure. This can be avoided by providing an antidote therapy containing a muscarinic cholinergic receptor antagonist (atropine sulfate) and an oxime (pralidoxime, asoxime, etc.) with a strong nucleophile to reactivate the OP-inhibited acetylcholinesterase (AChE). Anticonvulsant medications are also added to the antidote therapy to avoid epileptic seizures and subsequent brain damage [6,7]. However, animal models have revealed that even if the pharmaceutical-based intervention stabilizes the convulsions and improves survival, it is still not enough to avoid long-term neurologic deficits [8]. In addition, other animal models exposed to lower doses of OP have also exhibited long-term neuropsychological dysfunctions, specifically anxiety, depression and cognitive deficits [8–10]. These results correlate with the delayed neurological consequences observed in humans after acute OP intoxication, specifically regarding anxiety-related behavior and cognitive deficits [4,11–13]. Nevertheless, clarification is needed in animal models, as a single low-dose exposure to OP induces only transient behavioral changes [14] and long-term deficits require concurrent stress [10] or repeated exposure to NA [14–16].

Restrictions regulating CWA use and storage are a major restraint to the study of OP-based NA neurotoxicity. Recently, using the sarin surrogate 4-nitrophenyl isopropyl methylphosphonate (NIMP), we developed a murine model to facilitate investigation in CWA-unauthorized laboratories [17]. NIMP is less toxic than sarin, but it has proven itself to be highly effective at inhibiting ChE in vitro and in vivo [18–22]. Furthermore, it can reproduce several features of sarin intoxication, including seizure-like behavior, cortical and hippocampal neuropathologies, neuroinflammatory processes and memory impairment in rodents [17,18,20,22]. These findings thus offer great promise for the study of the consequences of longer-term NIMP exposure in order to better anticipate neurologic sequelae induced by NA exposure.

The aim of this study was to characterize the long-term effects of a single NIMP exposure in mice by evaluating two different sublethal OP doses (0.5 and 0.9 LD50) for over 6 months. For this, we evaluated cerebral and blood ChE inhibition, neuroinflammation, systemic inflammation, behavioral modifications, anatomical and diffusion magnetic resonance imaging (MRI) and gut microbiota impacts at different timepoints for up to 6 months post-exposure. Our results demonstrate that exposure to a low dose of NA has the potential to disrupt gut microbiota and immune homeostasis as well as alter emotional behavior in the long-term.

2. Materials and Methods

2.1. Animals

All experimental procedures were approved by the SSA animal ethics committee according to applicable French legislation (Directive 2010/63/UE, decret 2013-118). To avoid any potential sexually dimorphic effect, only male Swiss mice (Janvier Labs, Genest-Saint-Isle, France) aged 7 to 8 weeks were housed, with four per cage on a 12 h/12 h light/dark cycle with food and water ad libitum. After a 7-day acclimation period, the animals were randomly assigned to their dedicated experiment. Animal group housing was the same before and after the NIMP challenge. Nine-week-old animals were used for the OP challenge.

2.2. NIMP Exposure

4-nitrophenyl isopropyl methylphosphonate (NIMP), used as a sarin surrogate, was synthesized by Dr. Rachid Baati (Université Strasbourg, CNRS UMR 7199, Strasbourg, France) according to a previously reported procedure [21]. On day 0, mice received a single subcutaneous injection (10 mL/kg) of NIMP (LD50 = 0.63 mg/kg), freshly diluted in 0.9% NaCl at two different sublethal doses (0.5 or 0.9 LD50). LD50 was estimated using the improved method of Dixon's up-and-down procedure described by Rispin et al. [23]. Control mice (CTL) received a similar vehicle injection. Mouse weight was monitored before the intoxication, every day during the first week following NIMP exposure, twice per week for up to 3 months and then once per week up until 6 months post-intoxication.

2.3. Behavioral Observation

2.3.1. Intoxication Severity Scale

All behavioral changes observed in NIMP-exposed mice during the first hour post-intoxication compared to CTL were noted at their onset and considered as observable signs of intoxication. Thirteen intoxication levels (12 signs) were used, grading animals from normal (=0) to death (=12) (Figure 1a). For the entirety of this study, 165 mice were evaluated including 54 CTL mice, 55 mice exposed to 0.5 LD50 and 56 mice exposed to 0.9 LD50 of NIMP.

2.3.2. Anxiety-like Behavior Tests

Before and every month after NIMP exposure, anxiety behaviors were evaluated for each group of mice (CTL *n* = 14; 0.5 LD50 *n* = 15; 0.9 LD50 *n* = 16). Anxiety tests are based on the balance between the natural tendency of mice to explore novel environments and their apprehension for open and bright areas. This approach–avoidance conflict results in behaviors correlated with an increase in physiological stress indicators. A new anxiety test was performed every month to avoid test habituation and exploration diminution during the task. Tests were randomly selected. Due to the similarity of the elevated plus maze (EPM) and elevated zero maze (EZM) tests, these tests were assigned to the first and the last period of the study, respectively.

- Open-field test: square area or circular area
  Mice were placed for 5 min in either an empty square open-field (45 × 45 cm) or circular open-field (40-cm diameter) box surrounded by high walls to prevent escape. The test period was videorecorded, and the activity of the animal over time was analyzed using EthoVision XT software (Noldus, Wageningen, The Netherlands). The distance, speed and time spent walking around the outer edge of the box vs. the center (square or round area depending on the open-field shape) of the box were evaluated.

- EPM test or EZM test
  The EPM apparatus consists of a raised maze (80 cm off the floor) with four arms in a cross shape: two arms are exposed to the open air, and the other two arms are enclosed. EZM is a circular apparatus with dark enclosed sections alternating with open sections. Mice were placed in the maze for a 5-min period. The test period was videorecorded,

and the activity of the animal over time was analyzed using EthoVision XT software. The time spent in the open arms was evaluated.

- Staircase test
  The staircase contains six identical steps (2.5 cm high and 7.5 cm deep) enclosed between vertical walls (10 cm wide). Wall levels are constant along the staircase. Each mouse was placed individually at the bottom of the staircase for a 5-min observation period. The top step (5th step) is considered to be the most anxious for mice, as it is more elevated and brighter than the others. The number of rearings on the 5th step was recorded and used as the anxiety index.

- Dark–light box test
  The dark–light box apparatus is divided into two compartments: a light one (white, bright and without a lid) and a dark one (black, closed and covered). Mice were placed in the light compartment and videorecorded for a 5-min period. The time spent in both compartments was evaluated using the EthoVision XT software.

- Neophobia test
  This test is based on the appetence for sugar in mice and was adapted from the "Novelty Suppressed Feeding Test" without food restriction. Mice were placed in a bright box with chocolate cereal in the center of a platform and videorecorded for a 10-min period. The platform visit frequency was measured using the EthoVision XT software.

2.4. Cholinesterase Activity and Multiplex Biomarker Assays

At different timepoints (6 h, 24 h, 3 days, 7 days, 1 month and 6 months) after NIMP exposure, animals were deeply anesthetized with pentobarbital. Immediately afterwards, blood samples (800 μL) were collected by intracardial sampling with a syringe and mice were transcardially perfused with 15 mL of cold NaCl (0.9%). Blood samples were divided and prepared for two separate analyses, with 400 μL containing 60 μL of 1.6% EDTA to avoid platelet aggregation and 400 μL used for serum analyses. Blood samples with EDTA were used to count white blood cells (WBC). Mononuclear and polynuclear cell counts were carried out using an IDEXX ProCyte Dx Hematology analyzer.

Brains were quickly removed and dropped in cold saline buffer before being sliced in 2-mm-thick coronal sections. The piriform cortex and amygdala were dissected, collected in microtubes and frozen in dry ice. Samples were homogenized in 50 mM phosphate buffer (pH 7.4)/0.5% Tween using a bead mill homogenizer (OMNI International) with 1.4-mm ceramic beads and centrifuged at $10,000\times g$ (4 °C) for 10 min. The resulting supernatants were stored at -80 °C. Total protein concentrations were determined using the DC Protein Assay (Bio-Rad, Marnes-la-Coquette, France) according to the manufacturer's protocol.

2.4.1. Cholinesterase Inhibition Assay

Total ChE activity was determined using the Ellman method by adding 5 μL of piriform cortex and amygdala sample to 0.22 mM 5,5′-dithiobis-2-nitrobenzoic acid (DTNB, Sigma Aldrich, St-Quentin Falavier, France) in phosphate buffer (pH 7.4). In parallel, 0.1 mM ethopropazine hydrochloride (Sigma Aldrich) was added for AChE-specific activity analyses. After a 15-min baseline reading to account for thiols present in the samples, 1 mM acetylthiocholine (Sigma Aldrich) was added and the reaction between thiocholine and DTNB was monitored for 30 min at 412 nm and at 25 °C in a microplate reader (Spark 10, Tecan). All samples were assayed in duplicate. Activities of the piriform cortex and amygdala samples were normalized to total protein concentration for each sample. The final results were expressed as percentages of average CTL activity.

2.4.2. Milliplex Multiplex Assays

Cerebral and serum concentrations of KC, IL-1α, IL-10, IL-9, IL-17, GM-CSF, M-CSF and G-CSF were measured using the Milliplex MCYTMAG-70K-PX32 (Merck-Millipore, Burlington, MA, USA) according to the manufacturer's protocol.

2.5. Anatomical Examination

2.5.1. In Vivo MRI and Analysis

MRI scans were performed with a Bruker Biospec 70/30 (7T) preclinical scanner (Bruker Biospin MRI, Ettlingen, Germany). Images were acquired and reconstructed, and parametric maps were generated using Paravision 6.0.1 (Bruker Biospin MRI, Ettlingen, Germany). Mice were imaged at five timepoints (48 h prior to intoxication and 72 h, 7 days, 1 month and 6 months post-NIMP exposure). Immediately prior to imaging, animals were anesthetized with 5% isoflurane and thereafter maintained with 2% isoflurane in a 70%/30% mixture of NO2/O2. Body temperature was maintained at 37 °C, with a respiration rate of 50–70 breaths per min. High resolution T2 sequences were performed using the following parameters: repetition time (TR) = 3500 ms, effective echo time (TE) = 40 ms, field of view (FOV) = 17.92 × 17.92 mm2 (256 × 256 data matrix) and 18 slices with a 0.5-mm thickness. Diffusion-weighted images (DWI) were collected using the following parameters: TR = 2500 ms, TE = 22 ms, FOV = 20 × 20 mm2 (128 × 128 data matrix), eight slices with a 0.8-mm thickness, three diffusion-weighted orthogonal directions with b = 650 s/mm2 and a total acquisition time of 16 min. Apparent diffusion coefficient (ADC) maps were generated and volumes of interest (VOI) were manually traced on ADC parametric maps around the amygdala brain region.

2.5.2. Histology

Six months post-NIMP exposure, mice were deeply anaesthetized with pentobarbital and transcardially perfused with 10 mL of cold NaCl (0.9%), followed by 30 mL of 4% paraformaldehyde (PFA) in phosphate buffer.

- Microglia staining
  Brains were quickly removed and immersed in cold 4% PFA for overnight post-fixation. Brains were cryoprotected for 24–48 h in cold 20% sucrose and then frozen in −40 °C isopentane. All brains were sliced in 14-μm coronal sections using a cryostat, and slices were sequentially mounted on Superfrost + slides (VWR, Radnor, PA, USA). Immunofluorescent labeling with rabbit anti-IBA1 (1/1000, Wako, USA), detected by anti-rabbit 555 (1/500, Invitrogen, Waltham, MA, USA), was performed on one slide per animal to analyze microglial reactivity. Fluorescent labeling of the amygdala was imaged using an automated Leica DM6000 B research microscope (Leica Microsystems, Wetzlar, Germany), and the two hemispheres of three different slices were analyzed for each animal. All acquisitions were performed using the same acquisition setup. Stereotaxic consistency between animals was maintained with the help of a reference mouse brain atlas [24]. Analyses were performed using the ImageJ software. To quantify Iba1 staining, we performed an area fraction analysis. Briefly, after setting a threshold, the pixels in the image with values inside this range were converted to white, whereas pixels with values outside this range were converted to black. The threshold was determined to obtain a clear area representing IBA1 labeling in CTL animals. The same threshold was applied to each image of all animal groups. This analysis measured the area of labeling in regions of interest.

- Gut histology
  The Swiss-rolling technique was used to examine complete colonic sections. This technique helps in the histological assessment of the complete colonic sections examined. The result is an intestinal roll which allows for the scanning of a large part of the intestine. The intestinal villi and the epithelial lining remain intact despite being rolled up. Following PFA fixation, the organs were rinsed with distilled water

and dehydrated. Samples were cut at a thickness of 4 μm on a rotary microtome (LEICA®). Hematoxylin-eosin-saffron staining was performed on successive sections for structural and functional analysis of the colon. The length of villi, number of goblet cells and area of immune infiltrates were quantified manually using Histolab software (GT Vision, UK).

2.6. Gut Microbiota Modification

2.6.1. Gram+/Gram− Ratio Determination

To observe the bacterial microflora, a fecal smear of each animal was performed using the Gram staining technique. For this, the colon was cut lengthwise and the feces were directly removed, diluted in 500 μL of 1X PBS (Gibco) and spread on a slide. Differential staining of Gram+ bacteria from Gram− bacteria was performed using the Gram−Hucker R Kit (RAL Diagnostics). The Gram+/Gram− ratio was determined for the entire smear by microscopy using a 40× objective.

2.6.2. Assessment of Gut Microbiota Composition by High-Throughput Sequencing

DNA extraction, 16S rRNA gene amplification, 16S rDNA amplicon library preparation and sequencing were carried out by SMALTIS (http://www.smaltis.fr/, accessed on 26 September 2019). DNA was extracted from 200 mg of distal colonic luminal content using the QIAamp Fast DNA stool Mini Kit (Qiagen) according to the manufacturer's recommendations (isolation of DNA from stool for pathogen detection). The V3–V4 region of the 16S rRNA gene was amplified using AccuStart™ II PCR ToughMix (QuantaBio) and the following primers: V3F "CTTTCCCTACACGACGCTCTTCCGATC-TACGGRAGGCAGCAG" (344F) and V4R "GGAGTTCAGACGTGTGCTCTTCCGATCT-TACCAGGGTATCTAATCCT" (802R). The thermocycler was programmed with an initial DNA denaturing step at 95 °C for 2 min followed by 30 cycles at 95 °C for 1 min, 65 °C for 1 min, 72 °C for 1 min and a final extension step at 72 °C for 10 min. Ligation of MiSeq sequencing adapters was performed by PCR using the MTP™ Taq DNA Polymerase (Sigma) and the following thermocycler conditions: 94 °C for 1 min followed by 12 cycles at 95 °C for 1 min, 65 °C for 1 min, 72 °C for 1 min and a final extension step at 72 °C for 10 min. Sequencing was performed on a MiSeq device using the 2 × 250bp V3 kit. The remaining adapter/primer sequences were trimmed, and reads were checked for quality ($\geq$20) and length ($\geq$200 bp) using cutadapt [25]. Reads were further corrected for known sequencing errors using SPAdes [26] and then merged using PEAR [27]. Operational taxonomic units (OUTs) were identified using a Vsearch pipeline [28] set up to dereplicate (–derep_prefix –minuquesize 2), cluster (–unoise3) and chimera check (uchime3_denovo) the merged reads. OTU taxonomical classification was performed using a classifier from the RDPTools suit [29].

2.7. Statistics

2.7.1. General Statistics

Data were expressed as means $\pm$ standard deviation (SD) and analyzed using PRISM 7 software (GraphPad, San Diego, CA, USA). Statistical tests and sample sizes are indicated in the figure legends and figures, respectively. Graphs display the mean and error bars represent the SD. The significance between groups is denoted by * $p < 0.05$, ** $p < 0.01$, *** $p < 0.001$ or **** $p < 0.0001$.

2.7.2. Microbiota-Specific Statistics

Statistical analyses were run using the R programming language and software together with the gplots, gdata, vegan (http://cran.r-project.org/package=vegan, accessed on 3 March 2020), ade4, Hmisc, corrplot and phangorn packages. OTU counts were normalized via simple division to their sample size and then multiplied by the size of the smallest sample. α-diversity and richness were estimated using diversity and estimateR. The distance matrix for β-diversity analysis was computed using vegdist and the Bray–Curtis method. Principal coordinates analysis was computed on a distance matrix using dudi.pco. Associations between the microbiota composition at the genus level, anxiety-like behavior measurements and mononuclear cell percentages were assessed using rcorr. The Kruskal–Wallis rank sum test and post hoc Dunn's all-pairs rank test were used as required to detect differences between groups. *p*-values were adjusted as necessary using false discovery rate correction.

3. Results

3.1. General Effects of NIMP Intoxication

The severity of NIMP intoxication was evaluated based on our intoxication scale (Figure 1a). This scale includes 12 relevant behavioral changes, from no behavioral perturbation to death, the maximum on our scale. Almost all mice exposed to 0.5 LD50 of NIMP displayed fasciculation and face stereotypies (chewing, yawning), considered as light intoxication symptoms. On the other hand, the majority of the 0.9 LD50-exposed mice presented long-lasting convulsions. This dose was thus selected to be the high sublethal dose in our study. The mean intoxication scores were 2.5 ± 0.2 for 0.5 LD50 and 10.4 ± 0.2 for the 0.9 LD50-exposed mice (Figure 1b). Six animals (10.7%) exposed to the highest dose and none exposed to the lowest dose died in the first day following the NIMP challenge. No delayed death was observed.

As expected, measurements of cerebral ChE activities in mice exposed to 0.9 LD50 revealed a significant important total ChE inhibition 6 h after intoxication ($5.4 \pm 0.6\%$ of ChE activity). Late measurements showed a persisting significant inhibition until 1-month post-intoxication ($72.5 \pm 6.7\%$ of ChE activity), with recovery being complete 6 months post-exposure (Figure 1c). Exposure to 0.5 LD50 induced a large inhibition of ChE activity ($36.9 \pm 4.6\%$) at 6 h post-exposure. The recovery appeared incomplete at 1 month ($81.7 \pm 5\%$) and 6 months after exposure at this dose ($77 \pm 13\%$); however, it was not significant from CTL activity (Figure 1c). Interestingly, the profile of AChE activity was the same as the total ChE activity for both doses (Figure 1d), suggesting that in the CNS, the majority of ChE activity is due to AChE activity.

Finally, the evolution in weight gain during the 6 months post-exposure revealed a significant loss of weight for only the 0.9 LD50-exposed mice during the first week after intoxication. One month after exposure, the change in mean body weight of the exposed animals reached the change in body weight of the CTL group. Interestingly, the 0.5 LD50-exposed mice did not display any difference in weight gain at any time compared to the CTL group (Figure 1e).

3.2. Anxiety-like Behavior Modification

Anxiety-related behaviors were evaluated every month after NIMP exposure. To avoid any test habituation, a new anxiety test was performed for each measurement. The initial test, which was performed before any intoxication, showed no difference in terms of the time passed exploring the center zone in the square area open-field and was thus considered as the anxiety level baseline (Figure 2a). One month after NIMP intoxication, the time passed in the open arms of the EPM significantly decreased in the NIMP-intoxicated animals compared to the CTL (Figure 2b). The number of rearings on the 5th step of the staircase measured on the second month post-exposure did not show any significant difference between groups, even though this number was slightly decreased in both NIMP intoxication groups (Figure 2c). In contrast, the recovery was complete 3 months post-exposure, since the time passed in the light box during the dark–light box test was equivalent between groups (Figure 2d). Four months post-intoxication, the time passed exploring the center zone in the circular area open-field decreased drastically in the 0.9 LD50-exposed mice group, while the 0.5 LD50-exposed mice did not show any difference with the CTL group (Figure 2e). Finally, the two NIMP-intoxicated animal groups at 5 and 6 months post-NIMP exposure showed a significant increase in the anxiety index, as measured by the decreased frequency of food zone exploration in the neophobia test and the time passed in the open arms in the O-maze (Figure 2f,g).

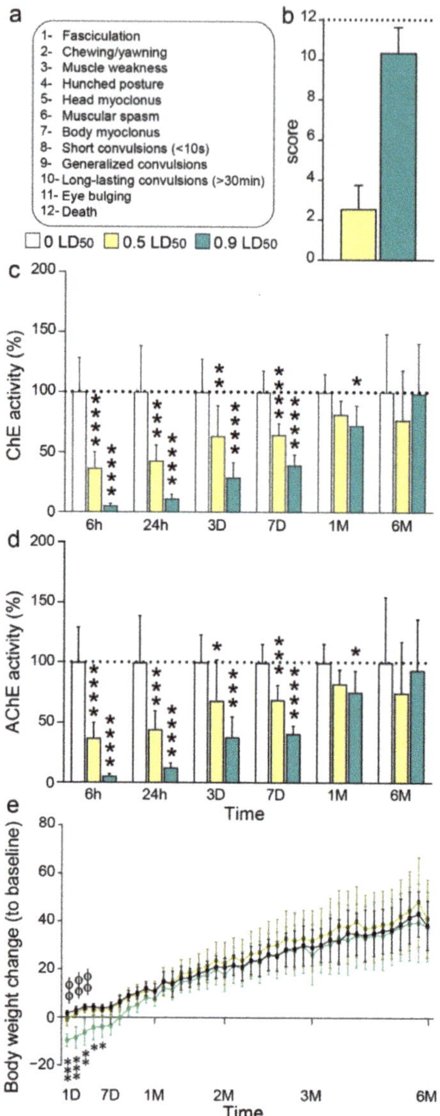

Figure 1. General effects of NIMP intoxication. (**a**) Intoxication scale based on the most relevant behavioral signs. (**b**) Average intoxication score induced by the two doses of interest: 2.55 ± 0.17 for 0.5 LD50 and 10.38 ± 0.16 for 0.9 LD50. (**c**) ChE activity at different timepoints post-NIMP exposure: 6 h (100 ± 11.5 for 0 LD50; 37.0 ± 4.6 for 0.5 LD50 and 5.4 ± 0.6 for 0.9 LD50); 24 h (100 ± 13.6 for 0 LD50; 43.0 ± 4.5 for 0.5 LD50 and 11.3 ± 1.3 for 0.9 LD50); 3 days (100 ± 9.6 for 0 LD50; 63.5 ± 8.8 for 0.5 LD50 and 29.0 ± 4.4 for 0.9 LD50); 7 days (100 ± 6.3 for 0 LD50; 64.8 ± 3.2 for 0.5 LD50 and 39.7 ± 3.3 for 0.9 LD50); 1 month (100 ± 6.7 for 0 LD50; 81.7 ± 5.0 for 0.5 LD50 and 72.5 ± 6.7 for 0.9 LD50) and 6 months (100 ± 17.2 for 0 LD50; 77.0 ± 13.7 for 0.5 LD50 and 98.9 ± 14.6 for 0.9 LD50); ($n = 8$ per group). Significant differences were determined by one-way ANOVA (F = 51.45, $p < 0.0001$ for 6H; F = 26.45, $p < 0.0001$ for 24H; F = 19.82, $p < 0.0001$ for 3D; F = 42.6, $p < 0.0001$ for 7D; F = 4.99, $p = 0.02$ for 1M and F = 0.76, $p = 0.48$ for 6M) with Dunnett's post hoc test compared to the CTL group

($****$ $p < 0.0001$; $***$ $p < 0.001$; $**$ $p < 0.01$; $*$ $p < 0.05$). (**d**) AChE activity at different timepoints post-NIMP exposure: 6 h (100 ± 11.9 for 0 LD50; 37.1 ± 4.5 for 0.5 LD50 and 5.3 ± 0.7 for 0.9 LD50); 24 h (100 ± 14.6 for 0 LD50; 44.3 ± 5.4 for 0.5 LD50 and 12.6 ± 1.3 for 0.9 LD50); 3 days (100 ± 8.7 for 0 LD50; 68.0 ± 12.0 for 0.5 LD50 and 37.9 ± 6.0 for 0.9 LD50); 7 days (100 ± 5.6 for 0 LD50; 68.9 ± 4.2 for 0.5 LD50 and 40.9 ± 2.3 for 0.9 LD50); 1 month (100 ± 6.8 for 0 LD50; 82.2 ± 5.2 for 0.5 LD50 and 75.3 ± 7.2 for 0.9 LD50) and 6 months (100 ± 19.3 for 0 LD50; 74.8 ± 14.1 for 0.5 LD50 and 93.5 ± 15.1 for 0.9 LD50); ($n = 8$ per group). Significant differences were determined by one-way ANOVA ($F = 48.94$, $p < 0.0001$ for 6H; $F = 24.4$, $p < 0.0001$ for 24H; $F = 10.86$, $p = 0.0006$ for 3D; $F = 45.38$, $p < 0.0001$ for 7D; $F = 3.7$, $p = 0.04$ for 1M and $F = 0.67$, $p = 0.52$ for 6M) with Dunnett's post hoc test compared to the CTL group ($****$ $p < 0.0001$; $***$ $p < 0.001$; $**$ $p < 0.01$; $*$ $p < 0.05$). (**e**) Body weight change relative to baseline ($n = 14$ for 0 LD50; $n = 15$ for 0.5 LD50 and $n = 16$ for 0.9 LD50 group). Significant differences were determined by two-way repeated measures ANOVA ($F_{Time} = 282.8$, $p < 0.0001$; $F_{Dose} = 2.3$, $p = 0.11$; $F_{TimexDose} = 1.67$, $p = 0.0009$) with Tukey's multiple comparisons test ($***$ $p < 0.001$; $**$ $p < 0.01$; $*$ $p < 0.05$ CTL vs. 0.9 LD50; $\Phi\Phi$ $p < 0.01$ 0.5 LD50 vs. 0.9 LD50).

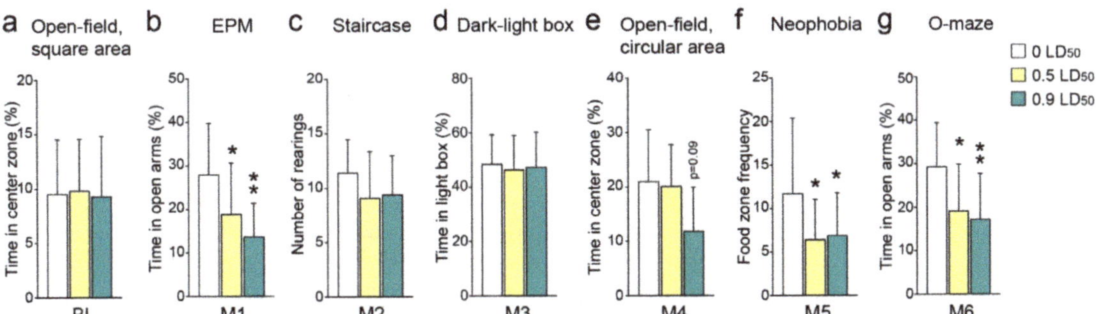

Figure 2. Development of anxiety-like behavior caused by NIMP intoxication. Anxiety-like behaviors were evaluated in NIMP-exposed and CTL animals using a different test every month ($n = 14$ for 0 LD50; $n = 15$ for 0.5 LD50 and $n = 16$ for 0.9 LD50 group). (**a**) Open-field, square area, at the baseline with the percentage of time spent in the center zone (9.6 ± 1.3 for 0 LD50; 9.9 ± 1.2 for 0.5 LD50 and 9.4 ± 1.4 for 0.9 LD50). (**b**) EPM at 1 month with the percentage of time spent in the open arms (28.1 ± 3.1 for 0 LD50; 18.9 ± 3.0 for 0.5 LD50 and 13.8 ± 1.9 for 0.9 LD50). (**c**) Staircase test, 2 months post-intoxication with the number of rearings on the top step (11.5 ± 0.8 for 0 LD50; 9.1 ± 1.1 for 0.5 LD50 and 9.4 ± 0.9 for 0.9 LD50). (**d**) Dark-light box at 3 months with the percentage of time spent in the light box (48.6 ± 2.8 for 0 LD50; 46.5 ± 3.2 for 0.5 LD50 and 47.5 ± 3.2 for 0.9 LD50). (**e**) Open-field, circular area, at 4 months with the percentage of time spent in the center zone (21.1 ± 3.8 for 0 LD50; 20.2 ± 2.7 for 0.5 LD50 and 11.9 ± 2.9 for 0.9 LD50). (**f**) Neophobia test at 5 months showing the frequency of visits to the center platform (11.8 ± 2.3 for 0 LD50; 6.5 ± 1.2 for 0.5 LD50 and 6.9 ± 1.2 for 0.9 LD50). (**g**) O-maze at 6 months post-exposure with the percentage of time spent in the open arms (29.4 ± 2.7 for 0 LD50; 19.2 ± 2.8 for 0.5 LD50 and 17.2 ± 2.6 for 0.9 LD50). Significant differences were determined by one-way ANOVA ($F = 0.04$, $p = 0.9$ for BL; $F = 7.1$, $p = 0.002$ for 1M; $F = 1.6$, $p = 0.2$ for 2M; $F = 0.11$, $p = 0.89$ for 3M; $F = 2.8$, $p = 0.08$ for 4M; $F = 3.2$, $p = 0.04$ for 5M and $F = 5.4$, $p = 0.008$ for 6M) with Dunnett's post hoc test compared to the CTL group ($**$ $p < 0.01$; $*$ $p < 0.05$).

3.3. Longitudinal Imaging Examination and Long-Term Induced Inflammation

To explain these behavioral long-term modifications, we conducted MRI in parallel on 0.5 and 0.9 LD50 NIMP-exposed mice for 6 months to evaluate morphological modifications and brain edema after exposure. The image analysis did not show significant neuroanatomical changes using T2-weighted images or edema formation using ADC mapping on DWI images (Figure 3).

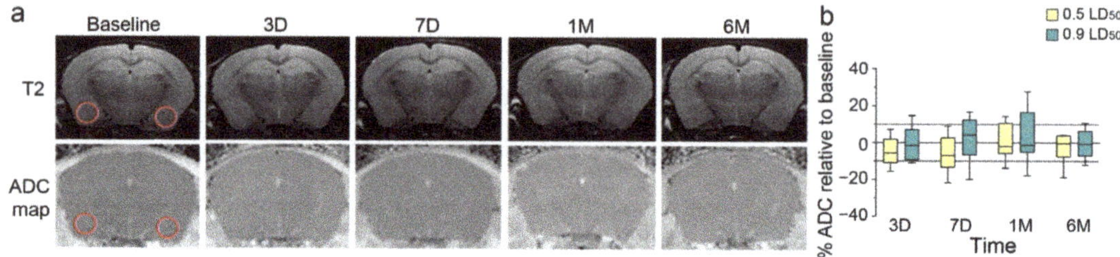

Figure 3. NIMP exposure does not induce morphological modifications or brain edema in the amygdala/piriform cortex. (**a**) T2 images with the apparent diffusion coefficient (ADC) map of one representative NIMP-exposed mouse (0.9 LD50) at different timepoints: before intoxication (baseline), and 3 days, 7 days, 1 month and 6 months after NIMP exposure. Regions of interest used for analyses are shown with red circles. (**b**) The percentage of ADC normalized to the baseline as a function of different timepoints for NIMP-exposed animal groups [(0.5 LD50: -5.1 ± 2.7 at 3D; -6.0 ± 3.5 at 7D; 1.7 ± 3.9 at 1M and -2.9 ± 3.1 at 6M) and (0.9 LD50: -0.8 ± 3.1 at 3D; 2.1 ± 4.3 at 7D; 2.9 ± 5.2 at 1M and -0.5 ± 2.7 at 6M)]. Statistical analyses were conducted by mixed-effects model (REML) analysis ($F_{Time} = 0.89$, $p = 0.44$; $F_{Dose} = 1.5$, $p = 0.24$; $FT_{imexDose} = 0.36$, $p = 0.78$) ($n = 7$ to 9 per group).

Next, cerebral cytokine levels were evaluated in the amygdala/pyriform cortex extracts using Luminex technology. A slight increase in several cytokine levels was observed in the NIMP-exposed mice group 6 months post-exposure (Figure 4a). Thus, sublethal doses of NIMP did not induce any robust long-term neuroinflammation. In addition, histological exploration showed no persistent microglia reactivity 6 months post-intoxication in the amygdala of exposed animals (Figure 4b,c), confirming the previous cytokine results.

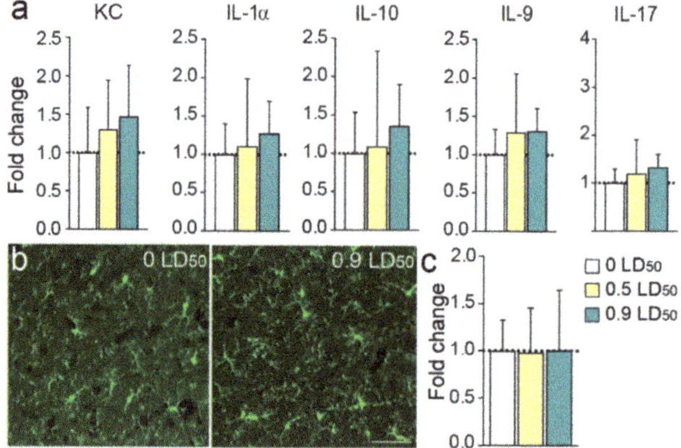

Figure 4. NIMP exposure does not induce long-term neuroinflammation in the amygdala/piriform cortex. (**a**) Expression levels of the cytokines KC, IL-1α, IL-10, IL-9 and IL-17 normalized to CTL values at 6 months post-intoxication [(KC: 1.0 ± 0.2 for 0 LD50; 1.3 ± 0.2 for 0.5 LD50 and 1.5 ± 0.2 for 0.9 LD50); (IL-1α: 1.0 ± 0.2 for 0 LD50; 1.1 ± 0.3 for 0.5 LD50 and 1.3 ± 0.2 for 0.9 LD50); (IL-10: 1.0 ± 0.2 for 0 LD50; 1.1 ± 0.5 for 0.5 LD50 and 1.4 ± 0.2 for 0.9 LD50); (IL-9: 1.0 ± 0.1 for 0 LD50; 1.3 ± 0.3 for 0.5 LD50 and 1.3 ± 0.1 for 0.9 LD50) and (IL-17: 1.0 ± 0.1 for 0 LD50; 1.2 ± 0.3 for 0.5 LD50 and 1.3 ± 0.1 for 0.9 LD50)]. Statistical analyses were conducted by one-way ANOVA ($n = 8$ per group). (**b**) IBA-1 labeling in the amygdala region of CTL (**left**) and 0.9 LD50 (**right**) mice. Scale bar = 50 µm. (**c**) Fold change of IBA1 labeling the optical density percentage normalized to CTL at 6 months post-intoxication: 1.0 ± 0.1 for 0 LD50; 0.98 ± 0.2 for 0.5 LD50 and 1.0 ± 0.2 for 0.9 LD50. Statistical analyses were conducted by Kruskal–Wallis test ($p = 0.83$; $n = 7$ per group).

Because the impact of a long-term systemic inflammation could alter the nervous system, we evaluated the evolution of WBC after NIMP exposure. A first phase could be observed, with significant WBC modifications in the 0.9 LD50-exposed mice group. Two timepoints, one day and 7 days after NIMP exposure, showed significant decreases in lymphocytes, monocytes and eosinophils in the highest dose-exposed group, whereas the lowest dose induced little or no modifications in most cell types (Figure 5). One-month post-intoxication, no difference was observed between the NIMP-exposed groups or in comparison to the CTL group. Surprisingly, a second phase was also observed between 3 and 6 months post-NIMP exposure, with significant modifications to WBC quantities in the 0.5 LD50-exposed mice group. With the exception of basophils and eosinophils, which were significantly decreased in the two exposed groups, the 0.5 LD50-exposed mice group presented significant long-term modification in the proportion of lymphocytes, monocytes and neutrophils compared to the 0.9 LD50-exposed and CTL groups (Figure 5).

Although serum cytokine levels were not significantly increased 6 months post-exposure, the expression of some cytokines was slightly enhanced in the 0.9 LD50-exposed mice group (+183% $\pm$ 146% for G-CSF; +46% $\pm$ 26% for KC; +66% $\pm$ 35% for IL-1α). Interestingly, the expression of granulocyte-macrophage colony-stimulating factor (GM-CSF or CSF2) as well as IL-17 was slightly elevated in the 0.5 LD50-exposed mice group (+184% $\pm$ 131% and 86% $\pm$ 58%, respectively) (Figure 6).

3.4. Gut Modification

The consequences of chronic systemic low-level inflammation could have a significant impact on multiple physiological systems. Among the possible targets, we investigated the impact on intestinal function. A gut morphological evaluation was thus conducted 6 months post-NIMP exposure. Histological modifications were observed for both NIMP sublethal doses (Figure 7a). The quantification of goblet cell numbers and the ratio of intestinal villi size revealed a significant decrease in the 0.5 and 0.9 LD50-exposed mice groups (goblet cell numbers compared to CTL: 77.2% $\pm$ 8.3% and 78.7% $\pm$ 3.5%, respectively; intestinal villi size compared to CTL: 75.8% $\pm$ 3% and 69.6% $\pm$ 2.7%, respectively) (Figure 7b,c). Moreover, lymphoid foci numbers were significantly increased in both exposed mice groups ($\times 3 \pm 1.1$ and $\times 5.5 \pm 1.3$ for 0.5 LD50 and 0.9 LD50, respectively) (Figure 7d). Finally, we conducted an evaluation of the Gram+/Gram− ratio. A dysbiosis of Gram− microbiota was observed in both NIMP-treated groups (Gram+/Gram− ratio: 0.55 $\pm$ 0.09 and 0.47 $\pm$ 0.11 for 0.5 LD50 and 0.9 LD50, respectively) (Figure 7e). This level of gut dysbiosis can trigger the innate immune response and chronic low-grade inflammation, leading to many age-related degenerative pathologies and unhealthy aging, which could in turn influence the gut microbiota composition [30].

3.5. Altered Microbiota Composition

A total of 788,900 16S rDNA reads was analyzed to establish the gut microbiota composition of the mice included in the present study. Our analyses led to the detection of 873 OTUs, evenly distributed among the mice groups (CTL: 551 $\pm$ 41, 0.5 LD50: 562 $\pm$ 54 and 0.9 LD50: 561 $\pm$ 44). No differences were detected regarding gut bacterial diversity and richness (Figure 8a) for mice challenged with NIMP 6 months post-exposure in comparison to the CTL group. As displayed in Figure 8b, nine predominant bacterial genera (Alistipes, Barnesiella, Bacteroides, Prevotella, Odoribacter, Alloprevotella, Clostridium_XIVa, Bilophila and Oscillibacter) shaped the gut microbiota composition of the mice, with highly similar abundances between groups. Furthermore, we observed a moderate shift in the gut microbiota composition of NIMP-exposed mice in comparison to the CTL group (Figure 8c). Two clusters consisting of CTL and NIMP-exposed mice (both 0.5 and 0.9 LD50 doses) were identified in a principal coordinate analysis (PCoA). This moderate shift can mostly be explained by higher abundances of Turicibacter (in both 0.5 and 0.9 LD50 groups) and Parabacteroides (0.9 LD50 group only) and a lower abundance of Coprococcus in NIMP-exposed mice compared to the CTL group (Figure 8d).

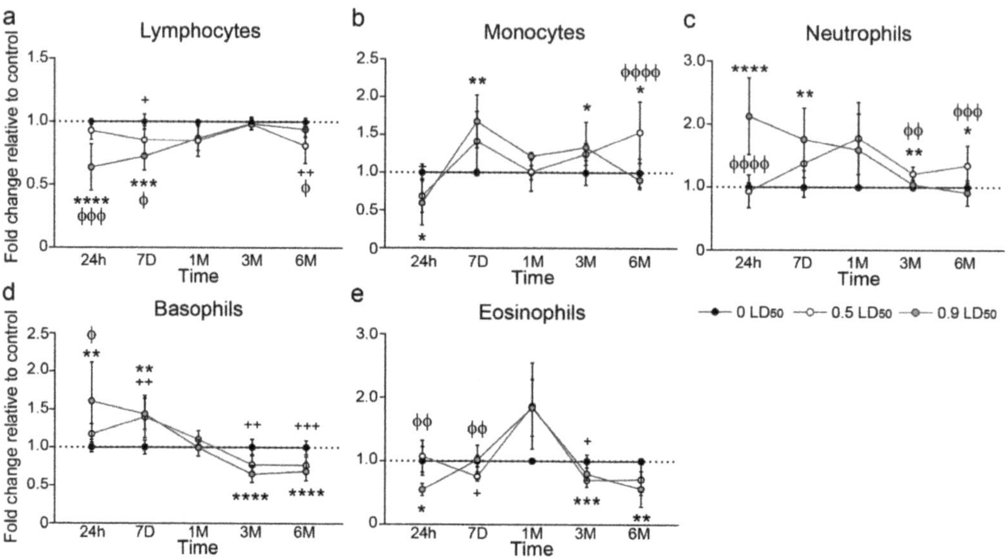

Figure 5. Evolution in white blood cell count after NIMP exposure. (**a**) Lymphocyte count at different timepoints after NIMP exposure relative to CTL: 24 h (1.0 ± 0.01 for 0 LD50; 0.93 ± 0.02 for 0.5 LD50 and 0.64 ± 0.06 for 0.9 LD50); 7 days (1.0 ± 0.02 for 0 LD50; 0.85 ± 0.03 for 0.5 LD50 and 0.72 ± 0.04 for 0.9 LD50); 1 month (1.0 ± 0.004 for 0 LD50; 0.85 ± 0.07 for 0.5 LD50 and 0.87 ± 0.05 for 0.9 LD50); 3 months (1.0 ± 0.003 for 0 LD50; 0.98 ± 0.01 for 0.5 LD50 and 0.99 ± 0.02 for 0.9 LD50) and 6 months (1.0 ± 0.12 for 0 LD50; 0.81 ± 0.05 for 0.5 LD50 and 0.94 ± 0.02 for 0.9 LD50); (*n* = 8 to 10 per group). Significant differences were determined by one-way ANOVA (F = 22.52, *p* < 0.0001 for 24H; F = 13.33, *p* = 0.0002 for 7D; F = 5.21, *p* = 0.02 for 1M; F = 0.32, *p* = 0.73 for 3M and F = 8.51, *p* = 0.0023 for 6M with Tukey's multiple comparisons test. (**b**) Monocyte count at different timepoints after NIMP exposure relative to CTL: 24 h (1.0 ± 0.04 for 0 LD50; 0.69 ± 0.08 for 0.5 LD50 and 0.60 ± 0.13 for 0.9 LD50); 7 days (1.0 ± 0.02 for 0 LD50; 1.41 ± 0.12 for 0.5 LD50 and 1.68 ± 0.12 for 0.9 LD50); 1 month (1.0 ± 0.04 for 0 LD50; 1.01 ± 0.12 for 0.5 LD50 and 1.22 ± 0.02 for 0.9 LD50); 3 months (1.0 ± 0.06 for 0 LD50; 1.24 ± 0.05 for 0.5 LD50 and 1.33 ± 0.11 for 0.9 LD50) and 6 months (1.0 ± 0.07 for 0 LD50; 1.53 ± 0.13 for 0.5 LD50 and 0.90 ± 0.03 for 0.9 LD50); (*n* = 8 to 10 per group). Significant differences were determined by one-way ANOVA (F = 5.54, *p* = 0.015 for 24H; F = 6.39, *p* = 0.0065 for 7D; F = 2.88, *p* = 0.095 for 1M; F = 4.46, *p* = 0.024 for 3M and F = 15.78, *p* < 0.0001 for 6M with Tukey's multiple comparisons test. (**c**) Neutrophil count at different timepoints after NIMP exposure relative to CTL: 24 h (1.0 ± 0.02 for 0 LD50; 0.93 ± 0.08 for 0.5 LD50 and 2.12 ± 0.23 for 0.9 LD50); 7 days (1.0 ± 0.06 for 0 LD50; 1.37 ± 0.11 for 0.5 LD50 and 1.76 ± 0.15 for 0.9 LD50); 1 month (1.0 ± 0.02 for 0 LD50; 1.77 ± 0.33 for 0.5 LD50 and 1.59 ± 0.29 for 0.9 LD50); 3 months (1.0 ± 0.01 for 0 LD50; 1.21 ± 0.04 for 0.5 LD50 and 1.04 ± 0.03 for 0.9 LD50) and 6 months (1.0 ± 0.03 for 0 LD50; 1.34 ± 0.11 for 0.5 LD50 and 0.92 ± 0.06 for 0.9 LD50); (*n* = 8 to 10 per group). Significant differences were determined by one-way ANOVA (F = 27.24, *p* < 0.0001 for 24H; F = 7.68, *p* = 0.0028 for 7D; F = 3.7, *p* = 0.067 for 1M; F = 12.19, *p* = 0.0003 for 3M and F = 9.50, *p* = 0.0009 for 6M with Tukey's multiple comparisons test. (**d**) Basophil count at different timepoints after NIMP exposure relative to CTL: 24 h (1.0 ± 0.03 for 0 LD50; 1.17 ± 0.04 for 0.5 LD50 and 1.60 ± 0.17 for 0.9 LD50); 7 days (1.0 ± 0.03 for 0 LD50; 1.40 ± 0.09 for 0.5 LD50 and 1.44 ± 0.08 for 0.9 LD50); 1 month (1.0 ± 0.01 for 0 LD50; 1.10 ± 0.06 for 0.5 LD50 and 0.99 ± 0.04 for 0.9 LD50); 3 months (1.0 ± 0.05 for 0 LD50; 0.77 ± 0.04 for 0.5 LD50 and 0.65 ± 0.03 for 0.9 LD50) and 6 months (1.0 ± 0.03 for 0 LD50; 0.77 ± 0.03 for 0.5 LD50 and 0.69 ± 0.03 for 0.9 LD50); (*n* = 8 to 10 per group). Significant differences were determined by one-way ANOVA (F = 7.218, *p* = 0.0041 for 24H; F = 9.08, *p* = 0.0016 for 7D; F = 1.84, *p* = 0.19 for 1M;

F = 14.12, $p < 0.0001$ for 3M and F = 20.04, $p < 0.0001$ for 6M with Tukey's multiple comparisons test. (**e**) Eosinophil count at different timepoints after NIMP exposure relative to CTL: 24 h (1.0 ± 0.09 for 0 LD50; 1.07 ± 0.09 for 0.5 LD50 and 0.54 ± 0.05 for 0.9 LD50); 7 days (1.0 ± 0.04 for 0 LD50; 0.76 ± 0.02 for 0.5 LD50 and 1.02 ± 0.08 for 0.9 LD50); 1 month (1.0 ± 0.02 for 0 LD50; 1.87 ± 0.30 for 0.5 LD50 and 1.84 ± 0.26 for 0.9 LD50); 3 months (1.0 ± 0.04 for 0 LD50; 0.70 ± 0.05 for 0.5 LD50 and 0.80 ± 0.04 for 0.9 LD50) and 6 months (1.0 ± 0.02 for 0 LD50; 0.71 ± 0.08 for 0.5 LD50 and 0.56 ± 0.11 for 0.9 LD50); (n = 8 to 10 per group). Significant differences were determined by one-way ANOVA (F = 8.17, p = 0.0045 for 24H; F = 7.52, p = 0.0042 for 7D; F = 3.95, p = 0.058 for 1M; F = 12.92, p = 0.0007 for 3M and F = 6.13, p = 0.0088 for 6M) with Tukey's multiple comparisons test. (**** $p < 0.0001$; *** $p < 0.001$; ** $p < 0.01$; * $p < 0.05$ CTL vs. 0.5 LD50; +++ $p < 0.001$; ++ $p < 0.01$; + $p < 0.05$ CTL vs. 0.9 LD50; ΦΦΦΦ $p < 0.01$; ΦΦΦ $p < 0.001$; ΦΦ $p < 0.01$; Φ $p < 0.05$ 0.5 LD50 vs. 0.9 LD50).

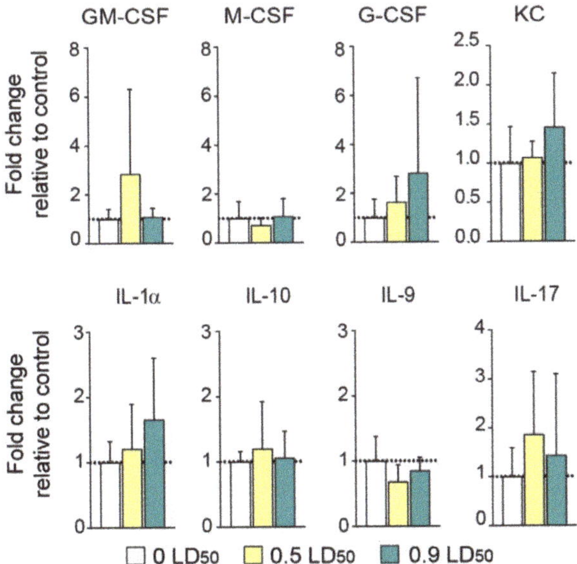

Figure 6. NIMP does not induce long-term systemic inflammation. Expression levels of the serum cytokines GM-CSF, M-CSF, G-CSF, KC, IL-1α, IL-10, IL-9 and IL-17 normalized to CTL values at 6 months post-intoxication [(GM-CSF: 1.0 ± 0.2 for 0 LD50; 2.8 ± 0.13 for 0.5 LD50 and 1.1 ± 0.01 for 0.9 LD50); (M-CSF: 1.0 ± 0.3 for 0 LD50; 0.72 ± 0.10 for 0.5 LD50 and 1.08 ± 0.3 for 0.9 LD50); (G-CSF: 1.0 ± 0.3 for 0 LD50; 1.63 ± 0.4 for 0.5 LD50 and 2.8 ± 1.46 for 0.9 LD50); (KC: 1.0 ± 0.2 for 0 LD50; 1.1± 0.08 for 0.5 LD50 and 1.46 ± 0.2 for 0.9 LD50); (IL-1α: 1.0 ± 0.1 for 0 LD50; 1.2 ± 0.3 for 0.5 LD50 and 1.7 ± 0.4 for 0.9 LD50); (IL-1α: 1.0 ± 0.07 for 0 LD50; 1.2 ± 0.3 for 0.5 LD50 and 1.06 ± 0.2 for 0.9 LD50); (IL-9: 1.0 ± 0.2 for 0 LD50; 0.68 ± 0.1 for 0.5 LD50 and 0.85 ± 0.07 for 0.9 LD50); (IL-17: 1.0 ± 0.3 for 0 LD50; 1.86 ± 0.6 for 0.5 LD50 and 1.43 ± 0.6 for 0.9 LD50)]. Statistical analyses were conducted by one-way ANOVA (n = 8 per group).

By combining other metadata related to anxiety-like behavior status and mononuclear blood cell percentages with gut microbiota composition at the genus level, we identified multiple significant associations linking the host with specific bacterial taxa (Figure 9). Notably, blood monocyte levels were positively correlated with the abundance of Aestuari-ispira (r = 0.62) and Parasutterella (r = 0.71) and negatively correlated with the levels of Bilophila (r = −0.61), Flavonifractor (r = −0.61), Gemmiger (r = −0.61), Hydrogenoanaer-obacterium (r = −0.69), Oscillibacter (r = −0.67) and Pseudoflavonifractor (r = −0.70). Neutrophil and lymphocyte percentages were positively and negatively associated with Ruminococcus abundance (r = 0.75 and −0.77, respectively). The Intestinimonas genera was associated with anxiety status parameters such as cumulative open area activity (r = 0.62).

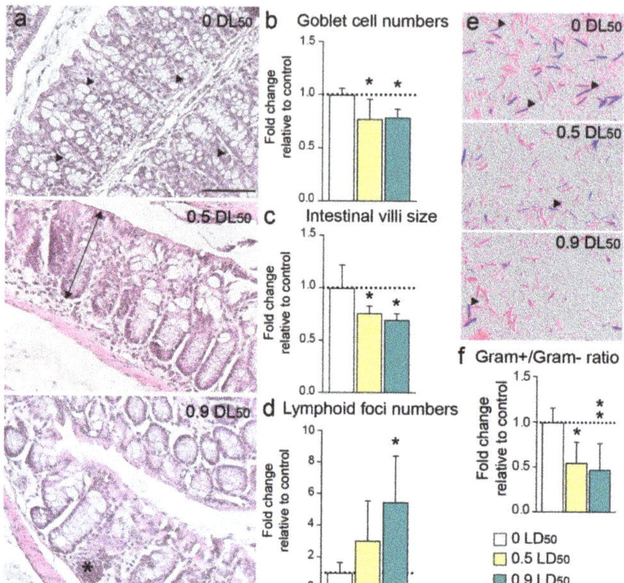

Figure 7. Long-term morphological changes in the large intestine induced by NIMP exposure. (**a**) Large intestine pictures of CTL (top panel), 0.5 LD50 (middle panel) and 0.9 LD50 (bottom panel) mice. Scale bar = 100 μm. Mucin-producing goblet cells (arrow heads) were identified in colon sections of CTL mice. The double arrow indicates an example of villus size in a 0.5 LD50 NIMP-exposed mouse; the asterisk indicates a lymphoid focus in a 0.9 LD50 NIMP-exposed mouse. Fold change in (**b**) goblet cell numbers (1.0 ± 0.03 for 0 LD50; 0.77 ± 0.08 for 0.5 LD50 and 0.79 ± 0.03 for 0.9 LD50), (**c**) villi size (1.0 ± 0.10 for 0 LD50; 0.76 ± 0.03 for 0.5 LD50 and 0.70 ± 0.03 for 0.9 LD50) and (**d**) lymphoid foci numbers (1.0 ± 0.28 for 0 LD50; 3.02 ± 1.12 for 0.5 LD50 and 5.46 ± 1.31 for 0.9 LD50) normalized to CTL values at 6 months post-intoxication. Significant differences were determined by one-way ANOVA (F = 5.5, $p = 0.02$ for goblet cell numbers; F = 6.87, $p = 0.01$ for villi size; F = 4.89, $p = 0.028$ for lymphoid foci numbers) with Tukey's multiple comparisons test. (* $p < 0.05$ CTL vs. NIMP-treated mice). (**e**) Gram−stained stool smears in the large intestine of CTL (top panel), 0.5 LD50 (middle panel) and 0.9 LD50 (bottom panel) mice. Arrow heads illustrate Gram+ bacteria stained blue, while Gram− bacteria are stained pink (X100). (**f**) The ratio of Gram+/Gram− bacteria normalized to CTL animals (1.0 ± 0.7 for 0 LD50; 0.55 ± 0.09 for 0.5 LD50 and 0.47 ± 0.11 for 0.9 LD50). Significant differences were determined by one-way ANOVA (F = 7.6, $p = 0.0047$) with Tukey's multiple comparisons test. (** $p < 0.01$; * $p < 0.05$ CTL vs. NIMP-treated mice).

4. Discussion

The long-term follow-up of NA poisoning highlights the occurrence of neurological sequelae, even if the exposed victims only show transient low symptoms [2,31]. Psychological consequences are the main mood disorder reported by NA victims, and they can still be present more than 10 years after the incident [2,32]. Most of these complications affect emotions with increased fear, event recollections, irritability and agitation, which reflect anxiety-like behaviors. A growing concern in animal studies is the ability to evaluate behavioral and physiological modifications in order to better assess persistent long-term effects. We therefore developed an animal model of sublethal OP exposure that reproduces the early effects of sarin poisoning [17], which appears to be suitable for the study of the long-term effects of NA exposure.

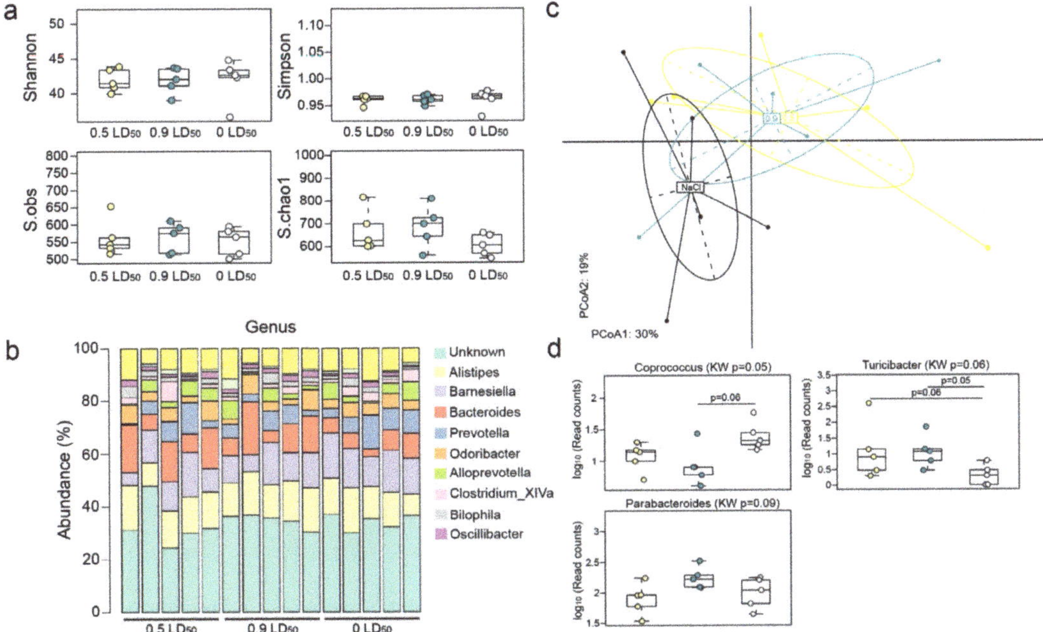

Figure 8. Mice exposed to NIMP harbor slight modifications in their gut microbiota composition 6 months after exposure. (**a**) Gut microbiota diversity was not affected by NIMP exposure. Microbial diversity and richness indexes were computed on the OTU abundance table. No difference was noted for these indexes between groups. (**b**) Overview of the gut microbiota composition at the genus level. Alistipes, Barnesiella, Bacteroides, Prevotella, Odoribacter, Alloprevotella, Clostridium_XIVa, Bilophila and Oscillibacter were the most abundant genera identified from mice gut microbiota. (**c**) NIMP exposure had a moderate impact on mice gut microbiota composition. Two-dimensional principal coordinates analysis was performed using the Bray–Curtis distances computed on the OTU abundance table. The total inertia explained (PCoA1 and PCoA2 axes) accounted for 49%. The CTL mice centroid (average microbiota profile) clustered away from both doses of NIMP-treated groups (0.5 LD50 in yellow and 0.9 LD50 in blue). This observation suggests that both doses of NIMP intoxication have a long-lasting effect on gut microbiota composition up to 6 months post-treatment. (**d**) Several gut bacterial genera display a shifted abundance following NIMP treatment. Differences in genus abundance between groups were assessed using the Kruskal–Wallis rank sum test ($p \leq 0.1$) followed by a post hoc Dunn's all-pairs rank comparison test. Three genera harbored a shifted abundance following NIMP exposure. In comparison to CTL mice, the 0.5 LD50 and 0.9 LD50 mice groups had a lower abundance of Coprococcus ($p \leq 0.05$). The Turicibacter level increased in both 0.5 LD50 and 0.9 LD50 mice groups in comparison to the CTL group ($p \leq 0.06$). Parabacteroides was slightly increased in the 0.9 LD50 group.

In the present study, the use of NIMP as a sarin analog provided a robust replication of OP exposure symptoms and brain ChE activity inhibition. Similar parameters of intoxication severity have been observed in animals at early timepoints in a dose-dependent manner [17]. Mice subjected to the highest NIMP dose (0.9 LD50) showed marked brain ChE activity inhibition, behavioral alteration, weight loss and modification of WBC levels during the first week post-intoxication. Taken together, all of these signs are representative of the expected clinical features of OP poisoning and could be used to define a typical toxidrome of NA exposure consequences in animal subjects.

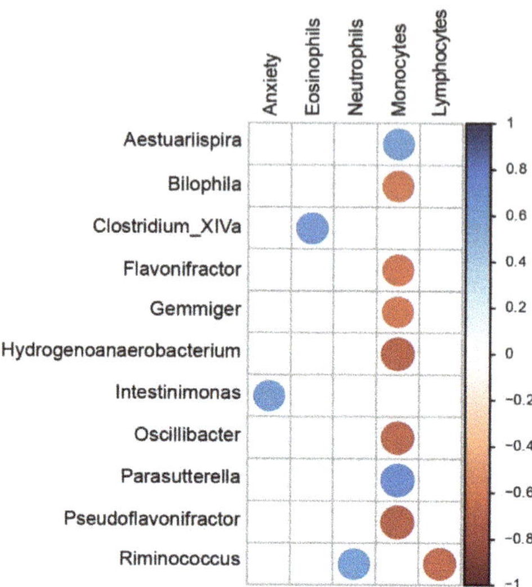

Figure 9. Pathological connection between modifications observed after NIMP exposure. Identification of bacterial genera found associated with host anxiety status and blood mononuclear cell abundance. Spearman correlations were computed using the genus abundance table and host metadata. Displayed correlations were all significant ($p \leq 0.05$) and lower than -0.6 or higher than 0.6. Anxiety: cumulative open area activity (%), Eosinophils (%), Neutrophils (%), Monocytes (%) and Lymphocytes (%).

Animals exposed to 0.5 LD50 NIMP did not have any strong behavior or weight consequences, despite a significant inhibition of cerebral ChE activity and significant modification in WBC (lymphocytes, basophils and eosinophils) levels that persisted during the first week. One month after exposure, animals in both NIMP-exposed groups showed similar recovery, with no observed differences in weight or daily behavior (food intake, socialization, fur and body maintenance) compared to the CTL group. In fact, only cerebral ChE activity was still decreased in both groups. However, a significant increase in anxious behavior as evaluated by EPM occurred in both NIMP-exposed groups.

Anxiety disorder is one of the most reported disorders in animal models in response to OP exposure [10,33–36]. Generally, such behavioral assessments are conducted in animals exposed to high doses of OP that show persistent mood disorder defects [33,34]. Similar results have also been observed for the development of anxiety-like behavior in animals exposed to sublethal dose of OP [10,37] and are consistent with reports in humans [2,38,39]. Interestingly, in our study, significant anxiety disorder was observed 30 days after NIMP exposure in both groups (0.5 and 0.9 LD50), but was compensated for after the first month, and no significant anxious behaviors were seen before the fifth month post-exposure. This anxiety recurrence affected both NIMP-exposed groups and persisted up to 6 months post-intoxication.

It is worth noting that most of the anxiety tests used in rodents are based on a balance between a fearful response to an aversive condition (open, brightly lit or elevated spaces) and the tendency of animals to engage in exploratory activity or social interaction. To maintain the test novelty and the curiosity of the mice, and thus avoid any habituation, we chose to change the anxiety tests every month. Compensation of behavioral and/or cognitive deficits following OP exposure has already been reported in different rodent models [37,40]. Indeed, mice presented a partial improvement in cognitive performance over time in a Morris water maze and T-maze 3 months after soman exposure [40]. Transient improvement was also observed in rats 4 months after sarin vapor exposure, but no recovery was observed at longer timepoints [37]. Hence, our results are clearly in line with previous animal models, demonstrating that NIMP could be a reliable OP compound to reproduce NA exposure in the long-term.

The amygdala is the primary brain region involved in anxiety behavior [41,42] and is one of the main brain regions to be particularly affected by NA exposure [43–45]. Neuronal hyperexcitability has been noted in the basolateral amygdala following exposure to a high dose of soman due to a decrease in GABAergic inhibition in this area [35,36]. A slow recovery in AChE activity coupled to a loss of GABAergic neurons may explain this hyperactivity and the development of anxiety-like behavior [35,36]. However, the low doses of NIMP used in our study did not induce any observable neuropathologies at any timepoint studied. This result is consistent with our previous study [17] and with other animal models exposed to a low dose of NA [15,46]. Furthermore, no anatomical volume modification of the amygdala was observed in our model at any timepoint studied, suggesting that neither cell loss nor swelling was induced after the low-dose exposure to NIMP. The amygdala is particularly sensitive to stress, which could be reflected by an increase in inflammatory cytokines leading to its enlargement [41,47].

In accordance with the lack of any architectural change in the amygdala, no neuroinflammation process (i.e., significant elevated cytokine levels, astrocytes or microglial activation) was observed in either group of NIMP-exposed animals 6 months post-intoxication. Therefore, we suspected that peripheral inflammation could be involved in the establishment of anxiety-like behavior observed in NIMP-exposed animals in the longer term. Indeed, elevated peripheral inflammatory cytokine levels (IL−6, IL-1beta and TNF-alpha) have been associated with mood disorder development [41,47]. However, our evaluation of serum cytokines did not reveal any significant modifications in the inflammatory cytokines under study in the two groups 6 months post-NIMP exposure. On the other hand, a significant decrease in circulating basophil and eosinophil counts was observed 3 months after NIMP exposure and persisted for up to 6 months. A decrease in venous blood basophil counts has already been observed in patients with major depression disorder displaying elevated anxiety [48], suggesting that the alteration of leukocytes may play a role in the development of anxious behavior. Basophils and eosinophils are leukocyte subtypes particularly involved in allergic responses, but they are also involved in gut homeostasis regulation [49,50]. Since basophils and eosinophils play a role in the maintenance of the protective mucosal barrier and contribute to immune modulation towards gut microbes, these decreasing basophil and eosinophil counts may participate in the gut morphological alteration and dysbiosis observed in NIMP-exposed animals. We therefore decided to investigate if NIMP exposure could affect mice microbiota.

The gastrointestinal system expresses several nicotinic receptors and is highly innervated by the cholinergic neurons of the parasympathetic and enteric systems. In addition, ACh is involved in regulating several functions such as gut motility, local blood flow, intestinal mucosal barrier permeability and inflammation regulation [51–53]. Furthermore, it has been shown that ChE inhibition by soman, neostigmine, PB or DFP alters intestinal functions [54–56]. The intestine also plays a crucial role in the elimination of NA [57], which should impact the microbiota homeostasis. Indeed, one previous study found that the administration of PB along with the insecticide permethrin to mice alters the gut microbiome, with the enrichment of several bacterial families and genera in the treated animals [58].

Interestingly, the animal model used in the aforementioned study reproduced some of the GW symptoms reported by deployed veterans and showed abundant Coprococcus and Turicibacter correlated with neuroinflammation and gut leaching [58]. Moreover, a previous study conducted in humans reported that Coprococcus abundance is also associated with depressive disorder [59]. In our study, a decrease in Coprococcus and a concomitant increase in Turicibacter were observed in both groups 6 months after NIMP exposure, regardless of the dose received. The difference in Coprococcus variation could be explained by the exposure protocol and the long-term evaluation period, but an alternative possibility is that a lack of Coprococcus could be associated with anxiety.

To our knowledge, our study is the first demonstration of altered microbiota after NA exposure observed over the long term. In fact, several bacterial populations were found to be modified in rats 3 days after soman exposure, and no long-term change was reported (75 days post-exposure) where the gut microbiota remained resilient [60]. Although we assessed the gut microbiota composition at 6 months after NIMP exposure in order to determine if this could explain the second wave of anxiety-like behavior observed in NIMP exposed animals, we could not determine the timepoints at which these changes were implemented. Nevertheless, our microbiota analysis revealed a positive correlation between Intestinimonas abundance and the anxiety-like behavior level measured in mice exposed to NIMP. It is noteworthy that a positive link between Intestinimonas and psychological stress was previously reported in a rat model, in association with intestinal and blood-brain barrier alterations [61]. Furthermore, several bacterial species modified by NIMP exposure were correlated with the WBC variation observed in 0.5 LD50-exposed animals, which may enhance immune complications in these animals. Together, our results demonstrate that a single acute NA exposure can lead to long-term gut microbiota modifications.

5. Conclusions

In conclusion, our study demonstrates that exposure to a low dose of NA, identified as barely symptomatic for the 0.5 LD50 NIMP-exposed mice, could disrupt gut microbiota and immune homeostasis and alter emotional behavior in the long-term (Graphical abstract). Further studies are needed in order to decipher the mechanisms linking microbiota, anxiety-like behavior and immune cell response, as well as to discriminate more specific middle timepoints. The association between Intestinimonas and anxiety-like behavior requires further investigations to be confirmed. Nevertheless, our results highlight the need to care for all NA victims, even those exposed to low-doses, and to disclose new biomarkers that could be useful for follow-up with NA victims.

Author Contributions: Conceptualization, G.D.B. and K.T.; methodology, G.D.B. and K.T.; validation, S.F., S.M., Q.G., C.O., G.D.B. and K.T.; formal analysis, S.M., Q.G. and K.T.; investigation, S.F., S.M., Q.G., R.B. (Rosalie Bel), J.K., A.B., S.C., G.D.B. and K.T.; resources, R.B. (Rachid Baati), C.O., G.D.B. and K.T.; writing—original draft preparation, S.F., S.M., Q.G., C.O., G.D.B. and K.T.; writing—review and editing, S.F., S.M., Q.G., R.B. (Rachid Baati), C.O., G.D.B. and K.T.; supervision, G.D.B. and K.T.; project administration, C.O., G.D.B. and K.T.; funding acquisition, G.D.B. and K.T.; visualization, G.D.B. and K.T. All authors have read and agreed to the published version of the manuscript.

Funding: This work was supported by the French Ministry of Armed Forces: Direction Générale de l'Armement (DGA) and the Service de Santé des Armées (SSA), grant #PDH-2-NRBC-4-C-4208. The funding sources had no involvement in study design, collection, data analysis or interpretation, or decision to publish.

Institutional Review Board Statement: The animal study protocol was approved by the SSA animal ethics committee according to applicable French legislation (Directive 2010/63/UE, decret 2013-118), authorization project AP 281/ARM/IGSSA/SP (25 November 2017) and decision 223/ARM/IGSSA/SP (19 September 2018).

Informed Consent Statement: Not applicable.

Data Availability Statement: Not applicable.

Acknowledgments: This work made use of the IRBA biological analysis platforms, supported by the Service de Santé des Armées. We particularly thank Francisca Fargeau and Christine Frederic for their technical support.

Conflicts of Interest: The authors declare no conflict of interest. The funders had no role in the design of the study; in the collection, analyses, or interpretation of data; in the writing of the manuscript, or in the decision to publish the results.

References

1. Da Fonseca Carvalho, L.M. Novichok(s) A Challenge to the Chemical Weapons Convention. *Port. J. Mil. Sci. (Rev. Ciências Mil.)* **2021**, *9*, 39–62.
2. Sugiyama, A.; Matsuoka, T.; Sakamune, K.; Akita, T.; Makita, R.; Kimura, S.; Kuroiwa, Y.; Nagao, M.; Tanaka, J. The Tokyo Subway Sarin Attack Has Long-Term Effects on Survivors: A 10-Year Study Started 5 Years after the Terrorist Incident. *PLoS ONE* **2020**, *15*, e0234967. [CrossRef] [PubMed]
3. Talabani, J.M.; Ali, A.I.; Kadir, A.M.; Rashid, R.; Samin, F.; Greenwood, D.; Hay, A. Long-Term Health Effects of Chemical Warfare Agents on Children Following a Single Heavy Exposure. *Hum. Exp. Toxicol.* **2018**, *37*, 836–847. [CrossRef] [PubMed]
4. Jeffrey, M.G.; Krengel, M.; Kibler, J.L.; Zundel, C.; Klimas, N.G.; Sullivan, K.; Craddock, T.J.A. Neuropsychological Findings in Gulf War Illness: A Review. *Front. Psychol.* **2019**, *10*, 2088. [CrossRef]
5. Abou-Donia, M.B.; Siracuse, B.; Gupta, N.; Sobel Sokol, A. Sarin (GB, O-Isopropyl Methylphosphonofluoridate) Neurotoxicity: Critical Review. *Crit. Rev. Toxicol.* **2016**, *46*, 845–875. [CrossRef]
6. Dorandeu, F.; Baille, V.; Mikler, J.; Testylier, G.; Lallement, G.; Sawyer, T.; Carpentier, P. Protective Effects of S(+) Ketamine and Atropine against Lethality and Brain Damage during Soman-Induced Status Epilepticus in Guinea-Pigs. *Toxicology* **2007**, *234*, 185–193. [CrossRef]
7. Reddy, S.D.; Reddy, D.S. Midazolam as an Anticonvulsant Antidote for Organophosphate Intoxication-A Pharmacotherapeutic Appraisal. *Epilepsia* **2015**, *56*, 813–821. [CrossRef]
8. Pereira, E.F.R.; Aracava, Y.; DeTolla, L.J.; Beecham, E.J.; Basinger, G.W.; Wakayama, E.J.; Albuquerque, E.X. Animal Models That Best Reproduce the Clinical Manifestations of Human Intoxication with Organophosphorus Compounds. *J. Pharmacol. Exp. Ther.* **2014**, *350*, 313–321. [CrossRef]
9. Allon, N.; Chapman, S.; Egoz, I.; Rabinovitz, I.; Kapon, J.; Weissman, B.A.; Yacov, G.; Bloch-Shilderman, E.; Grauer, E. Deterioration in Brain and Heart Functions Following a Single Sub-Lethal (0.8 LCt50) Inhalation Exposure of Rats to Sarin Vapor. *Toxicol. Appl. Pharmacol.* **2011**, *253*, 31–37. [CrossRef]
10. Mamczarz, J.; Pereira, E.F.R.; Aracava, Y.; Adler, M.; Albuquerque, E.X. An Acute Exposure to a Sub-Lethal Dose of Soman Triggers Anxiety-Related Behavior in Guinea Pigs: Interactions with Acute Restraint. *Neurotoxicology* **2010**, *31*, 77–84. [CrossRef]
11. Loh, Y.; Swanberg, M.M.; Ingram, M.V.; Newmark, J. Case Report: Long-Term Cognitive Sequelae of Sarin Exposure. *Neurotoxicology* **2010**, *31*, 244–246. [CrossRef] [PubMed]
12. Sullivan, K.; Krengel, M.; Bradford, W.; Stone, C.; Thompson, T.A.; Heeren, T.; White, R.F. Neuropsychological Functioning in Military Pesticide Applicators from the Gulf War: Effects on Information Processing Speed, Attention and Visual Memory. *Neurotoxicol. Teratol.* **2018**, *65*, 1–13. [CrossRef] [PubMed]
13. White, R.F.; Steele, L.; O'Callaghan, J.P.; Sullivan, K.; Binns, J.H.; Golomb, B.A.; Bloom, F.E.; Bunker, J.A.; Crawford, F.; Graves, J.C.; et al. Recent Research on Gulf War Illness and Other Health Problems in Veterans of the 1991 Gulf War: Effects of Toxicant Exposures during Deployment. *Cortex* **2016**, *74*, 449–475. [CrossRef] [PubMed]
14. Kassa, J.; Koupilova, M.; Vachek, J. The Influence of Low-Level Sarin Inhalation Exposure on Spatial Memory in Rats. *Pharmacol. Biochem. Behav.* **2001**, *70*, 175–179. [CrossRef]
15. Bloch-Shilderman, E.; Rabinovitz, I.; Egoz, I.; Raveh, L.; Allon, N.; Grauer, E.; Gilat, E.; Weissman, B.A. Subchronic Exposure to Low-Doses of the Nerve Agent VX: Physiological, Behavioral, Histopathological and Neurochemical Studies. *Toxicol. Appl. Pharmacol.* **2008**, *231*, 17–23. [CrossRef]
16. Shih, T.-M.; Hulet, S.W.; McDonough, J.H. The Effects of Repeated Low-Dose Sarin Exposure. *Toxicol. Appl. Pharmacol.* **2006**, *215*, 119–134. [CrossRef]
17. Angrand, L.; Takillah, S.; Malissin, I.; Berriche, A.; Cervera, C.; Bel, R.; Gerard, Q.; Knoertzer, J.; Baati, R.; Kononchik, J.P.; et al. Persistent Brainwave Disruption and Cognitive Impairment Induced by Acute Sarin Surrogate Sub-Lethal Dose Exposure. *Toxicology* **2021**, *456*, 152787. [CrossRef]
18. Chambers, J.E.; Meek, E.C.; Bennett, J.P.; Bennett, W.S.; Chambers, H.W.; Leach, C.A.; Pringle, R.B.; Wills, R.W. Novel Substituted Phenoxyalkyl Pyridinium Oximes Enhance Survival and Attenuate Seizure-like Behavior of Rats Receiving Lethal Levels of Nerve Agent Surrogates. *Toxicology* **2016**, *339*, 51–57. [CrossRef]
19. Chambers, J.E.; Meek, E.C. Novel Centrally Active Oxime Reactivators of Acetylcholinesterase Inhibited by Surrogates of Sarin and VX. *Neurobiol. Dis.* **2020**, *133*, 104487. [CrossRef]
20. Dail, M.B.; Leach, C.A.; Meek, E.C.; Olivier, A.K.; Pringle, R.B.; Green, C.E.; Chambers, J.E. Novel Brain-Penetrating Oxime Acetylcholinesterase Reactivators Attenuate Organophosphate-Induced Neuropathology in the Rat Hippocampus. *Toxicol. Sci.* **2019**, *169*, 465–474. [CrossRef]
21. Meek, E.C.; Chambers, H.W.; Coban, A.; Funck, K.E.; Pringle, R.B.; Ross, M.K.; Chambers, J.E. Synthesis and In Vitro and In Vivo Inhibition Potencies of Highly Relevant Nerve Agent Surrogates. *Toxicol. Sci.* **2012**, *126*, 525–533. [CrossRef] [PubMed]
22. Pringle, R.B.; Meek, E.C.; Chambers, H.W.; Chambers, J.E. Neuroprotection from Organophosphate-Induced Damage by Novel Phenoxyalkyl Pyridinium Oximes in Rat Brain. *Toxicol. Sci.* **2018**, *166*, 420–427. [CrossRef] [PubMed]
23. Rispin, A.; Farrar, D.; Margosches, E.; Gupta, K.; Stitzel, K.; Carr, G.; Greene, M.; Meyer, W.; McCall, D. Alternative Methods for the Median Lethal Dose (LD(50)) Test: The up-and-down Procedure for Acute Oral Toxicity. *ILAR J.* **2002**, *43*, 233–243. [CrossRef] [PubMed]
24. Franklin, K.B.J.; Paxinos, G. *The Mouse Brain in Stereotaxic Coordinates*, 3rd ed.; Elsevier, AP: Amsterdam, The Netherlands, 2008.

25. Martin, M. Cutadapt Removes Adapter Sequences from High-Throughput Sequencing Reads. *EMBnet. J.* **2011**, *17*, 10–12. [CrossRef]

26. Bankevich, A.; Nurk, S.; Antipov, D.; Gurevich, A.A.; Dvorkin, M.; Kulikov, A.S.; Lesin, V.M.; Nikolenko, S.I.; Pham, S.; Prjibelski, A.D.; et al. SPAdes: A New Genome Assembly Algorithm and Its Applications to Single-Cell Sequencing. *J. Comput. Biol.* **2012**, *19*, 455–477. [CrossRef] [PubMed]

27. Zhang, J.; Kobert, K.; Flouri, T.; Stamatakis, A. PEAR: A Fast and Accurate Illumina Paired-End ReAd MergeR. *Bioinformatics* **2014**, *30*, 614–620. [CrossRef]

28. Rognes, T.; Flouri, T.; Nichols, B.; Quince, C.; Mahé, F. VSEARCH: A Versatile Open Source Tool for Metagenomics. *PeerJ* **2016**, *4*, e2584. [CrossRef]

29. Cole, J.R.; Wang, Q.; Fish, J.A.; Chai, B.; McGarrell, D.M.; Sun, Y.; Brown, C.T.; Porras-Alfaro, A.; Kuske, C.R.; Tiedje, J.M. Ribosomal Database Project: Data and Tools for High Throughput RRNA Analysis. *Nucleic Acids Res.* **2014**, *42*, D633–D642. [CrossRef]

30. Kim, S.; Jazwinski, S.M. The Gut Microbiota and Healthy Aging. *Gerontology* **2018**, *64*, 513–520. [CrossRef]

31. Ohbu, S.; Yamashina, A.; Takasu, N.; Yamaguchi, T.; Murai, T.; Nakano, K.; Matsui, Y.; Mikami, R.; Sakurai, K.; Hinohara, S. Sarin Poisoning on Tokyo Subway. *South. Med. J.* **1997**, *90*, 587–593. [CrossRef]

32. Yanagisawa, N.; Morita, H.; Nakajima, T. Sarin Experiences in Japan: Acute Toxicity and Long-Term Effects. *J. Neurol. Sci.* **2006**, *249*, 76–85. [CrossRef] [PubMed]

33. Calsbeek, J.J.; González, E.A.; Bruun, D.A.; Guignet, M.A.; Copping, N.; Dawson, M.E.; Yu, A.J.; MacMahon, J.A.; Saito, N.H.; Harvey, D.J.; et al. Persistent Neuropathology and Behavioral Deficits in a Mouse Model of Status Epilepticus Induced by Acute Intoxication with Diisopropylfluorophosphate. *Neurotoxicology* **2021**, *87*, 106–119. [CrossRef] [PubMed]

34. Coubard, S.; Béracochéa, D.; Collombet, J.-M.; Philippin, J.-N.; Krazem, A.; Liscia, P.; Lallement, G.; Piérard, C. Long-Term Consequences of Soman Poisoning in Mice. *Behav. Brain Res.* **2008**, *191*, 95–103. [CrossRef] [PubMed]

35. Prager, E.M.; Pidoplichko, V.I.; Aroniadou-Anderjaska, V.; Apland, J.P.; Braga, M.F.M. Pathophysiological Mechanisms Underlying Increased Anxiety after Soman Exposure: Reduced GABAergic Inhibition in the Basolateral Amygdala. *NeuroToxicology* **2014**, *44*, 335–343. [CrossRef] [PubMed]

36. Prager, E.M.; Aroniadou-Anderjaska, V.; Almeida-Suhett, C.P.; Figueiredo, T.H.; Apland, J.P.; Rossetti, F.; Olsen, C.H.; Braga, M.F.M. The Recovery of Acetylcholinesterase Activity and the Progression of Neuropathological and Pathophysiological Alterations in the Rat Basolateral Amygdala after Soman-Induced Status Epilepticus: Relation to Anxiety-like Behavior. *Neuropharmacology* **2014**, *81*, 64–74. [CrossRef]

37. Grauer, E.; Chapman, S.; Rabinovitz, I.; Raveh, L.; Weissman, B.-A.; Kadar, T.; Allon, N. Single Whole-Body Exposure to Sarin Vapor in Rats: Long-Term Neuronal and Behavioral Deficits. *Toxicol. Appl. Pharmacol.* **2008**, *227*, 265–274. [CrossRef]

38. Harrison, V.; Mackenzie Ross, S. Anxiety and Depression Following Cumulative Low-Level Exposure to Organophosphate Pesticides. *Environ. Res.* **2016**, *151*, 528–536. [CrossRef]

39. Levin, H.S.; Rodnitzky, R.L.; Mick, D.L. Anxiety Associated with Exposure to Organophosphate Compounds. *Arch. Gen. Psychiatry* **1976**, *33*, 225–228. [CrossRef]

40. Filliat, P.; Coubard, S.; Pierard, C.; Liscia, P.; Beracochea, D.; Four, E.; Baubichon, D.; Masqueliez, C.; Lallement, G.; Collombet, J.-M. Long-Term Behavioral Consequences of Soman Poisoning in Mice. *NeuroToxicology* **2007**, *28*, 508–519. [CrossRef]

41. Łoś, K.; Waszkiewicz, N. Biological Markers in Anxiety Disorders. *J. Clin. Med.* **2021**, *10*, 1744. [CrossRef]

42. Shin, L.M.; Liberzon, I. The Neurocircuitry of Fear, Stress, and Anxiety Disorders. *Neuropsychopharmacology* **2010**, *35*, 169–191. [CrossRef] [PubMed]

43. Aroniadou-Anderjaska, V.; Figueiredo, T.H.; Apland, J.P.; Prager, E.M.; Pidoplichko, V.I.; Miller, S.L.; Braga, M.F.M. Long-Term Neuropathological and Behavioral Impairments after Exposure to Nerve Agents: Long-Term Health Effects of Nerve Agent Exposure. *Ann. N. Y. Acad. Sci.* **2016**, *1374*, 17–28. [CrossRef] [PubMed]

44. Collombet, J.-M. Nerve Agent Intoxication: Recent Neuropathophysiological Findings and Subsequent Impact on Medical Management Prospects. *Toxicol. Appl. Pharmacol.* **2011**, *255*, 229–241. [CrossRef] [PubMed]

45. Myhrer, T. Identification of Neuronal Target Areas for Nerve Agents and Specification of Receptors for Pharmacological Treatment. *NeuroToxicology* **2010**, *31*, 629–638. [CrossRef] [PubMed]

46. Oswal, D.P.; Garrett, T.L.; Morris, M.; Lucot, J.B. Low-Dose Sarin Exposure Produces Long Term Changes in Brain Neurochemistry of Mice. *Neurochem. Res.* **2013**, *38*, 108–116. [CrossRef] [PubMed]

47. Felger, J.C. Imaging the Role of Inflammation in Mood and Anxiety-Related Disorders. *Curr. Neuropharmacol.* **2018**, *16*, 533–558. [CrossRef]

48. Baek, J.H.; Kim, H.-J.; Fava, M.; Mischoulon, D.; Papakostas, G.I.; Nierenberg, A.; Heo, J.-Y.; Jeon, H.J. Reduced Venous Blood Basophil Count and Anxious Depression in Patients with Major Depressive Disorder. *Psychiatry Investig.* **2016**, *13*, 321. [CrossRef]

49. Kim, S.; Prout, M.; Ramshaw, H.; Lopez, A.F.; LeGros, G.; Min, B. Basophils Are Transiently Recruited into the Draining Lymph Nodes during Helminth Infection via IL-3 but Infection-Induced Th2 Immunity Can Develop without Basophil Lymph Node Recruitment or IL-3. *J. Immunol.* **2010**, *184*, 1143–1147. [CrossRef]

50. Loktionov, A. Eosinophils in the Gastrointestinal Tract and Their Role in the Pathogenesis of Major Colorectal Disorders. *World J. Gastroenterol.* **2019**, *25*, 3503–3526. [CrossRef]

51. Furness, J.B. The Enteric Nervous System: Normal Functions and Enteric Neuropathies. *Neurogastroenterol. Motil.* **2008**, *20* (Suppl. 1), 32–38. [CrossRef]
52. Matteoli, G.; Gomez-Pinilla, P.J.; Nemethova, A.; Di Giovangiulio, M.; Cailotto, C.; van Bree, S.H.; Michel, K.; Tracey, K.J.; Schemann, M.; Boesmans, W.; et al. A Distinct Vagal Anti-Inflammatory Pathway Modulates Intestinal Muscularis Resident Macrophages Independent of the Spleen. *Gut* **2014**, *63*, 938–948. [CrossRef] [PubMed]
53. Rueda Ruzafa, L.; Cedillo, J.L.; Hone, A.J. Nicotinic Acetylcholine Receptor Involvement in Inflammatory Bowel Disease and Interactions with Gut Microbiota. *Int. J. Environ. Res. Public. Health* **2021**, *18*, 1189. [CrossRef] [PubMed]
54. Brezenoff, H.E.; McGee, J.; Hymowitz, N. Inhibition of Acetylcholinesterase in the Gut Inhibits Schedule-Controlled Behavior in the Rat. *Life Sci.* **1985**, *37*, 49–54. [CrossRef]
55. Hernandez, S.; Fried, D.E.; Grubišić, V.; McClain, J.L.; Gulbransen, B.D. Gastrointestinal Neuroimmune Disruption in a Mouse Model of Gulf War Illness. *FASEB J.* **2019**, *33*, 6168–6184. [CrossRef]
56. Seth, R.K.; Kimono, D.; Alhasson, F.; Sarkar, S.; Albadrani, M.; Lasley, S.K.; Horner, R.; Janulewicz, P.; Nagarkatti, M.; Nagarkatti, P.; et al. Increased Butyrate Priming in the Gut Stalls Microbiome Associated-Gastrointestinal Inflammation and Hepatic Metabolic Reprogramming in a Mouse Model of Gulf War Illness. *Toxicol. Appl. Pharmacol.* **2018**, *350*, 64–77. [CrossRef]
57. Kadar, T.; Raveh, L.; Cohen, G.; Oz, N.; Baranes, I.; Balan, A.; Ashani, Y.; Shapira, S. Distribution of 3H-Soman in Mice. *Arch. Toxicol.* **1985**, *58*, 45–49. [CrossRef]
58. Alhasson, F.; Das, S.; Seth, R.; Dattaroy, D.; Chandrashekaran, V.; Ryan, C.N.; Chan, L.S.; Testerman, T.; Burch, J.; Hofseth, L.J.; et al. Altered Gut Microbiome in a Mouse Model of Gulf War Illness Causes Neuroinflammation and Intestinal Injury via Leaky Gut and TLR4 Activation. *PLoS ONE* **2017**, *12*, e0172914. [CrossRef]
59. Valles-Colomer, M.; Falony, G.; Darzi, Y.; Tigchelaar, E.F.; Wang, J.; Tito, R.Y.; Schiweck, C.; Kurilshikov, A.; Joossens, M.; Wijmenga, C.; et al. The Neuroactive Potential of the Human Gut Microbiota in Quality of Life and Depression. *Nat. Microbiol.* **2019**, *4*, 623–632. [CrossRef]
60. Getnet, D.; Gautam, A.; Kumar, R.; Hoke, A.; Cheema, A.K.; Rossetti, F.; Schultz, C.R.; Hammamieh, R.; Lumley, L.A.; Jett, M. Poisoning with Soman, an Organophosphorus Nerve Agent, Alters Fecal Bacterial Biota and Urine Metabolites: A Case for Novel Signatures for Asymptomatic Nerve Agent Exposure. *Appl. Environ. Microbiol.* **2018**, *84*, e00978-18. [CrossRef]
61. Geng, S.; Yang, L.; Cheng, F.; Zhang, Z.; Li, J.; Liu, W.; Li, Y.; Chen, Y.; Bao, Y.; Chen, L.; et al. Gut Microbiota Are Associated with Psychological Stress-Induced Defections in Intestinal and Blood–Brain Barriers. *Front. Microbiol.* **2020**, *10*, 3067. [CrossRef]

 *biomedicines*

Review

Exogenous Antioxidants in Remyelination and Skeletal Muscle Recovery

Ricardo Julián Cabezas Perez [1], Marco Fidel Ávila Rodríguez [2] and Doris Haydee Rosero Salazar [3,*]

[1]   School of Medicine, Faculty of Health, Universidad Antonio Nariño, Bogota 110111, Colombia
[2]   Department of Clinical Sciences, Universidad del Tolima, Ibagué 730007, Colombia
[3]   School of Dentistry, University of Washington, Seattle, WA 98105, USA
*   Correspondence: dhross@uw.edu

Abstract: Inflammatory, oxidative, and autoimmune responses cause severe damage to the nervous system inducing loss of myelin layers or demyelination. Even though demyelination is not considered a direct cause of skeletal muscle disease there is extensive damage in skeletal muscles following demyelination and impaired innervation. In vitro and in vivo evidence using exogenous antioxidants in models of demyelination is showing improvements in myelin formation alongside skeletal muscle recovery. For instance, exogenous antioxidants such as EGCG stimulate nerve structure maintenance, activation of glial cells, and reduction of oxidative stress. Consequently, this evidence is also showing structural and functional recovery of impaired skeletal muscles due to demyelination. Exogenous antioxidants mostly target inflammatory pathways and stimulate remyelinating mechanisms that seem to induce skeletal muscle regeneration. Therefore, the aim of this review is to describe recent evidence related to the molecular mechanisms in nerve and skeletal muscle regeneration induced by exogenous antioxidants. This will be relevant to identifying further targets to improve treatments of neuromuscular demyelinating diseases.

Keywords: antioxidants; inflammation; demyelination; nerve; skeletal muscle; regeneration; remyelination; myelin; oxidative stress

Citation: Cabezas Perez, R.J.; Ávila Rodríguez, M.F.; Rosero Salazar, D.H. Exogenous Antioxidants in Remyelination and Skeletal Muscle Recovery. *Biomedicines* **2022**, *10*, 2557. https://doi.org/10.3390/biomedicines10102557

Academic Editor: Marie-Louise Bang

Received: 1 May 2022
Accepted: 25 July 2022
Published: 13 October 2022

Publisher's Note: MDPI stays neutral with regard to jurisdictional claims in published maps and institutional affiliations.

1. Introduction

Inflammation and tissue disorder are outcomes of the imbalance in the production of reactive oxygen species (ROS) accompanied by insufficient activity of endogenous antioxidants that boost oxidative stress [1,2]. The increase of ROS such as superoxide and hydroxyl radicals may cause DNA lesions, protein structural damage, oxidize lipids, and mitochondrial dysfunction ending in apoptosis [3–6]. These events are associated with neurodegenerative and demyelinating disorders including Alzheimer's disease, Parkinson's disease, and multiple sclerosis [1]. In addition, low levels of exogenous antioxidants exacerbate ROS effects and subsequent degenerative effects [1,2,5].

Myelin is a sphingolipid-based multilayer produced by oligodendrocytes and Schwann cells [7,8]. This multilayer by wrapping the nerve axons forms the isolation to improve neuronal action potential maintaining the speed of nerve impulses [7,8]. Autoimmune and chronic inflammatory diseases in the nervous system exhibit removal of the myelin sheath, or demyelination, in central and peripheral nerves [8,9]. Demyelinating diseases outline a wide group of clinical conditions with heterogeneous clinical outcomes [9]. These outcomes include severe nerve injury resulting in myopathy-like damage, increased collagen production, or fibrosis, and loss of muscle fibers [10–12]. Early mechanisms related to these events were observed in rat models of nerve injuries showing massive ROS production in the oxidative stage [11]. This occurs along with increased protein kinases activity and hypoxia-inducible factor-1 (HIF-1) during cellular adaptation as compensatory mechanisms [11]. All this shows that nerve damage and demyelination impair skeletal muscle arrangement.

Demyelination is not described as a direct cause of myopathies, however, there is extensive muscle damage derived from impaired innervation as occurs in multiple sclerosis and other demyelinating diseases [13]. This is similar to what is observed in autoimmune myopathies as dermatomyositis with severe inflammation and comparable to the damage observed in chronic inflammatory demyelinating polyneuropathy [11,14,15]. The neuromuscular junctions also show impairment due to the oxidative and inflammatory reactions in nerves and muscle fibers [13,14]. Hence, autoimmune disorders either of the nervous system or skeletal muscle result in inflammation, increased oxidative stress, demyelination, and skeletal muscle damage.

The endogenous antioxidant system regulates oxidative balance by reducing ROS damage through specific mechanisms such as ROS scavenging and restriction of ROS generation [16]. These mechanisms neutralize ROS production by binding metal ions and decreasing oxidative effects [16]. Antioxidant molecules include superoxide dismutase (SOD), catalase, peroxidase, and glutathione, all able to detoxify ROS and keep balanced aerobic events [16]. Endogenous antioxidants seem to promote remyelination in vitro and in vivo for what stimulating production and activity through exogenous mechanisms might be promising in nerve and skeletal muscle regeneration [5,17–19]. Exogenous antioxidants such as polyphenols and others are showing protective effects in nerve and skeletal muscle cells by reducing free radicals [20]. Further research on the mechanisms of exogenous antioxidants in recovering demyelinated nerves and derived impaired skeletal muscles will feature targets to stimulate remyelination. Therefore, the aim of this review is to provide an overview of the recent evidence on the molecular mechanisms in nerve and skeletal muscle regeneration promoted by exogenous antioxidants. This will be relevant to identifying adjuvant targets to enhance the treatments of inflammatory and autoimmune neuromuscular demyelinating diseases.

2. Autoimmune and Inflammatory Demyelination

Myelination is the process performed by the specialized cells oligodendrocytes and Schwann cells to produce sphingolipids and form myelin layers [7]. The sphingolipids produced by oligodendrocytes including sphingomyelin, cerebrosides, and gangliosides, exhibit key roles in proliferation, migration, apoptosis, and remyelination [7,21]. Hence, failed sphingolipid synthesis may induce pathological and inflammatory reactions.

2.1. Role of Sphingolipids and ROS

The brain has the highest sphingolipid content and therefore requires a crucial balance in the synthesis, degradation, and removal of lipids and sphingolipids [22–24]. Changes in the lipid levels may trigger pathogenic reactions including neuroinflammation and oxidative damage [22–25]. For instance, defective sphingolipid metabolism may cause neurodegenerative diseases such as Alzheimer's and Parkinson's disease [22–25]. In a study including 30 patients diagnosed with multiple sclerosis was observed increased oxidative damage, lipid alteration, and DNA lesion [25]. In addition, these patients showed plaque formation containing oligodendrocytes, immunoproteasome activity, and astrocytes similar to brain biopsies from previous cases [21,25]. It was also observed inflammatory CD3-positive T-cells and oxidative stress markers in blood and cerebrospinal fluid including high concentrations of malondialdehyde and lipid hydroperoxide [6,25]. All this indicates oxidative effects and neuroinflammation associated with sphingolipid alteration and ROS production.

Also, oligodendrocytes under in vitro inflammatory conditions fail to differentiate and undergo apoptosis [21]. In vivo, a mice model for multiple sclerosis showed a reduction in oligodendrocyte proliferation, increased immune cell activation, immunoproteasome activity, increased demyelination, and decreased remyelination [21]. Immunoproteasome pathways are observed in both oligodendrocytes and astrocytes surrounded by a chronic inflammatory environment [21,26]. Interestingly, these pathways in astrocytes induce protective effects against oxidation and damage, while in oligodendrocytes accelerate apoptosis and tissue injury [21,26]. The immunoproteasome activity is one of the molecular mechanisms involved in demyelination and impaired remyelination yet to be understood.

The crosstalk between the sphingolipid-mediated pathway and the eicosanoid pathway may also impair myelination [7,27]. The sphingolipid-mediated pathway is associated with tissue development while the eicosanoid pathway is linked to pathological events such as the production of inflammatory mediators [7,27]. Orm1-like3 (ORMDL3) is an endoplasmic reticulum protein involved in leukotrienes synthesis regulation and sphingolipid homeostasis [28]. In vitro, mast cells isolated from wild-type mice and knockout-ORMDL3 mice showed higher production of sphingolipids and inflammatory mediators such as leukotrienes [28]. Low ORMDL3 expression induces failed sphingolipids production along with inflammatory reactions by leukotrienes [28]. In vivo, ORMDL1, ORMDL2, and ORMDL3 knockout mice showed altered sphingolipid synthesis during myelination resulting in severe demyelination in the sciatic nerve [29]. It seems that ORMDLs might be target proteins crucial for the stability in sphingolipid production and myelination.

Similarly, ceramide, ceramide1-phosphate (C1P), and sphingosine1-phosphate (S1P) may induce inflammatory reactions [7,30]. These intermediates are synthesized in the sphingomyelin cycle by sphingomyelinases [7,31]. Some studies show the critical balance of sphingolipids to regulate apoptotic or anti-apoptotic signals [21,22]. For instance, increasing ceramide levels may reduce S1P levels activating cell death pathways [21,22]. In vitro and in vivo using models of multiple sclerosis and inflammatory demyelinating diseases show that ceramides accumulate and activate astrocytes [30]. Sphingosine increases while S1P decreases, and both may be reestablished once remyelination occurs [30]. Ceramide, C1P, and S1P participate in the activation of protein kinase C isoforms and cytosolic phospholipase A (cPLA) [7,22,27]. This latter enzyme is associated with inflammation via the arachidonic acid pathway and activation of EDG-sphingolipid receptors. This leads to the activation of phospholipase C (PLC) and intracellular calcium release (Figure 1) [7,22,27].

Additionally, it is considered a synergistic action of C1P with S1P that induces COX-2 and activates PLA2 [7,21,32]. C1P contributes to inflammation via the production of prostaglandin E2 (PGE2) and stimulates mast cell degranulation via $Ca^{+}2$ dependent pathway associated with calmodulin [7,21,32,33]. Similarly, in the activated mast cells the sphingosine kinase produces S1P via the IgE receptor [7,21,32]. These events also occur in different cell types such as lung fibroblasts and epithelial cells. S1P stimulation via tumoral necrosis factor (TNF) induces COX-2 expression and triggers PGE2 production ending up in severe inflammation [7,21,32,34]. In a few words, the therapeutic regulation of the synergistic effects of C1P-S1P may reduce inflammatory reactions in other tissues preventing additional clinical conditions.

Furthermore, a metabolic syndrome involving insulin resistance is associated with a high risk of peripheral neuropathy affecting small unmyelinated axons described in recent reviews [35,36]. In brief, the neuropathy associated with type 2 diabetes includes an increase of long chain fatty acids, neuroinflammation, and neuronal oxidative stress [35]. These mechanisms induce a cascade of pro-inflammatory cytokines along with NF-kB activation and further production of COX-2 and TNF-α [35]. Interestingly, type 2 diabetes neuropathy shares COX-2 mechanisms with C1P and S1P lipid inductors [35]. Cytokine cascade induced by metabolic syndrome activates neutral SMase and SK1 with the concomitant activation of PLA2, PLC, S1P, and COX-2 mediated inflammatory effects [35,36]. Later, peripheral innervation is impaired with subsequent skeletal muscle damage.

Therefore, sphingolipids upon certain conditions prompt inflammatory responses including TNF-α induction of sphingomyelinase associated with increased ceramide levels and activation of the NF-κB pathway [21,37,38]. This induces more than 150 genes increasing cytokine and chemokine levels [21,37,38]. NF-κB activation encodes the production of interleukin-1β, IL-6, IL-8, and pro-inflammatory enzymes, such as COX-2 [21,37,38]. In murine models, ceramides stimulate in astrocytes the expression of proinflammatory mediators (TNF, IL-1B, and IL-6) and activation of NF-κB [7,22,27]. Interestingly, ceramides also increase the production of the c/EBP factor associated with the induction of TNF, IL-6, and IL-1β similar to NF-κB [7,21,32,34]. Thus, sphingolipid pathways represent potential targets to prevent inflammation, demyelination, and induce remyelination.

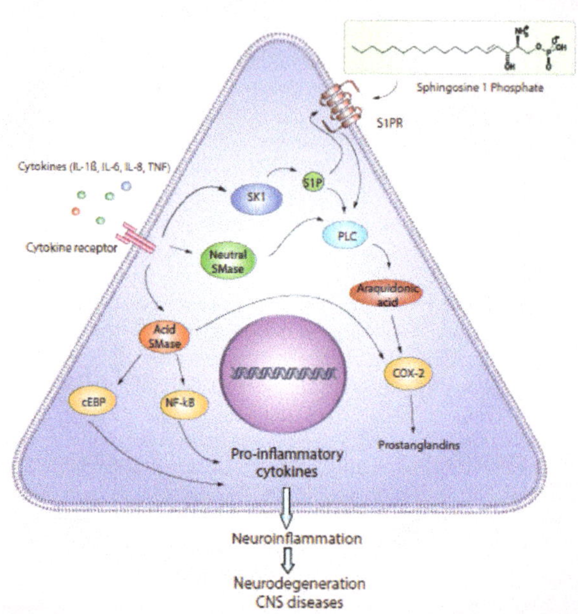

Figure 1. Mechanism of neuroinflammation associated with sphingolipids. Oligodendrocyte. Sphingolipids as sphingosine 1 phosphate cause the transduction of the PLC pathway with the ongoing activation of the araquidonic acid pathway. Araquinodate induces the activation of COX-2-producing prostaglandins that act as inflammatory mediators. Enzymes such as Acid sphingomyelinase SMase and Neutral sphingomyelinase are also capable of active araquidonic, and COX-2 paths. Interestingly, extracellular inflammatory mediators, including cytokines (IL-6, IL-1β, IL-8, and TNFα) may activate sphingomyelinase enzymes, mainly, acid sphingomyelinase with the subsequent activation of cEBP and NF-kB trans-activators that lead the production of pro-inflammatory cytokines. The inflammatory mediators and pro-inflammatory cytokines are associated with the development of neuroinflammation, demyelination, and degenerative diseases.

All this evidence suggests that sphingolipid derivatives including ceramide, C1P, and S1P, are associated with inflammatory mechanisms in the nervous system [7,22,23]. Ceramide is a key regulator of critical neuronal mechanisms, including cell differentiation, senescence, and cell death [7,22,23]. The increased levels of ceramide may induce cell death via caspase 3 activation, reactive oxygen species production, and mitochondrial dysfunction [7,22,27]. Sphingolipid metabolites constitute key mediators during inflammatory reactions being associated with neurodegenerative effects [37–39]. Neuroinflammatory events in neurodegenerative diseases include activation of both microglia and astrocytes showing overexpression of inflammatory mediators [37,39–41]. Microglia and astrocytes continuously produce pro-inflammatory cytokines conducting to chronic inflammation and the progression of neurodegeneration linked with neuroinflammation [37,39–41]. Such mechanisms are primary the pathophysiological foundations of neurodegenerative diseases.

2.2. Inflammation in Demyelinating Diseases

Overall, autoimmune and inflammatory diseases such as multiple sclerosis show increased ROS production rates [42]. Preclinical encephalomyelitis models using rodents showed higher levels of ROS produced by microglia and infiltrated macrophages [25,43].

This stimulates increasing pro-inflammatory mediators and oxidizing radicals, including superoxide, hydroxyl radicals, hydrogen peroxide, and nitric oxide causing oxidative damage [25,43]. Simultaneously, depletion of endogenous antioxidants such as catalase, SOD, and glutathione is also observed in neuroinflammation and neurodegeneration [16,44]. SOD converts superoxide radical anions into hydrogen peroxide in two forms of this enzyme: MnSOD and Cu/ZnSOD both expressed in the brain [16,44]. Nuclear factor E2-related factor (Nrf2-ARE) induces expression of glutathione and thioredoxin-antioxidant system along with peroxiredoxins, superoxide dismutase, catalases, and heme oxygenase upon inflammatory reactions and neurodegeneration [44]. Nrf2 deficiency causes in mice loss of oligodendrocyte, demyelination, neuroinflammation, and axonal damage [45,46]. Similarly, peripheral nerves from Nrf2 knockout mice exhibited reduced axonal remyelination, functional impairment, and damaged neuromuscular junctions [47].

Sensory and motor nerves are susceptible to degeneration and immune-mediated demyelination as seen in multiple sclerosis, chronic inflammatory demyelinating polyneuropathy, and Guillain-Barre syndrome [9,22,31,40]. Multiple sclerosis constitutes one of the major demyelinating diseases with effects on skeletal muscles [9,38]. Demyelination in the central nervous system is a complex process involving blood-brain barrier disturbance, immune cell recruitment, B-lymphocytes and T-Lymphocytes activation, secretion of pro-inflammatory mediators, activation of microglia, and macrophage stimulation [23,31,38]. Altogether, destabilizes structural myelin proteins including basic myelin-protein, myelin oligodendrocyte glycoprotein, myelin acidic protein, contactin1, and neurofascin [38]. The activity of the innate and adaptive immune system produces reactive oxygen species and nitrogen reactive species causing myelin damage and oligodendrocytes decreasing. These mechanisms also involve macrophage stimulation, IgM antibody-mediated damage against myelin, and demyelination [38,48,49].

Demyelination in multiple sclerosis is the final phase of the disease showing myelin sheath degradation by immune reactive cells [23,31,41]. Different types of lipids as phospholipids and sphingomyelin form the myelin sheath [23,31,41]. Myelin is also comprised of immunoglobulin-nature proteins such as myelin-associated glycoproteins (MAG), myelin oligodendrocyte glycoprotein (MOG), and cell adhesion proteins [23,31,41]. These components may induce the activation of autoimmune or immune-derived responses that causes myelin degradation [21,31,41]. First evidence shows a direct antibody-antigen interaction in multiple sclerosis in preclinical autoimmune encephalomyelitis models against MOG [21,31,41]. These models showed oxidative damage and plaque formation that, to some extent, are caused by reduced arterial perfusion and low tissue oxygen supply [23,38,41]. Then, demyelination follows the increased oxidative damage and plaques deteriorating the tissue area as seen in multiple sclerosis.

On the other hand, diseases such as neuromyelitis optica show significant immunoglobulin deposition and characteristic immunoreactivity against astrocytic aquaporin AQP4 [23,27,41]. This demyelination in the optic nerve may cover the brain stem and rarely, the brain tissue [23,27,41]. Histopathological findings show loss of axons, low AQP4 and AQP1, and the production of glial acidic fibrillary protein GFAP, a key landmark of glial reactivity [23,41]. Eosinophiles and granulocytes infiltrate the lesion causing an increased inflammatory process with the ongoing activation of macrophages and lymphocytes [23,38].

Acute disseminated encephalomyelitis (ADEM) is generally preceded by an infection or rarely by vaccination and it is associated with an immunoreactive process against MOG antigens [23,38,41]. ADEM usually affects children and young adults, and some variants are multiphasic (MDEM) [23,32,38]. The differentiation of such conditions with multiple sclerosis represents clinical challenges. ADEM and MDEM show macrophages reactivity, foamy macrophages that increase demyelination throughout the central nervous system or are restricted to a single location [23,32,38]. MOG seems to play a key role as a surface receptor and adhesion molecule as seen in demyelinating encephalomyelitis models [22,23,38]. MOG may be used as a diagnostic marker and possible target in diseases including ADEM, MDEM, and rarely multiple sclerosis in which IgG1 antibody production against MOG is found [22,23,38].

In peripheral nerves, Schwann cells produce myelin and keep a balanced nerve environment [34]. Interestingly, Schwann cells are capable to sense damaging signals and they may be involved in developing the inflammatory process via the expression of macrophage colony-stimulating factor (CSF-1) [34]. CSF-1 upregulates the production of TNF-α, and several cytokines such as IL-1α and IL-1β, to stimulate the recruitment of macrophages [34]. Trias et al. demonstrated in an amyotrophic lateral sclerosis model that Schwann cells might change their phenotype to induce nerve repair [50,51]. This change promotes proliferation and secretion of pro-inflammatory factors including CSF-1 and IL-34 from distal to proximal, as Wallerian degeneration, increasing nerve injury [50,51]. Schwann cells also remove myelin debris, ensure nerve homeostasis, and promote axonal regeneration and reinnervation [50,51]. Therefore, Schwann cells are critical and double-edge cellular components to regulate neuro-regeneration or neurodegeneration.

In demyelinating disorders involving Schwann cells, it is observed oxidative stress that impairs their mitochondrial metabolism [33,52]. For example, Friedreich's ataxia disease occurs with loss of dorsal root ganglion seen in clinical studies along with oxidative stress that diminishes the expression of frataxin [33,52]. Frataxin seems to regulate iron molecules for proper protein functioning. The precise mechanism and downregulation of frataxin are not yet well understood. Studies in knockout models showed increasing selective inflammatory toxicity when Schwann cells are devoid of frataxin [33,52]. In fact, the absence of frataxin in Schwann cells may turn the phenotype from normal to inflammatory [33,52]. This shows a suitable target to promote peripheral nerve regeneration.

Similarly, other evidence suggests a role of Schwann cells in the balanced expression of TNF-α to promote axonal regeneration [53]. Kato et al. demonstrated via the inhibition of TNFα using etanercept-a TNFα-antagonist, improvement in axonal regeneration after nerve injury [53]. The adaptive immune response uses a major histocompatibility-II (MHC-II) complex to coordinate the T-cell activation [54]. MHC-II is expressed by antigen-presenting cells (APC) including B cells, dendritic cells, and macrophages [54]. However, inflammatory conditions may also induce the non-immune cells as endothelial or muscle cells to upregulate the expression of MHC-II [54]. Hartlehnert et al. showed that Schwann cells may upregulate MHC-II under inflammatory conditions after injury in models of neuropathic pain [54]. Intriguingly, Schwann cells may function as conditional antigen-presenting cells under inflammatory disorders [54]. All this evidence shows the role of Schwann cells in the regulation of inflammatory and immune responses following nerve damage.

To sum up, molecular mechanisms in demyelination acting as potential targets for nerve regeneration seem to be ORMDL proteins, immune proteasomes showing protective activity in astrocytes, destruction of oligodendrocytes, regulation of ceramides, sphingolipid pathways and endogenous antioxidants. Molecular events in peripheral nerves include the regulation of TNF-α, frataxin, and the ability of Schwann cells to change their phenotype turning into inflammatory. All these molecular mechanisms might be involved in skeletal muscle damage and would certainly be relevant targets to induce remyelination and nerve regeneration.

3. Skeletal Muscle Damage and Oxidative Stress Demyelination Related

Striated skeletal muscle tissue is formed by aligned fibers associated with the intramuscular extracellular matrix arranged into endomysium, perimysium, and epimysium [55]. The endomysium is the connective tissue surrounding each muscle fiber containing capillaries and nerve terminals [55–58]. The perimysium is a stronger connective tissue that arranges muscle fibers into small and large myobundles [55–58]. This arrangement provides mechanical stability and supports muscle contraction and force production [55–58]. The epimysium surrounds the entire muscle being related to tendons and fasciae [55–58]. The regenerative ability of skeletal muscles relies on their satellite cells, the quiescent stem cells located in the proximity of the muscle fibers [59–61]. Upon injury, these satellite cells activate, proliferate, fuse, and differentiate into multinucleated fibers to replace or repair the injured ones [59–61]. However, multiple conditions trigger the destruction of muscle fibers, defeat their regenerative capacity, and induce loss of muscle arrangement [59,62,63]. Severe conditions such as acquired or inherited diseases, demyelination and trauma exhibit such effects decreasing the muscle mass [59,62–64]. All this hampers the regenerative ability of skeletal muscles, increases the connective tissue, and forms fibrosis ending up in functional impairment [59,62,63].

Inherited myopathies such as dystrophies are caused by genetic factors while acquired myopathies including infectious and inflammatory diseases are linked to systemic factors [62,63]. These types of myopathies show that ROS triggering skeletal muscle damage are SOD, hydrogen peroxide, hydroxyl radical, neuronal nitric oxide, endothelial nitric oxide, and peroxynitrite [3]. Hydroxyl radical is a highly reactive oxidant that causes DNA and protein damage as occur in impaired nerves [3,11]. ROS in nerve injury induce inflammatory stages that in turn induces proteolysis in muscle fibers causing skeletal muscle impairment [11]. This damage induces the activation of TNF, TGF-β, and JAK-STAT, among other inflammatory molecules [11]. Simultaneously, decreasing of proteins involved in the regulation of oxidative stress such as Nrf2, may cause myopathy and dysfunction as recently observed in knockout mice [65]. These events induce structural changes in the intramuscular extracellular matrix causing reduction of the muscle mass and atrophy [11,66]. The permanent loss of muscle mass causes long-term functional impairment altogether defined as volumetric muscle loss [67].

Volumetric muscle loss also impairs neuromuscular junctions and induces secondary denervation as shown in rat models [67]. This secondary denervation results in chronic functional impairment [67]. In mice, passive rehabilitation after a massive muscle loss induces adaptation and minor functional recovery [68]. In general, this incomplete muscle recovery indicates persistent damage for what stimulating nerve regeneration alongside neuromuscular junction and muscle fiber may conduct an overall recovery.

As mentioned earlier in this review, the role of immunoproteasomes in nerve impairment is not well understood. In skeletal muscles, proteasomes seem to have key roles in recovery after exercise, minor injuries, and muscle atrophy [69]. In vitro, proteasomes 20S and 26S in C2C12 myoblasts and human myoblasts increase during muscle cell differentiation [69]. The inhibition of these proteasomes causes protein oxidation and stimulates proapoptotic pathways [69]. Similarly, the proteasome mechanism in atrophy is related to the regulation of the ubiquitin-proteasome system by transcription factors including Nrf1 and Nrf2 [70–72]. In a model of oculopharyngeal muscular dystrophy using Drosophila the increased activity of the ubiquitin-proteasome system conducts to muscle damage and impaired function [71]. The gene therapy to stimulate the production of inhibitors improved muscle weakness and reduced degeneration [71]. In other clinical conditions causing skeletal muscle atrophy, there was also increased activity of the ubiquitin-proteasome system and muscle protein degradation [72]. Intriguingly, proteasome activity seems to have effects on myoblast differentiation and muscle atrophy.

Sphingolipids and ceramides also have a role in muscle inflammation and degeneration. Recent in vitro findings showed that the impairment of ceramide kinase might be one of the molecular mechanisms causing these effects [73]. In a different in vitro study, high levels of TNF-α increased ceramides synthesis, harmed protein metabolism and induced sphingolipid accumulation impairing myofiber formation [74]. This might be related to a damaged endoplasmic reticulum involved in protein and lipid synthesis [75]. In vitro, lipotoxicity and abnormal protein folding induce the production of long-chain ceramide signals, metabolic disease, inflammation, and skeletal muscle degeneration [75]. To our knowledge, the role of ORMDL-proteins in skeletal muscle damage has not been yet described. Surely, there might be mechanisms related to ORMDL-proteins occurring during muscle degeneration as seen in nerve impairment.

In summary, molecular mechanisms inducing demyelination seem to increase skeletal muscle damage. All of this represents targets to improve regeneration in nerves and muscles. In the next section, we will discuss the potential effects of exogenous antioxidants on specific molecular mechanisms to improve remyelination and muscle recovery.

4. Skeletal Muscle Regeneration along Antioxidant Induced Remyelination

Remyelination involves oligodendrocytes and Schwann cell differentiation for nerve recovery [76]. In rats and mice, the inhibition of sphingomyelinase 2 (SMase2) restored myelin production and improved ceramide content in the remyelinated nerves [77]. In mice, remyelination occurs after contusion injury with the activation of the myelin regulatory factor (Myrf) [78]. Myrf produced by oligodendrocytes and Schwann cells induce myelin recovery in the impaired nerves [78]. Interestingly, Schwann cells in Myrf knockout mice continue myelin production showing expression of specific markers such as myelin protein zero (P0) [78]. Myelination by Schwann cells allowed partial recovery of the injured nerves [78]. This indicates the critical adaptation of specialized myelinating cells to sustain myelination upon injury.

Exogenous antioxidants are expected to activate molecular and anti-inflammatory mechanisms to prevent degeneration and/or induce regeneration. Thus, therapeutical antioxidant options and further research ought to target oxidative imbalance and altered cellular pathways that cause neurodegeneration and skeletal muscle damage (Figure 2). In this section, specific mechanisms of exogenous antioxidants observed in nerve and skeletal muscle will be discussed.

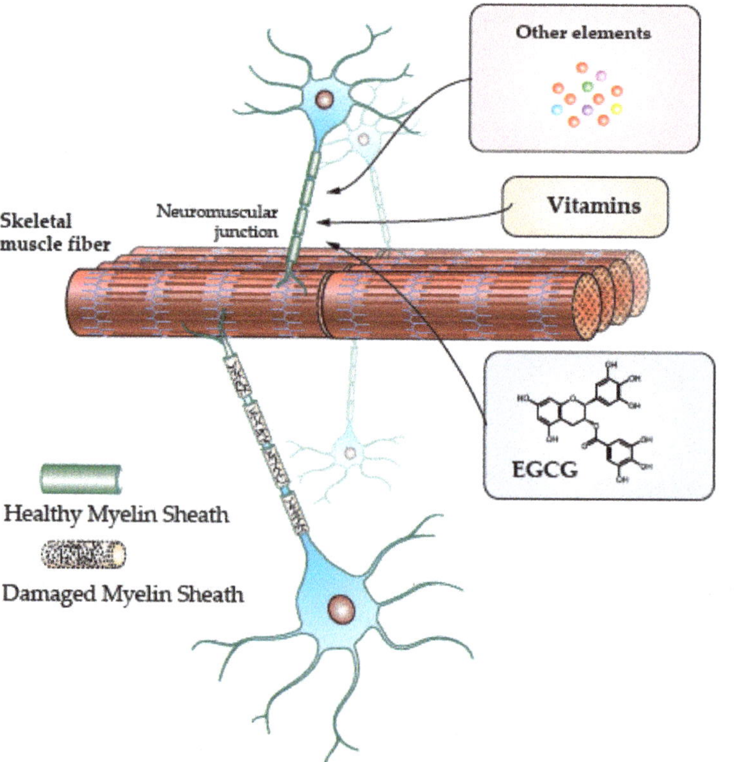

Figure 2. Exogenous antioxidants in regeneration. Damaged myelin impairs innervation and neuromuscular junctions end in skeletal muscle fibers damage. Remyelination is one of the potential effects of exogenous antioxidants that activate mechanisms to induce regenerative effects in skeletal muscles including their functional recovery.

Major endogenous antioxidants in nerves and skeletal muscles are superoxide dismutases (SOD1, SOD2, SOD3), catalase, and glutathione [3,16,44]. Along with exogenous antioxidants such as curcumin, EGCG, vitamins, and others, endogenous antioxidants may induce regeneration by increasing protection against peroxidation [3,16,44]. Similarly, the potential effects of polyphenols such as EGCG and other natural exogenous antioxidants exhibit ROS reduction, decreasing inflammation, and myofiber formation [19,79]. The effects of exogenous antioxidants on nerve regeneration and muscle recovery are summarized in Table 1.

4.1. Curcumin

Curcumin is a phenolic compound with antibacterial, anti-inflammatory, and antioxidant properties with potential effects on the recovery of neurodegenerative disorders [19]. In vitro, oligodendrocyte progenitor cells treated with curcumin showed improvement in mitochondrial activity, and increased cell differentiation via PPAR-γ activation, ERK1/2 phosphorylation, and increased PGC1-α expression [80]. In vivo, a sciatic nerve crushed injury model in rats was treated using curcumin [81]. This curcumin was administered through a subcutaneously implanted pump at the site of injury [81]. There was functional and morphological muscle recovery along with reduction of oxidative markers and increased Nrf2 antioxidant in the treated groups [81]. This model shows promising results, and it may consider delivering curcumin using a less invasive strategy. In general, Nrf2 seems to be relevant in muscle recovery and nerve regeneration by reducing demyelination via exogenous antioxidant stimulation.

4.2. Flavonoids

Quercetin is a flavonoid found in multiple vegetables showing protective mitochondrial mechanisms, antiinflammation, and neuroprotection [19]. The flavonoid-derived medication Baicalin was recently analyzed in vitro using oligodendrocytes and in vivo in mice models of demyelination [82]. In vitro, Baicalin induced oligodendrocyte proliferation, differentiation, and a reduction in the number of astrocytes [82]. In vivo, the anti-inflammatory effect of this medication inhibited demyelination, promoted remyelination, and enhanced coordinated movement [82]. Additional neuroprotective effects of Baicalin and similar flavonoids include mechanisms to reduce neurotoxicity caused by aminochromes [83]. An aminochrome is a quinone formed during dopamine oxidation known to induce mitochondrial dysfunction and subsequent neuroinflammation in Parkinson's disease [83]. It seems that flavonoids stimulate neuroprotection by preventing lysosomal dysfunction and protecting against oxidative damage, even though all these aspects are still under research [83]. Similarly, in a mice model of cuprizone, a model of toxic demyelination, the treatment with the flavonoid Icariin increased myelin restoration, APC$^+$/Olig2$^+$ mature oligodendrocytes, and brain-derived neurotrophic factor production [84]. In general, flavonoids seem to induce neuroprotection by reducing neurotoxicity showing possible therapeutical approaches, for instance, to improve treatments of Parkinson's disease.

4.3. EGCG

Polyphenols such as epigallocatechin-3-gallate (EGCG) derived from green tea are showing potential regenerative effects. In vitro, C2C12 myoblasts exposed to polyphenolic EGCG showed a higher number of myotubes with increasing length [85,86]. EGCG activates the transcriptional coactivator TAZ that in turn increases myogenin, myoblasts fusion, and myotubes formation [85]. TAZ is related to tissue homeostasis, regeneration, organ development, and myogenic differentiation that occurs through the stimulation of the myogenic differentiation factor (MyoD) [85,87]. Similarly, EGCG stimulates miRNA-486-5p expression and reduces myostatin promoting C2C12 myoblasts differentiation [86]. Myostatin is a member of the transcription growth factor β (TGFβ) family that inhibits satellite cell proliferation, myoblasts differentiation, and in vivo induces fibrosis in severely injured muscles [88,89].

In a study using senescent mice was provided an enriched-EGCG diet for up to eight weeks [86]. Their results showed a gradual increasing of muscle mass in this group than in the group without this diet [86]. A similar result was found in a mice model of dystrophy that showed improvements in force production in the treated mice for up to eight weeks with EGCG than in the untreated group [90]. In the same study, the dystrophic diaphragm and soleus showed a higher area of skeletal muscle and lower area of connective tissue in time [90]. Aged rats were used in a hindlimb muscle atrophy disuse model [91]. The treatment with EGCG induced reduction of pro-apoptotic pathways, satellite cell proliferation, cell differentiation, and force production in the treated group [91]. In this group, oxidative markers were lower while SOD antioxidant markers were higher [91]. In a different study, the crushed nerve injury in the hindlimbs of rats was treated with EGCG intraperitoneally for up to 8 weeks [92]. There were increasing myelin sheath thickness, nociceptive recovery, hindlimb reflex, and posture improvements such as standing on the injured limb in the treated groups [92]. This study did not describe the functional or histological analysis of hindlimb skeletal muscles. However, the recovery in posture and coordinated movements indicate potential recovery in the injured muscles. All this evidence suggests the activation of regenerative mechanisms in nerves and skeletal muscles induced by exogenous antioxidant effects.

Another molecular mechanism involved in skeletal muscle regeneration induced by EGCG is similar to the effects observed in insulin-like growth factor (IGF-1). In vitro, IGF-1 enhances myotube formation in aligned scaffolds showing increasing length and higher number of nuclei per myotube [93]. In vivo, IGF-1 delivered in aligned scaffolds implanted in large excisional wounds showed myofiber formation and lower areas of collagen in time [93]. In recent in vitro studies, skeletal muscle fibers isolated from mice were stimulated to express Forehead box (Foxo)-O1 and Foxo-O3 [94,95]. These proteins are implicated in muscle atrophy mainly by increasing protein degradation via E3 ubiquitin ligase expression [94,95]. A decreased nuclear Foxo expression was observed in muscle fibers exposed to IGF-1 and EGCG but not in the untreated cells [95]. The addition of ROS such as H_2O_2 to the EGCG group cells did not increase nuclear Foxo activity [95]. All this indicates that EGCG might promote additional protective effects in skeletal muscle fibers comparable to those that occur upon IGF-1 stimulation.

In situ, in a model of severe nerve and muscle radiation injury in the rat hindlimb was delivered EGCG in a biodegradable synthetic hydrogel scaffold [96]. It was observed three months later increased antioxidant markers Nrf2 and MnSOD alongside reduced nNOS and TNF-α [96]. The area of the regenerated nerve was higher in the EGCG-hydrogel group than in the untreated group. The myelin sheath thickness and the number and diameter of axons were also higher in the treated groups with EGCG-hydrogel [96]. In the same study, the area of muscle fibers in the fibrotic muscles due to radiation exposition showed higher recovery in the EGCG-hydrogel group than in the untreated group. Similarly, strength and muscle mass were higher in the treated group [96]. The in vitro component in this study included the effects of the scaffold EGCG-loaded using Schwann cell and skeletal muscle fibers isolated from rats [96]. Their results one week after culture showed higher proliferative ability and differentiation of both cell types in EGCG-hydrogel than in control cultures [96]. All this preclinical evidence indicates that EGCG stimulates nerve regeneration and potential recovery of atrophic muscles. All this is promising for therapeutic approaches in nerve and skeletal muscle regeneration.

In a clinical study, 60 years-old patients were treated with EGCG for 12 weeks [97]. Their main results include increases in antioxidant activity, reduction of myostatin, and grip strength [97]. The autoimmune and inflammatory effects seen in multiple sclerosis disease are mostly demyelination and skeletal muscle damage [98]. A daily dose of 600mg of EGCG was provided for up to 12 weeks to patients upon treatment for multiple sclerosis [98]. After moderate exercise, improvements in muscle metabolism such as lactate reduction, and stable carbohydrate oxidation were found over time [98]. Further analysis beyond this study may include specific neural and muscular parameters for regeneration in these patients receiving EGCG along with their treatment. This clinical evidence shows regenerative effects of EGCG in the recovery of severely impaired skeletal muscle due to demyelination. In summary, EGCG seems to stimulate molecular mechanisms for satellite cell differentiation, skeletal muscle growth, and regeneration.

4.4. Vitamins and Other Elements

Regarding vitamins and other remyelinating components, vitamin E promotes myelin maintenance and prevents cell damage as shown in mice [99]. The lack of this vitamin causes axonopathy and demyelination [99]. The administration of vitamin E and vitamin D3 in a rat model of multiple sclerosis showed antioxidant effects, reduced apoptosis, and increased remyelination [100]. The treatment of the sciatic nerve injury in rats with vitamin E mixed with pyrroloquinoline-quinone, a small molecule with antioxidant effects, showed improvement in nerve function, increased muscle mass, and motor functional recovery [101]. Vitamin C, or ascorbic acid, is also involved in peripheral nerve development and maintenance for collagen synthesis and lipid protection showing antioxidant effects [17].

Selenium is a trace element considered an essential micronutrient for eukaryotes and procaryotes through the function of selenoproteins (Se1P). Knockout mice for this protein showed problems in learning during training and impaired motor coordination [102]. Se1P incorporates selenium in the form of selenocysteine and selenium-methionine residues [103]. In the CNS, selenium acts as a cofactor for glutathione peroxidase types I–IV and VI [104].

Table 1. Exogenous antioxidant effects on muscle and nerve regeneration.

Compound	Type	Effect on Muscle	Effect on Brain/Neurons	References
Vitamin C	Ascorbic acid	Antioxidant	Peripheral nerve development Antioxidant	[17]
Quercetin	Flavonoid	Recovery of neuromuscular function. Reduction in skeletal muscle atrophy.	Antiinflammatory neuroprotection	[19]
Baicalin	Flavone glicoside	Reduce skeletal muscle damage	Oligodendrocyte proliferation, differentiation. Improve remyelination, activation of PGC1a	[82,83]
Icariin	Flavonoid	Angiogenesis, tendon bone healing. Increase myotubes number and length	Myelin restoration, oligodendrocyte maturation, BDNF increase.	[84]
EGCG	Polyphenol	Myogenic differentiation Increase muscle area Increase antioxidative production. Reduction lactate and stable carbohydrate oxidation	Increase myelin sheath thickness Nociceptive recovery Improvements in posture	[85–98]
Vitamin E/	Tocopherol/	Antioxidant Muscle recovery	Antioxidant effects reduced apoptosis, increase remyelination. Improvements in nerve and motor function.	[99–101]
Vitamin D3	Cholecalciferol		Antioxidant effects Reduced apoptosis, increase remyelination	
Selenium			Antioxidant, microglial inhibition, increased remyelination	[102–106]

Selenium is also important in the function of SelP, a selenium-rich protein, present in glial cells, the choroid plexus, and cerebral spinal fluid [104]. SELENOP1 is a metal binding protein with antioxidant functions able to regulate the concentration of tau phosphorylated protein and synaptic Zn^2 in a mice model [105]. Finally, 18β-glycyrrhetinic acid is a compound found in licorice roots shown to suppress the proinflammatory chemokines CCL2, CCL3, CCL5, CXCL10, and CCL20 in mice models of encephalomyelitis [106]. This study showed microglial inhibition and increased remyelination possibly via the expression of brain-derived neurotrophic factor (BDNF) [106]. All these vitamins and elements are showing potential remyelinating effects worthy of deeper research. In general, the evidence in this review is showing promising effects in remyelination and muscle regeneration mostly induced by exogenous antioxidants.

Further strategies in sync with pharmacological treatment and antioxidant approaches include physical activity. Exercise is another therapeutical intervention to reduce oxidative mechanisms following demyelination to recover muscle mass and function [107,108]. The genetic approach in mice models also aims to stimulate and accelerate myelin production reducing inflammatory responses [64]. One of the mechanisms of accelerated remyelination is the activation of peroxisome proliferator-activated receptor gamma co-activator 1-alpha (PGC1a) that increases during exercise in mice [109]. It seems that accelerating remyelination after autoimmune inflammation protects the axon structure and prevents axonal loss in the central nervous system.

To last, a growing worldwide disorder triggering neurodegeneration is obesity [35,110]. This condition increases nerve damage and shows impaired nerve regeneration also in absence of diabetes [35,110]. In obese rats, the recovery of mechanical injuries of the sciatic nerve showed a lower number of myelinated axons and thinner myelin sheath than the normal weight group [111]. Similarly, hyperlipidemia, or high levels of lipids in the blood, constitutes a risk that may induce neuropathy and peripheral nerve dysfunction [112]. Lowering weight and preventing hyperlipidemia may reduce the risk and peripheral neuropathy effects [113]. Obesity may also cause muscle damage and impair skeletal muscle regeneration [114]. Satellite cells from limb muscles in obese mice showed reduced activity of AMP-activated protein kinaseα [114]. This enzyme stimulates myogenin expression and myoblasts fusion for its inhibition decreases the regenerative ability in skeletal muscles [114]. Oxidative stress and exacerbation of multiple inflammatory responses in obesity induce the inhibition of AMP-activated protein kinaseα [114,115]. Then, based on the evidence in this review, exogenous antioxidants such as flavonoids, EGCG, and vitamin E may reduce oxidative damage and inflammation improving the regenerative ability of nerves and skeletal muscles.

To conclude, autoimmune and inflammatory mechanisms cause demyelination and nerve damage that harms skeletal muscles causing myopathy-like destruction. Oligodendrocytes and Schwann cells show adaptive responses upon injury and incomplete remyelination. Approaching mechanisms such as Nrf2, ORMDL, proteasomes, and others, might improve remyelinating therapies. Additional targets with potential pharmacological interest include the metabolic regulation of ceramides and their association with inflammatory mechanisms. Preclinical evidence using antioxidants such as curcumin, EGCG, and other exogenous antioxidants showed regenerative effects. These effects include stimulation of oligodendrocyte and Schwann cell differentiation, increased myelin production, reduction of oxidative stress, increased endogenous antioxidant activity, recovery of the neuromuscular junctions, and skeletal muscle functional recovery. Then, inducing simultaneous regeneration of nerves, skeletal muscle, and neuromuscular junctions along with current approaches will enhance the overall recovery discussed in this review.

Funding: The APC was funded by the Central Office of Research, Universidad del Tolima, grant number 20520.

Institutional Review Board Statement: Not applicable.

Informed Consent Statement: Not applicable.

Data Availability Statement: Not applicable.

Conflicts of Interest: The authors declare no conflict of interest.

Abbreviations

ADEM	Acute disseminated encephalomyelitis
AQP4	Aquaporin 4
BDNF	Brain-derived 590 neurotrophic factor
CCL	Chemokine
CD3	Cluster of differentiation 3
C1P	Ceramide1-phosphate
COX-2	Cyclooxygenase 2
CSF-1	Colony stimulating factor
EGCG	Epigallocatechin Gallate
ERK1/2	Extracellular signal-regulated kinase 1/2
GFAP	Glial acidic fibrillary protein
HIF-1	Hypoxia inducible factor-1
IGF-1	Insulin-like growth factor 1
IL	Interleukin
JAK-STAT	JAK (Janus Kinase)-STAT signaling pathway
MAG	Myelin-associated glycoproteins
MDEM	Multiphasic disseminated encephalomyelitis
MHC-II	Major 301 histocompatibility-II Complex
Myrf	Myelin regulatory factor
MOG	Myelin oligodendrocyte glycoprotein
NF-kB	Nuclear Factor type KB
Nrf2	Nuclear factor E2-related factor
ORMDL3	ORMDL (Orm1-like3) sphingolipid biosynthesis Regulator 3
PLC	Phospholipase C
PGE2	Prostaglandin E2
PGC1a	Peroxisome proliferator-activated receptor gamma co-activator 1-alpha
PPAR-γ	Peroxisome proliferator-activated receptor gamma
ROS	Reactive Oxygen Species
SC	Schwann cells
SMase	Sphingomyelinase
S1P	Sphingosine1-phosphate
SOD	Superoxide dismutase
TNF	Tumoral necrosis factor
TGF-β	Transforming growth factor beta

References

1. Sayre, L.M.; Perry, G.; Smith, M.A. Oxidative stress and neurotoxicity. *Chem. Res. Toxicol.* **2008**, *21*, 172–188. [CrossRef] [PubMed]
2. Campbell, G.R.; Mahad, D.J. Mitochondrial changes associated with demyelination: Consequences for axonal integrity. *Mitochondrion* **2012**, *12*, 173–179. [CrossRef] [PubMed]
3. Le Moal, E.; Pialoux, V.; Juban, G.; Groussard, C.; Zouhal, H.; Chazaud, B.; Mounier, R. Redox Control of Skeletal Muscle Regeneration. *Antioxid. Redox Signal.* **2017**, *27*, 276–310. [CrossRef]
4. Pizzino, G.; Irrera, N.; Cucinotta, M.; Pallio, G.; Mannino, F.; Arcoraci, V.; Squadrito, F.; Altavilla, D.; Bitto, A. Oxidative Stress: Harms and Benefits for Human Health. *Oxid. Med. Cell. Longev.* **2017**, *2017*, 8416763. [CrossRef]
5. Uttara, B.; Singh, A.V.; Zamboni, P.; Mahajan, R.T. Oxidative stress and neurodegenerative diseases: A review of upstream and downstream antioxidant therapeutic options. *Curr. Neuropharmacol.* **2009**, *7*, 65–74. [CrossRef]
6. Zhang, S.Y.; Gui, L.N.; Liu, Y.Y.; Shi, S.; Cheng, Y. Oxidative Stress Marker Aberrations in Multiple Sclerosis: A Meta-Analysis Study. *Front. Neurosci.* **2020**, *14*, 823. [CrossRef] [PubMed]
7. Nixon, G.F. Sphingolipids in inflammation: Pathological implications and potential therapeutic targets. *Br. J. Pharmacol.* **2009**, *158*, 982–993. [CrossRef] [PubMed]
8. Alizadeh, A.; Dyck, S.M.; Karimi-Abdolrezaee, S. Myelin damage and repair in pathologic CNS: Challenges and prospects. *Front. Mol. Neurosci.* **2015**, *8*, 35. [CrossRef]
9. Yin, L.; Li, N.; Jia, W.; Wang, N.; Liang, M.; Yang, X.; Du, G. Skeletal muscle atrophy: From mechanisms to treatments. *Pharmacol. Res.* **2021**, *172*, 105807. [CrossRef] [PubMed]
10. Mahdy, M.A.A. Skeletal muscle fibrosis: An overview. *Cell Tissue Res.* **2019**, *375*, 575–588. [CrossRef] [PubMed]
11. Shen, Y.; Zhang, R.; Xu, L.; Wan, Q.; Zhu, J.; Gu, J.; Huang, Z.; Ma, W.; Shen, M.; Ding, F.; et al. Microarray Analysis of Gene Expression Provides New Insights Into Denervation-Induced Skeletal Muscle Atrophy. *Front. Physiol.* **2019**, *10*, 1298. [CrossRef]

12. Ahmad, K.; Shaikh, S.; Ahmad, S.S.; Lee, E.J.; Choi, I. Cross-Talk Between Extracellular Matrix and Skeletal Muscle: Implications for Myopathies. *Front. Pharmacol.* **2020**, *11*, 142. [CrossRef]
13. Bhagavati, S. Autoimmune Disorders of the Nervous System: Pathophysiology, Clinical Features, and Therapy. *Front. Neurol.* **2021**, *12*, 664664. [CrossRef]
14. Chu, X.L.; Song, X.Z.; Li, Q.; Li, Y.R.; He, F.; Gu, X.S.; Ming, D. Basic mechanisms of peripheral nerve injury and treatment via electrical stimulation. *Neural. Regen. Res.* **2022**, *17*, 2185–2193. [CrossRef]
15. Simon, J.P.; Marie, I.; Jouen, F.; Boyer, O.; Martinet, J. Autoimmune Myopathies: Where Do We Stand? *Front. Immunol.* **2016**, *7*, 234. [CrossRef]
16. He, L.; He, T.; Farrar, S.; Ji, L.; Liu, T.; Ma, X. Antioxidants Maintain Cellular Redox Homeostasis by Elimination of Reactive Oxygen Species. *Cell. Physiol. Biochem.* **2017**, *44*, 532–553. [CrossRef]
17. Podratz, J.L.; Rodriguez, E.H.; Windebank, A.J. Antioxidants are necessary for myelination of dorsal root ganglion neurons, in vitro. *Glia* **2004**, *45*, 54–58. [CrossRef]
18. Dong, L.; Li, R.; Li, D.; Wang, B.; Lu, Y.; Li, P.; Yu, F.; Jin, Y.; Ni, X.; Wu, Y.; et al. FGF10 Enhances Peripheral Nerve Regeneration via the Preactivation of the PI3K/Akt Signaling-Mediated Antioxidant Response. *Front. Pharm.* **2019**, *10*, 1224. [CrossRef] [PubMed]
19. Simioni, C.; Zauli, G.; Martelli, A.M.; Vitale, M.; Sacchetti, G.; Gonelli, A.; Neri, L.M. Oxidative stress: Role of physical exercise and antioxidant nutraceuticals in adulthood and aging. *Oncotarget* **2018**, *9*, 17181–17198. [CrossRef] [PubMed]
20. Cui, X.; Lin, Q.; Liang, Y. Plant-Derived Antioxidants Protect the Nervous System From Aging by Inhibiting Oxidative Stress. *Front. Aging Neurosci.* **2020**, *12*, 209. [CrossRef] [PubMed]
21. Kirby, L.; Jin, J.; Cardona, J.G.; Smith, M.D.; Martin, K.A.; Wang, J.; Strasburger, H.; Herbst, L.; Alexis, M.; Karnell, J.; et al. Oligodendrocyte precursor cells present antigen and are cytotoxic targets in inflammatory demyelination. *Nat. Commun.* **2019**, *10*, 3887. [CrossRef]
22. Lee, J.Y.; Jin, H.K.; Bae, J.S. Sphingolipids in neuroinflammation: A potential target for diagnosis and therapy. *BMB Rep.* **2020**, *53*, 28–34. [CrossRef]
23. Latov, C.B.T.H.B.W.T.N. Chronic Inflammatory Demyelinating Polineuropathy. *Neuromusc. Disord.* **1996**, *6*, 311–325.
24. Tidball, J.G. Mechanisms of muscle injury, repair, and regeneration. *Compr. Physiol.* **2011**, *1*, 2029–2062. [CrossRef] [PubMed]
25. Haider, L.; Fischer, M.T.; Frischer, J.M.; Bauer, J.; Hoftberger, R.; Botond, G.; Esterbauer, H.; Binder, C.J.; Witztum, J.L.; Lassmann, H. Oxidative damage in multiple sclerosis lesions. *Brain* **2011**, *134*, 1914–1924. [CrossRef] [PubMed]
26. Smith, B.C.; Sinyuk, M.; Jenkins, J.E., 3rd; Psenicka, M.W.; Williams, J.L. The impact of regional astrocyte interferon-gamma signaling during chronic autoimmunity: A novel role for the immunoproteasome. *J. Neuroinflammation* **2020**, *17*, 184. [CrossRef] [PubMed]
27. Pettus, B.J.; Chalfant, C.E.; Hannun, Y.A. Sphingolipids in inflammation: Roles and implications. *Curr. Mol. Med.* **2004**, *4*, 405–418. [CrossRef]
28. Bugajev, V.; Paulenda, T.; Utekal, P.; Mrkacek, M.; Halova, I.; Kuchar, L.; Kuda, O.; Vavrova, P.; Schuster, B.; Fuentes-Liso, S.; et al. Crosstalk between ORMDL3, serine palmitoyltransferase, and 5-lipoxygenase in the sphingolipid and eicosanoid metabolic pathways. *J. Lipid Res.* **2021**, *62*, 100121. [CrossRef] [PubMed]
29. Clarke, B.A.; Majumder, S.; Zhu, H.; Lee, Y.T.; Kono, M.; Li, C.; Khanna, C.; Blain, H.; Schwartz, R.; Huso, V.L.; et al. The Ormdl genes regulate the sphingolipid synthesis pathway to ensure proper myelination and neurologic function in mice. *eLife* **2019**, *8*, e51067. [CrossRef]
30. Kim, S.; Steelman, A.J.; Zhang, Y.; Kinney, H.C.; Li, J. Aberrant upregulation of astroglial ceramide potentiates oligodendrocyte injury. *Brain Pathol.* **2012**, *22*, 41–57. [CrossRef]
31. Rodriguez, Y.; Vatti, N.; Ramirez-Santana, C.; Chang, C.; Mancera-Paez, O.; Gershwin, M.E.; Anaya, J.M. Chronic inflammatory demyelinating polyneuropathy as an autoimmune disease. *J. Autoimmun.* **2019**, *102*, 8–37. [CrossRef]
32. Constantinescu, C.S.; Farooqi, N.; O'Brien, K.; Gran, B. Experimental autoimmune encephalomyelitis (EAE) as a model for multiple sclerosis (MS). *Br. J. Pharmacol.* **2011**, *164*, 1079–1106. [CrossRef]
33. Park, H.T.; Kim, Y.H.; Lee, K.E.; Kim, J.K. Behind the pathology of macrophage-associated demyelination in inflammatory neuropathies: Demyelinating Schwann cells. *Cell. Mol. Life Sci. CMLS* **2020**, *77*, 2497–2506. [CrossRef]
34. Yadav, A.; Huang, T.C.; Chen, S.H.; Ramasamy, T.S.; Hsueh, Y.Y.; Lin, S.P.; Lu, F.I.; Liu, Y.H.; Wu, C.C. Sodium phenyl-butyrate inhibits Schwann cell inflammation via HDAC and NFkappaB to promote axonal regeneration and remyelination. *J. Neuroinflammation* **2021**, *18*, 238. [CrossRef]
35. Kazamel, M.; Stino, A.M.; Smith, A.G. Metabolic syndrome and peripheral neuropathy. *Muscle Nerve* **2021**, *63*, 285–293. [CrossRef]
36. Zilliox, L.A. Diabetes and Peripheral Nerve Disease. *Clin. Geriatr. Med.* **2021**, *37*, 253–267. [CrossRef]
37. Perandini, L.A.; Chimin, P.; Lutkemeyer, D.D.S.; Camara, N.O.S. Chronic inflammation in skeletal muscle impairs satellite cells function during regeneration: Can physical exercise restore the satellite cell niche? *FEBS J.* **2018**, *285*, 1973–1984. [CrossRef]
38. Hoftberger, R.; Lassmann, H. Inflammatory demyelinating diseases of the central nervous system. *Handb. Clin. Neurol.* **2017**, *145*, 263–283. [CrossRef]
39. Dufresne, S.S.; Frenette, J.; Dumont, N.A. Inflammation and muscle regeneration, a double-edged sword. *Med. Sci. M/S* **2016**, *32*, 591–597. [CrossRef]

40. Superstein, R.J.B.D.S. Guillain-BarrC Syndrome and Chronic Inflammatory Demyelinating Polyneuropathy. *Semin. Neurol.* **1998**, *18*, 1–13.
41. Mayo, L.; Quintana, F.J.; Weiner, H.L. The innate immune system in demyelinating disease. *Immunol. Rev.* **2012**, *248*, 170–187. [CrossRef]
42. Ohl, K.; Tenbrock, K.; Kipp, M. Oxidative stress in multiple sclerosis: Central and peripheral mode of action. *Exp. Neurol.* **2016**, *277*, 58–67. [CrossRef]
43. Lassmann, H.; van Horssen, J.; Mahad, D. Progressive multiple sclerosis: Pathology and pathogenesis. *Nat. Rev. Neurol.* **2012**, *8*, 647–656. [CrossRef]
44. Mirshafiey, A.; Mohsenzadegan, M. Antioxidant therapy in multiple sclerosis. *Immunopharmacol. Immunotoxicol.* **2009**, *31*, 13–29. [CrossRef]
45. Draheim, T.; Liessem, A.; Scheld, M.; Wilms, F.; Weissflog, M.; Denecke, B.; Kensler, T.W.; Zendedel, A.; Beyer, C.; Kipp, M.; et al. Activation of the astrocytic Nrf2/ARE system ameliorates the formation of demyelinating lesions in a multiple sclerosis animal model. *Glia* **2016**, *64*, 2219–2230. [CrossRef]
46. Nellessen, A.; Nyamoya, S.; Zendedel, A.; Slowik, A.; Wruck, C.; Beyer, C.; Fragoulis, A.; Clarner, T. Nrf2 deficiency increases oligodendrocyte loss, demyelination, neuroinflammation and axonal damage in an MS animal model. *Metab. Brain Dis.* **2020**, *35*, 353–362. [CrossRef] [PubMed]
47. Zhang, L.; Johnson, D.; Johnson, J.A. Deletion of Nrf2 impairs functional recovery, reduces clearance of myelin debris and decreases axonal remyelination after peripheral nerve injury. *Neurobiol. Dis.* **2013**, *54*, 329–338. [CrossRef]
48. Hu, W.; Lucchinetti, C.F. The pathological spectrum of CNS inflammatory demyelinating diseases. *Semin. Immunopathol.* **2009**, *31*, 439–453. [CrossRef]
49. Hardy, T.A.; Reddel, S.W.; Barnett, M.H.; Palace, J.; Lucchinetti, C.F.; Weinshenker, B.G. Atypical inflammatory demyelinating syndromes of the CNS. *Lancet Neurol.* **2016**, *15*, 967–981. [CrossRef]
50. Zhang, J.; Chen, H.; Duan, Z.; Chen, K.; Liu, Z.; Zhang, L.; Yao, D.; Li, B. The Effects of Co-transplantation of Olfactory Ensheathing Cells and Schwann Cells on Local Inflammation Environment in the Contused Spinal Cord of Rats. *Mol. Neurobiol.* **2017**, *54*, 943–953. [CrossRef]
51. Trias, E.; Kovacs, M.; King, P.H.; Si, Y.; Kwon, Y.; Varela, V.; Ibarburu, S.; Moura, I.C.; Hermine, O.; Beckman, J.S.; et al. Schwann cells orchestrate peripheral nerve inflammation through the expression of CSF1, IL-34, and SCF in amyotrophic lateral sclerosis. *Glia* **2020**, *68*, 1165–1181. [CrossRef] [PubMed]
52. Lu, C.; Schoenfeld, R.; Shan, Y.; Tsai, H.J.; Hammock, B.; Cortopassi, G. Frataxin deficiency induces Schwann cell inflammation and death. *Biochim. Et Biophys. Acta* **2009**, *1792*, 1052–1061. [CrossRef] [PubMed]
53. Kato, K.; Liu, H.; Kikuchi, S.; Myers, R.R.; Shubayev, V.I. Immediate anti-tumor necrosis factor-alpha (etanercept) therapy enhances axonal regeneration after sciatic nerve crush. *J. Neurosci. Res.* **2010**, *88*, 360–368. [CrossRef] [PubMed]
54. Hartlehnert, M.; Derksen, A.; Hagenacker, T.; Kindermann, D.; Schafers, M.; Pawlak, M.; Kieseier, B.C.; Meyer Zu Horste, G. Schwann cells promote post-traumatic nerve inflammation and neuropathic pain through MHC class II. *Sci. Rep.* **2017**, *7*, 12518. [CrossRef]
55. Zhang, W.; Liu, Y.; Zhang, H. Extracellular matrix: An important regulator of cell functions and skeletal muscle development. *Cell Biosci.* **2021**, *11*, 65. [CrossRef]
56. Rosero Salazar, D.H.; Elvira Flórez, L.J. Image analysis in Gomori's trichrome stain of skeletal muscles subjected to ischemia and reperfusion injury. *Explor. Anim. Med. Res.* **2016**, *6*, 15–25.
57. Salazar, D.H.R.; Monsalve, L.S. Digital image analysis of striated skeletal muscle tissue injury during reperfusion after induced ischemia. *Proc. SPIE* **2015**, *9287*, 92870I. [CrossRef]
58. Purslow, P.P. Muscle fascia and force transmission. *J. Bodyw. Mov.* **2010**, *14*, 411–417. [CrossRef] [PubMed]
59. Rosero Salazar, D.H.; Carvajal Monroy, P.L.; Wagener, F.; Von den Hoff, J.W. Orofacial Muscles: Embryonic Development and Regeneration after Injury. *J. Dent. Res.* **2020**, *99*, 125–132. [CrossRef] [PubMed]
60. Dumont, N.A.; Bentzinger, C.F.; Sincennes, M.C.; Rudnicki, M.A. Satellite Cells and Skeletal Muscle Regeneration. *Compr. Physiol.* **2015**, *5*, 1027–1059. [CrossRef]
61. Pruller, J.; Mannhardt, I.; Eschenhagen, T.; Zammit, P.S.; Figeac, N. Satellite cells delivered in their niche efficiently generate functional myotubes in three-dimensional cell culture. *PLoS ONE* **2018**, *13*, e0202574. [CrossRef]
62. Chawla, J. Stepwise approach to myopathy in systemic disease. *Front. Neurol.* **2011**, *2*, 49. [CrossRef] [PubMed]
63. Nagy, H.; Veerapaneni, K.D. Myopathy. In *StatPearls*; StatPearls Publishing: Treasure Island, FL, USA, 2022. [PubMed]
64. Mei, F.; Lehmann-Horn, K.; Shen, Y.A.; Rankin, K.A.; Stebbins, K.J.; Lorrain, D.S.; Pekarek, K.; S, A.S.; Xiao, L.; Teuscher, C.; et al. Accelerated remyelination during inflammatory demyelination prevents axonal loss and improves functional recovery. *eLife* **2016**, *5*, e18246. [CrossRef] [PubMed]
65. Abu, R.; Yu, L.; Kumar, A.; Gao, L.; Kumar, V. A Quantitative Proteomics Approach to Gain Insight into NRF2-KEAP1 Skeletal Muscle System and Its Cysteine Redox Regulation. *Genes* **2021**, *12*, 1655. [CrossRef]
66. Vallee, A.; Lecarpentier, Y. TGF-beta in fibrosis by acting as a conductor for contractile properties of myofibroblasts. *Cell Biosci.* **2019**, *9*, 1–15. [CrossRef] [PubMed]
67. Sorensen, J.R.; Hoffman, D.B.; Corona, B.T.; Greising, S.M. Secondary denervation is a chronic pathophysiologic sequela of volumetric muscle loss. *J. Appl. Physiol. (1985)* **2021**, *130*, 1614–1625. [CrossRef]

68. Greising, S.M.; Warren, G.L.; Southern, W.M.; Nichenko, A.S.; Qualls, A.E.; Corona, B.T.; Call, J.A. Early rehabilitation for volumetric muscle loss injury augments endogenous regenerative aspects of muscle strength and oxidative capacity. *BMC Musculoskelet. Disord.* **2018**, *19*, 173. [CrossRef] [PubMed]

69. Cui, Z.; Hwang, S.M.; Gomes, A.V. Identification of the immunoproteasome as a novel regulator of skeletal muscle differentiation. *Mol. Cell. Biol.* **2014**, *34*, 96–109. [CrossRef] [PubMed]

70. Kitajima, Y.; Yoshioka, K.; Suzuki, N. The ubiquitin-proteasome system in regulation of the skeletal muscle homeostasis and atrophy: From basic science to disorders. *J. Physiol. Sci.* **2020**, *70*, 40. [CrossRef]

71. Ribot, C.; Soler, C.; Chartier, A.; Al Hayek, S.; Nait-Saidi, R.; Barbezier, N.; Coux, O.; Simonelig, M. Activation of the ubiquitin-proteasome system contributes to oculopharyngeal muscular dystrophy through muscle atrophy. *PLoS Genet.* **2022**, *18*, e1010015. [CrossRef]

72. Haberecht-Muller, S.; Kruger, E.; Fielitz, J. Out of Control: The Role of the Ubiquitin Proteasome System in Skeletal Muscle during Inflammation. *Biomolecules* **2021**, *11*, 1327. [CrossRef] [PubMed]

73. Pierucci, F.; Frati, A.; Battistini, C.; Penna, F.; Costelli, P.; Meacci, E. Control of Skeletal Muscle Atrophy Associated to Cancer or Corticosteroids by Ceramide Kinase. *Cancers* **2021**, *13*, 3285. [CrossRef] [PubMed]

74. De Larichaudy, J.; Zufferli, A.; Serra, F.; Isidori, A.M.; Naro, F.; Dessalle, K.; Desgeorges, M.; Piraud, M.; Cheillan, D.; Vidal, H.; et al. TNF-alpha- and tumor-induced skeletal muscle atrophy involves sphingolipid metabolism. *Skelet. Muscle* **2012**, *2*, 2. [CrossRef] [PubMed]

75. McNally, B.D.; Ashley, D.F.; Hanschke, L.; Daou, H.N.; Watt, N.T.; Murfitt, S.A.; MacCannell, A.D.V.; Whitehead, A.; Bowen, T.S.; Sanders, F.W.B.; et al. Long-chain ceramides are cell non-autonomous signals linking lipotoxicity to endoplasmic reticulum stress in skeletal muscle. *Nat. Commun.* **2022**, *13*, 1748. [CrossRef]

76. Chen, C.Z.; Neumann, B.; Forster, S.; Franklin, R.J.M. Schwann cell remyelination of the central nervous system: Why does it happen and what are the benefits? *Open Biol.* **2021**, *11*, 200352. [CrossRef] [PubMed]

77. Yoo, S.W.; Agarwal, A.; Smith, M.D.; Khuder, S.S.; Baxi, E.G.; Thomas, A.G.; Rojas, C.; Moniruzzaman, M.; Slusher, B.S.; Bergles, D.E.; et al. Inhibition of neutral sphingomyelinase 2 promotes remyelination. *Sci. Adv.* **2020**, *6*, eaba5210. [CrossRef] [PubMed]

78. Duncan, G.J.; Manesh, S.B.; Hilton, B.J.; Assinck, P.; Liu, J.; Moulson, A.; Plemel, J.R.; Tetzlaff, W. Locomotor recovery following contusive spinal cord injury does not require oligodendrocyte remyelination. *Nat. Commun.* **2018**, *9*, 3066. [CrossRef] [PubMed]

79. Luk, H.Y.; Appell, C.; Chyu, M.C.; Chen, C.H.; Wang, C.Y.; Yang, R.S.; Shen, C.L. Impacts of Green Tea on Joint and Skeletal Muscle Health: Prospects of Translational Nutrition. *Antioxidants* **2020**, *9*, 1050. [CrossRef] [PubMed]

80. Bernardo, A.; Plumitallo, C.; De Nuccio, C.; Visentin, S.; Minghetti, L. Curcumin promotes oligodendrocyte differentiation and their protection against TNF-alpha through the activation of the nuclear receptor PPAR-gamma. *Sci. Rep.* **2021**, *11*, 4952. [CrossRef]

81. Caillaud, M.; Chantemargue, B.; Richard, L.; Vignaud, L.; Favreau, F.; Faye, P.A.; Vignoles, P.; Sturtz, F.; Trouillas, P.; Vallat, J.M.; et al. Local low dose curcumin treatment improves functional recovery and remyelination in a rat model of sciatic nerve crush through inhibition of oxidative stress. *Neuropharmacology* **2018**, *139*, 98–116. [CrossRef]

82. Ai, R.S.; Xing, K.; Deng, X.; Han, J.J.; Hao, D.X.; Qi, W.H.; Han, B.; Yang, Y.N.; Li, X.; Zhang, Y. Baicalin Promotes CNS Remyelination via PPARgamma Signal Pathway. *Neurol. Neuroimmunol. Neuroinflamm.* **2022**, *9*, e1142. [CrossRef] [PubMed]

83. Silva, V.; Segura-Aguilar, J. State and perspectives on flavonoid neuroprotection against aminochrome-induced neurotoxicity. *Neural. Regen. Res.* **2021**, *16*, 1797–1798. [CrossRef] [PubMed]

84. Zhang, Y.; Yin, L.; Zheng, N.; Zhang, L.; Liu, J.; Liang, W.; Wang, Q. Icariin enhances remyelination process after acute demyelination induced by cuprizone exposure. *Brain Res. Bull.* **2017**, *130*, 180–187. [CrossRef] [PubMed]

85. Kim, A.R.; Kim, K.M.; Byun, M.R.; Hwang, J.H.; Park, J.I.; Oh, H.T.; Jeong, M.G.; Hwang, E.S.; Hong, J.H. (-)-Epigallocatechin-3-gallate stimulates myogenic differentiation through TAZ activation. *Biochem. Biophys. Res. Commun.* **2017**, *486*, 378–384. [CrossRef] [PubMed]

86. Chang, Y.C.; Liu, H.W.; Chan, Y.C.; Hu, S.H.; Liu, M.Y.; Chang, S.J. The green tea polyphenol epigallocatechin-3-gallate attenuates age-associated muscle loss via regulation of miR-486-5p and myostatin. *Arch. Biochem. Biophys.* **2020**, *692*, 108511. [CrossRef]

87. Jeong, M.G.; Kim, H.K.; Hwang, E.S. The essential role of TAZ in normal tissue homeostasis. *Arch. Pharm. Res.* **2021**, *44*, 253–262. [CrossRef]

88. Wang, Q.; McPherron, A.C. Myostatin inhibition induces muscle fibre hypertrophy prior to satellite cell activation. *J. Physiol.* **2012**, *590*, 2151–2165. [CrossRef]

89. Li, Z.B.; Kollias, H.D.; Wagner, K.R. Myostatin directly regulates skeletal muscle fibrosis. *J. Biol. Chem.* **2008**, *283*, 19371–19378. [CrossRef]

90. Nakae, Y.; Hirasaka, K.; Goto, J.; Nikawa, T.; Shono, M.; Yoshida, M.; Stoward, P.J. Subcutaneous injection, from birth, of epigallocatechin-3-gallate, a component of green tea, limits the onset of muscular dystrophy in mdx mice: A quantitative histological, immunohistochemical and electrophysiological study. *Histochem Cell Biol.* **2008**, *129*, 489–501. [CrossRef]

91. Alway, S.E.; Bennett, B.T.; Wilson, J.C.; Sperringer, J.; Mohamed, J.S.; Edens, N.K.; Pereira, S.L. Green tea extract attenuates muscle loss and improves muscle function during disuse, but fails to improve muscle recovery following unloading in aged rats. *J. Appl. Physiol. (1985)* **2015**, *118*, 319–330. [CrossRef]

92. Renno, W.M.; Benov, L.; Khan, K.M. Possible role of antioxidative capacity of (-)-epigallocatechin-3-gallate treatment in morphological and neurobehavioral recovery after sciatic nerve crush injury. *J. Neurosurg. Spine* **2017**, *27*, 593–613. [CrossRef] [PubMed]

93. Alcazar, C.A.; Hu, C.; Rando, T.A.; Huang, N.F.; Nakayama, K.H. Transplantation of insulin-like growth factor-1 laden scaffolds combined with exercise promotes neuroregeneration and angiogenesis in a preclinical muscle injury model. *Biomater. Sci.* **2020**, *8*, 5376–5389. [CrossRef]

94. Kang, S.H.; Lee, H.A.; Kim, M.; Lee, E.; Sohn, U.D.; Kim, I. Forkhead box O3 plays a role in skeletal muscle atrophy through expression of E3 ubiquitin ligases MuRF-1 and atrogin-1 in Cushing's syndrome. *Am. J. Physiol. Endocrinol. Metab.* **2017**, *312*, E495–E507. [CrossRef]

95. Wimmer, R.J.; Russell, S.J.; Schneider, M.F. Green tea component EGCG, insulin and IGF-1 promote nuclear efflux of atrophy-associated transcription factor Foxo1 in skeletal muscle fibers. *J. Nutr. Biochem.* **2015**, *26*, 1559–1567. [CrossRef]

96. Qian, Y.; Yao, Z.; Wang, X.; Cheng, Y.; Fang, Z.; Yuan, W.E.; Fan, C.; Ouyang, Y. (-)-Epigallocatechin gallate-loaded polycaprolactone scaffolds fabricated using a 3D integrated moulding method alleviate immune stress and induce neurogenesis. *Cell Prolif.* **2020**, *53*, e12730. [CrossRef] [PubMed]

97. Seo, H.; Lee, S.H.; Park, Y.; Lee, H.S.; Hong, J.S.; Lim, C.Y.; Kim, D.H.; Park, S.S.; Suh, H.J.; Hong, K.B. (-)-Epicatechin-Enriched Extract from Camellia sinensis Improves Regulation of Muscle Mass and Function: Results from a Randomized Controlled Trial. *Antioxidants* **2021**, *10*, 1026. [CrossRef] [PubMed]

98. Mahler, A.; Steiniger, J.; Bock, M.; Klug, L.; Parreidt, N.; Lorenz, M.; Zimmermann, B.F.; Krannich, A.; Paul, F.; Boschmann, M. Metabolic response to epigallocatechin-3-gallate in relapsing-remitting multiple sclerosis: A randomized clinical trial. *Am. J. Clin. Nutr.* **2015**, *101*, 487–495. [CrossRef]

99. Ulatowski, L.; Parker, R.; Warrier, G.; Sultana, R.; Butterfield, D.A.; Manor, D. Vitamin E is essential for Purkinje neuron integrity. *Neuroscience* **2014**, *260*, 120–129. [CrossRef] [PubMed]

100. Goudarzvand, M.; Javan, M.; Mirnajafi-Zadeh, J.; Mozafari, S.; Tiraihi, T. Vitamins E and D3 attenuate demyelination and potentiate remyelination processes of hippocampal formation of rats following local injection of ethidium bromide. *Cell. Mol. Neurobiol.* **2010**, *30*, 289–299. [CrossRef]

101. Azizi, A.; Azizi, S.; Heshmatian, B.; Amini, K. Improvement of functional recovery of transected peripheral nerve by means of chitosan grafts filled with vitamin E, pyrroloquinoline quinone and their combination. *Int. J. Surg.* **2014**, *12*, 76–82. [CrossRef] [PubMed]

102. Glaser, V.; Moritz, B.; Schmitz, A.; Dafre, A.L.; Nazari, E.M.; Rauh Muller, Y.M.; Feksa, L.; Straliottoa, M.R.; de Bem, A.F.; Farina, M.; et al. Protective effects of diphenyl diselenide in a mouse model of brain toxicity. *Chem. Biol. Interact.* **2013**, *206*, 18–26. [CrossRef] [PubMed]

103. Moghadaszadeh, B.; Beggs, A.H. Selenoproteins and their impact on human health through diverse physiological pathways. *Physiology (Bethesda)* **2006**, *21*, 307–315. [CrossRef]

104. Solovyev, N.D. Importance of selenium and selenoprotein for brain function: From antioxidant protection to neuronal signalling. *J. Inorg. Biochem.* **2015**, *153*, 1–12. [CrossRef]

105. Kiyohara, A.C.P.; Torres, D.J.; Hagiwara, A.; Pak, J.; Rueli, R.; Shuttleworth, C.W.R.; Bellinger, F.P. Selenoprotein P Regulates Synaptic Zinc and Reduces Tau Phosphorylation. *Front. Nutr.* **2021**, *8*, 683154. [CrossRef] [PubMed]

106. Zhou, J.; Cai, W.; Jin, M.; Xu, J.; Wang, Y.; Xiao, Y.; Hao, L.; Wang, B.; Zhang, Y.; Han, J.; et al. 18beta-glycyrrhetinic acid suppresses experimental autoimmune encephalomyelitis through inhibition of microglia activation and promotion of remyelination. *Sci. Rep.* **2015**, *5*, 13713. [CrossRef] [PubMed]

107. Lozinski, B.M.; de Almeida, L.G.N.; Silva, C.; Dong, Y.; Brown, D.; Chopra, S.; Yong, V.W.; Dufour, A. Exercise rapidly alters proteomes in mice following spinal cord demyelination. *Sci. Rep.* **2021**, *11*, 7239. [CrossRef]

108. Saito, Y.; Chikenji, T.S.; Matsumura, T.; Nakano, M.; Fujimiya, M. Exercise enhances skeletal muscle regeneration by promoting senescence in fibro-adipogenic progenitors. *Nat. Commun.* **2020**, *11*, 889. [CrossRef] [PubMed]

109. Jensen, S.K.; Michaels, N.J.; Ilyntskyy, S.; Keough, M.B.; Kovalchuk, O.; Yong, V.W. Multimodal Enhancement of Remyelination by Exercise with a Pivotal Role for Oligodendroglial PGC1alpha. *Cell Rep.* **2018**, *24*, 3167–3179. [CrossRef] [PubMed]

110. Callaghan, B.C.; Reynolds, E.; Banerjee, M.; Chant, E.; Villegas-Umana, E.; Feldman, E.L. Central Obesity is Associated with Neuropathy in the Severely Obese. *Mayo Clin. Proc.* **2020**, *95*, 1342–1353. [CrossRef]

111. Bekar, E.; Altunkaynak, B.Z.; Balci, K.; Aslan, G.; Ayyildiz, M.; Kaplan, S. Effects of high fat diet induced obesity on peripheral nerve regeneration and levels of GAP 43 and TGF-beta in rats. *Biotech. Histochem.* **2014**, *89*, 446–456. [CrossRef] [PubMed]

112. Liu, P.; Peng, J.; Han, G.H.; Ding, X.; Wei, S.; Gao, G.; Huang, K.; Chang, F.; Wang, Y. Role of macrophages in peripheral nerve injury and repair. *Neural Regen. Res.* **2019**, *14*, 1335–1342. [CrossRef]

113. Iqbal, Z.; Bashir, B.; Ferdousi, M.; Kalteniece, A.; Alam, U.; Malik, R.A.; Soran, H. Lipids and peripheral neuropathy. *Curr. Opin Lipidol.* **2021**, *32*, 249–257. [CrossRef] [PubMed]

114. Fu, X.; Zhu, M.; Zhang, S.; Foretz, M.; Viollet, B.; Du, M. Obesity Impairs Skeletal Muscle Regeneration Through Inhibition of AMPK. *Diabetes* **2016**, *65*, 188–200. [CrossRef] [PubMed]

115. Perez-Torres, I.; Castrejon-Tellez, V.; Soto, M.E.; Rubio-Ruiz, M.E.; Manzano-Pech, L.; Guarner-Lans, V. Oxidative Stress, Plant Natural Antioxidants, and Obesity. *Int. J. Mol. Sci.* **2021**, *22*, 1786. [CrossRef] [PubMed]

Review

Creating an Optimal In Vivo Environment to Enhance Outcomes Using Cell Therapy to Repair/Regenerate Injured Tissues of the Musculoskeletal System

David A. Hart [1,2,*] **and Norimasa Nakamura** [3]

1 Department of Surgery, Faculty of Kinesiology, McCaig Institute for Bone & Joint Health, University of Calgary, Calgary, AB T2N 4N1, Canada
2 Bone & Joint Health Strategic Clinical Network, Alberta Health Services, Edmonton, AB T5J 3E4, Canada
3 Institute of Medical Science in Sport, Osaka Health Science University, 1-9-27 Tenma, Kita-ku, Osaka 530-0043, Japan; norimasa.nakamura@ohsu.ac.jp
* Correspondence: hartd@ucalgary.ca

Abstract: Following most injuries to a musculoskeletal tissue which function in unique mechanical environments, an inflammatory response occurs to facilitate endogenous repair. This is a process that usually yields functionally inferior scar tissue. In the case of such injuries occurring in adults, the injury environment no longer expresses the anabolic processes that contributed to growth and maturation. An injury can also contribute to the development of a degenerative process, such as osteoarthritis. Over the past several years, researchers have attempted to use cellular therapies to enhance the repair and regeneration of injured tissues, including Platelet-rich Plasma and mesenchymal stem/medicinal signaling cells (MSC) from a variety of tissue sources, either as free MSC or incorporated into tissue engineered constructs, to facilitate regeneration of such damaged tissues. The use of free MSC can sometimes affect pain symptoms associated with conditions such as OA, but regeneration of damaged tissues has been challenging, particularly as some of these tissues have very complex structures. Therefore, implanting MSC or engineered constructs into an inflammatory environment in an adult may compromise the potential of the cells to facilitate regeneration, and neutralizing the inflammatory environment and enhancing the anabolic environment may be required for MSC-based interventions to fulfill their potential. Thus, success may depend on first eliminating negative influences (e.g., inflammation) in an environment, and secondly, implanting optimally cultured MSC or tissue engineered constructs into an anabolic environment to achieve the best outcomes. Furthermore, such interventions should be considered early rather than later on in a disease process, at a time when sufficient endogenous cells remain to serve as a template for repair and regeneration. This review discusses how the interface between inflammation and cell-based regeneration of damaged tissues may be at odds, and outlines approaches to improve outcomes. In addition, other variables that could contribute to the success of cell therapies are discussed. Thus, there may be a need to adopt a Precision Medicine approach to optimize tissue repair and regeneration following injury to these important tissues.

Keywords: musculoskeletal repair; mesenchymal stem cells; inflammation; tissue engineering; tissue regeneration

Citation: Hart, D.A.; Nakamura, N. Creating an Optimal In Vivo Environment to Enhance Outcomes Using Cell Therapy to Repair/Regenerate Injured Tissues of the Musculoskeletal System. *Biomedicines* **2022**, *10*, 1570. https://doi.org/10.3390/biomedicines10071570

Academic Editors: Krisztina Nikovics and Christian Dani

Received: 24 March 2022
Accepted: 27 June 2022
Published: 1 July 2022

Publisher's Note: MDPI stays neutral with regard to jurisdictional claims in published maps and institutional affiliations.

1. Introduction

1.1. Purpose of the Review

The purpose of this review is to present the current state of cellular therapy uses in enhancing the repair of injured or damaged tissues of the musculoskeletal system (MSK), and then propose approaches to improve the application of such approaches. The review focuses on soft tissues of the MSK that function in unique biomechanical environments.

The organization is to initially present the scope of the problem and then address why inflammatory processes are relevant to the topic, and how they will likely influence outcomes of cellular therapy interventions. Subsequently, the discussion focuses on the way forward to improve outcomes regarding the healing of these tissues when injured. The content of the review and the perspectives presented are based on an assessment of articles in the PubMed database from the past 30 years with representative articles cited.

1.2. Characteristics of Musculoskeletal (MSK) Tissues

Tissues of the musculoskeletal system (MSK) (i.e., cartilage, menisci, ligaments, tendons, muscles, bone) are designed to provide a mechanical function to aid mobility. At the time of birth, many of them are "cell rich and matrix poor" but become more "matrix rich and cell poor" during growth and maturation discussed in [1,2]. This occurs as the mechanical demands increase and the cell density declines due to matrix deposition. In addition, many of these tissues have a low density of a microvasculature and innervation [3–5], with mature articular cartilage devoid of both. Some tissues, such as ligaments, also contain a few immune-related cells that could initiate an endogenous inflammatory response [6,7].

Many of the tissues of the MSK system have very complex organizations either at the macro or micro level. Tendons in different locations (i.e., Achilles tendon, patellar tendon, supraspinatus tendon of the shoulder) have different properties but all have an insertion site at the bone interface enthesis [8–10], a mid-substance and a myotendinous junction where it inserts into the muscle as discussed in [11,12]. The menisci of the knee have a central region devoid of a microvasculature and innervation that is collagen II-rich and a peripheral region that has a microvasculature and some innervation, discussed in [13]. The complexity of the matrix organization in the two areas are very different and are quite sophisticated in their complexity [14–16]. The template for the organization of this complexity is likely laid down during development and then expanded upon during growth and maturation. As this complexity was likely optimized during evolution to address the required function in the mechanical environment, it would be difficult to replicate many of them in a tissue-engineered construct. The exceptions to this conclusion are bone which heals quite well when injured in most locations, and muscle which also heals well in most circumstances.

Tissues of the MSK system are complex at multiple levels (macro structure and matrix organization) and vary in different mechanical environments. They are dynamic and can respond to changes in the mechanical environment. Bone can respond to changes in loading and become strengthened [17–19]. In contrast, several MSK tissues can undergo atrophy if not used appropriately and become "deconditioned" [20]. Following an injury, MSK tissues often requires immobilization (e.g., putting a leg in a cast when a bone is broken) which is a "deconditioning" environment for healing to initially take place.

This scenario poses some questions regarding signaling for healing and responsiveness to pro-healing signals/mediators. If one immobilizes the leg of a very young animal, such as a rabbit, the medial collateral ligament of the knee almost immediately ceases to grow in this anabolic environment [21]. The cells in what is normally a biomechanically active environment do not receive the signals regulating growth and growth stops in such a "deconditioning' environment. After skeletal maturity, when the presence of anabolic mediators contributing to growth and maturation are diminished in expression, immobilization and deprivation of normal loading can lead to atrophy, potentially by the de-repression of a cassette of catabolic genes that includes some pro-inflammatory genes, such as IL-1 [22].

The characteristics of the various MSK tissues described above must be considered when attempting to repair or regenerate these tissues following injury.

1.3. The Inflammatory Response

Fundamentally, the inflammatory response is one that is designed to enhance survival after an injury (i.e., a cut in the skin) or exposure to environmental threats, such as microorganisms. After an injury to a tissue, the inflammatory response can also facilitate repair via clearance of damaged tissue components or microorganisms, and then initiate the fibrotic process with formation of a scar tissue, a process that is central to wound healing success as reviewed in [23]. Thus, scar formation in response to the injury of some tissues offers a survival advantage, but for MSK tissues that function in defined mechanical environments, scar tissue is not adequate for the optimal functioning of tissues, such as tendons, ligaments and menisci.

Because of its potential to also cause harm, inflammatory responses are regulated in a very complex manner, depending in part on the initiating factors, the extent of the response that is needed (i.e., the size and location of the wound) and the ability to initiate the downregulation of the response via both removal of the inciting events and elaboration of anti-inflammatory molecules, such as resolvins and related molecules [24–27]. In "normal" circumstances, the inflammatory response can be acute in nature followed by a resolution. However, if the inciting stimulus is not removed, this can lead to a state of chronic inflammation and fibrosis, with the development of pathology and loss of function of the affected tissues (i.e., pulmonary fibrosis, liver cirrhosis). After the transition from an acute inflammatory state to a more chronic state, such inflammation can become more difficult to control, possibly due in part to epigenetic alterations in the site of the response [28–30]. As some epigenetic alterations may be reversible [29,31,32], this may be an effective approach to enhance the potential for regeneration of damaged MSK tissues where a chronic inflammatory state is evident.

The timing of the onset of the ability to mount an inflammatory response during development has provided some interesting insights into the relationship(s) between tissue development and inflammation. Thus, the organization template of tissues and organs appears prior to an ability to mount an effective inflammatory response; however, some aspects of this relationship are still controversial as discussed in [23,33,34]. Injuries in some locations, such as cutaneous wounds, heal by regeneration if incurred early in fetal life, but they heal with the formation of scar tissue after the onset of effective inflammatory response capabilities are in place. Further study of scarless versus scar-forming wound healing may provide new clues to how to regulate the responses in adults to further attempts for successful tissue regeneration as discussed in [35].

Inflammatory responses are also influenced by sex steroids, and thus the response pattern would be influenced by both puberty in males and females, and in females after menopause, reviewed in [36,37]. It is well known that inflammatory processes can decline in the elderly, and this appears to involve estrogens [38,39].

2. Interactions between Inflammation and Injured Connective Tissues of the MSK System-Loss of Function

After skeletal maturity, injury to load bearing connective tissues, such as some ligaments, tendons and menisci can lead to the host attempting to initiate repair using the wound healing apparatus, or a failure to initiate such a response as in the case of complete rupture of the anterior cruciate ligament (ACL) of the knee, or after generation of a defect in articular cartilage. In the case of the ruptured ACL, the two ends of the ligament cannot find each other and thus a reconstruction with an autologous tissue section of a tendon [40,41] or an allogeneic tissue is required. In this circumstance, the reconstruction operation is another inflammatory stimulus with additional tissue damage due to the drilling of bone and cutting to gain access to the interior of the joint. In the case of articular cartilage defects due to the lack of innervation and a microvasculature, they are believed to contribute to a lack of healing unless some interventions are initiated, discussed in [42].

For those connective tissues that evoke an inflammatory response when injured, the result is the formation of scar tissue (outlined in Figure 1). In the case of the injured medial collateral ligament (MCL) of the knee, the injured tissue becomes rapidly "healed" with an early scar tissue that is comprised of a ~50–50 mixture of collagen II and collagen I [43,44] (the normal ligament is ~90% collagen). In contrast to the normal MCL, the collagen fibrils in the early scar tissue exhibit a random orientation that gradually becomes aligned along the length of the tissue over time [43]. However, even when aligned, the collagen fibrils are of a small diameter compared to the biphasic size distribution in the normal tissue. By two years post-injury, some large collagen fibrils appear but the ligament is much weaker than normal for much of this time period [44]. For the MCL, this weakness does not compromise function as it is a stabilizing ligament that operates normally in the toe region of its stress–strain curve. However, this model does show that in the context of an inflammatory response, the healing process does not lead to a regeneration of the normal tissue even after a protracted period of time post-injury or post-surgery. The process does, however, lead to a partially functional tissue.

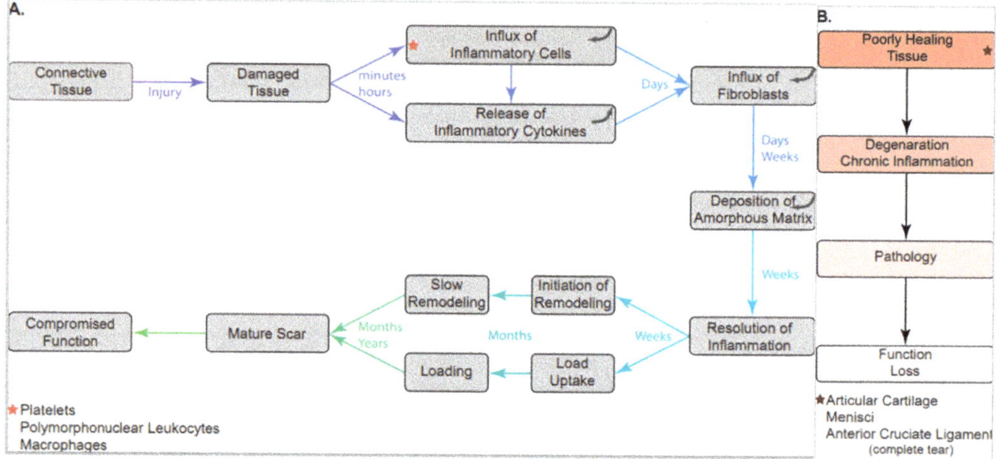

Figure 1. Sequence of events following injury to soft tissues of the MSK system. Following injury to most soft tissues of the MSK system, a sequence of events leading to scar tissue formation and maturation occurs (**A**). This sequence of events involves an inflammatory response that resolves over time as the scar tissue forms and matures. In some tissues, these events do not lead to a healing response and involve an inflammatory response that may become chronic with development of pathology and loss of function (**B**).

The situation in a reconstructed ACL is somewhat different, but again shows that an inflammatory process can likely lead to a functional compromise of the tissue. When implanted, a patellar tendon or hamstring tendon graft is likely stronger than the original ACL but has some properties that differ from the ACL as discussed in [45]. Using an allogeneic ACL graft should provide at least something of equivalent strength. However, over time, such reconstructions begin to undergo creep and stretch out [46–48], with the implanted material becoming more scar-like. The basis for this response pattern could be in part, due to damage to the grafts while preparing for the reconstruction, and/or due to the inflammation associated with the operation contributing to a local inflammatory environment that persists. Interestingly, treatment of the graft environment with an anti-inflammatory glucocorticoid can prevent or diminish this effect [49,50]. In contrast to the MCL, the ACL operates in a high stress environment and so the functional compromise over time that is observed following reconstruction can lead to dysfunction of the knee as discussed in [46].

The consequences following an injury to the menisci of the knee are likely more complex than those for the ligament injuries. The mechanical environment for a meniscus is complex, with the central area exposed to compressive loads, while the periphery is more ligament-like and is subjected to hoop stresses, discussed in [16]. Injuries to the periphery often do heal, depending on the extent and type of injury, perhaps due to its nascent blood supply and innervation. However, being in a high stress environment often leads to repair failure, and if the tissue has to be surgically removed, it can lead to an increased risk for OA as reviewed in [51]. Of relevance to this discussion is the fact that some injuries to menisci of the knee in young individuals do heal with some interventions [52], likely via the more anabolic environment operative during active growth and maturation.

Injuries to articular cartilage do not heal, and even surgical interventions to promote repair either do not work well or are only temporary in outcome. Initiation of surgical interventions to stimulate repair by microfracture of the bone beneath the cartilage leads to fibrocartilage and not hyaline cartilage [53,54], and if left alone, many defects will progress to overt OA over time. Transplantation of chondrocytes from non-loading parts of the joint to defects arising in highly loaded areas of the cartilage do offer some repair potential as reviewed in [55,56].

The above discussion leads to two important conclusions: (1) inflammation associated with injury or surgical interventions to mechanically active connective tissues of the MSK system serve an important function in relation to scar formation, but need to be controlled as scar tissue compromises the function of many of these tissues; and (2) a return to function of injured connective tissues in mechanically active environments is often compromised and new approaches are needed to facilitate regeneration and return to function (outlined in Figure 1). However, even the use of newer interventions with tissue engineered constructs to regenerate compromised tissues will need to consider effective control of inflammatory processes which could compromise the effectiveness of such approaches over the long term.

3. Factors That Can Complicate Post-Injury Processes and Inflammation

The repair of connective tissues of the MSK system damaged by injury and/or disease, particularly in skeletally mature subject or older individuals where the natural healing process has been compromised during the aging process, may be an influencing factor. As individuals age, the immune and inflammatory processes can diminish, as reviewed in [36]. In females, inflammatory responses may be altered after menopause, possibly via the effects of loss of estrogen on macrophage functioning [38,39]. Inflammatory responses in females can be altered during pregnancy as discussed in [57,58]. Therefore, females may generally regulate inflammatory processes differently than males. Therefore, sex and stage of life are important variables to be aware of in planned studies.

While normal healthy young adults can usually heal without complications, there are factors that can potentially interfere with healing processes in addition to those mentioned above. These include the presence of diseases, such as diabetes, inflammatory autoimmune diseases and obesity, reviewed in [59,60]. In a preclinical rat model, the presence of induced type 2 diabetes led to altered healing of an injury to the Achilles tendon [61–63], and it is well known that, in humans, those that are diabetics often do not heal well or exhibit delayed healing, reviewed in [60]. Obesity can lead to development of metabolic syndrome, with an on-going low level of inflammation, reviewed in [64]. Obesity can also lead to altered structure of tendons, such as the Achilles tendon [65,66], which may increase risk of injury. Based on these considerations, co-morbidities can likely also impact the healing of injured connective tissues, such as tendons, reviewed in [60], and others.

Obviously, the use of some medications to treat co-morbidities could also influence outcomes depending on the type and dosages being used. Treatment with anti-inflammatory medications, such as NSAIDS [67–69] or high or continuous doses of glucocorticoids [70,71], could potentially adversely affect outcomes after interventions with cell therapy approaches. Therefore, the presence of co-morbidities and their treatment, as well age and sex, should be recognized and addressed before initiating interventions to enhance the repair/regeneration of injured connective tissues of the MSK system. Failure to do so could complicate the results of clinical trials using experimental interventions, such as cellular therapies.

4. The Role of Mechanics in Connective Tissue Repair and Regeneration: The Interface with Biology Is Critical

While many tissues function in the context of a mechanical environment, such as the lung, skin, and the cardiovasculature system, tissues of the musculoskeletal system function in a variety of compressive, shear or tensile environments, and their complex structures reflect such requirements in unique and specialized environments. While the composition and functioning of tissues, such as tendons, ligaments, menisci and articular cartilage can undergo changes with aging [2,72–74], it is not known in detail if they have evolved to their optimum to last for >80–90 years. It is clear that structure and function relationships are critical to the performance of functional activities. The prevailing wisdom is that engineering of artificial MSK tissues should lead to a construct that mimics the original that developed in utero and during growth and maturation, discussed in [75].

There are various approaches to address the issue of the biomechanical environment with regard to the development of constructs to facilitate the repair and regeneration of damaged or diseased MSK tissues. The first is to use an artificial scaffold containing cells, with the scaffold supplying a somewhat rigid but biodegradable template after implantation to allow the cells to adapt to the in vivo conditions in a mechanically active environment [76,77]. A second approach is to condition the in vitro generated scaffold containing matrix molecules, such as collagen I to which cells have been added, in a biomechanically active environment in vitro, prior to implantation [78]. These two approaches allow for the cells in the constructs to adapt to early loading and respond with enhanced secretion of extracellular matrix components, as well as adapt to loading and make adjustments to the cellular apparatus to allow survival and functioning when implanted. While some of the studies have used cells derived from tissues, others have shown that undifferentiated mesenchymal stem cells can also respond to mechanical loading in vitro in unique manners [79–82].

In studies with a tissue engineered construct [TEC] containing synovium-derived MSC and the matrix generated by these cells in culture as discussed in [83], the TEC were generated in the absence of loading in vitro. Following implantation, the cells appear to respond to the in vivo loading environment and form a hyaline-like repair tissue based on in vivo cues and the cells that are presented [84]. In this model, it would likely not be advantageous to expose the cells to mechanical loading regimens in vivo as the self-aggregation of the in vitro generated TEC is critical for the subsequent in vivo implantation. Exposure to in vitro loading may interfere with the post-implantation process. For other applications focused on the repair of MSK tissues functioning in tensile-loading environments, in vitro loading could be beneficial.

While there is some variation in the use of mechanical loading of constructs for the repair and regeneration of damaged MSK tissues, their use depends on the type of tissue being repaired (i.e., ligament, tendon or cartilage) and the in vivo loading environment that the constructs will be subjected to following implantation. In some circumstances, in vitro loading prior to implantation could lead to the generation of a construct with a better organized matrix and increased mechanical integrity, but in other situations, such loading could compromise some of the attributes of the construct. Therefore, the use of in vitro loading depends on the applications a construct will be used for in vivo.

5. Enhancing Repair/Regeneration of Injured Connective Tissues of the MSK System

Given the limitations or variables affecting healing after an injury, much research has focused on developing new approaches to enhance repair, often with the goal of tissue regeneration. One approach involves the use of Platelet-rich Plasma (PRP) [85–87], while others have used growth factors [88] and molecular blocking approaches (anti-sense, specific antibodies, enzyme inhibitors) as reviewed in [89–93].

PRP is usually derived from autologous blood and then injected into the site of a wound such as that in a tendon [94,95], ligament [96–98] or meniscus [99]. As platelets contain a number of growth factors and other relevant molecules, by injecting the PRP into the wound site at the time of surgical repair after degranulation the platelets can release their contents and create an enhanced anabolic environment. Such preparations have also been used in the conservative treatment of OA of the knee where the PRP is believed to alleviate the symptoms of pain and perhaps exert an anti-inflammatory effect in the intra-articular space reviewed in [85–87]. Unfortunately, in the latter scenario the efficacy of the PRP can be variable and not all patients respond positively. Whether this variability is due to the platelets and their content, or the method of preparation is not well defined. There are several methods that have been used for the preparation of PRP [85] as well as host variables that could impact the efficacy (e.g., co-morbidities, medications,). Some applications of PRP for the treatment of MSK diseases, tissues and conditions are summarized in Table 1. As this is a large literature, the citations in Table 1 are representative of the field in general.

Table 1. Applications of Platelet-Rich Plasma in Cell Therapy for Connective Tissue Repair and Regeneration.

Tissue	Species	Condition	Article Type	Year	Citation
Sports Injuries	Humans	Several	Review	2022	Herdea et al. [100]
	Humans	Several	Review	2009	Sanchez et al. [101]
MSK	Canine	Several	Review	2021	Sharun et al. [102]
Cartilage	Human	OA	Review	2022	Cash et al. [103]
	Human	Knee OA	Review	2022	Sax et al. [104]
	Human	OA	Review	2022	Trams et al. [105]
	Human	OA	Review	2020	Kydd & Hart [86]
	Human	Defects	Trial	2022	Venosa et al. [106]
Tendons	Human	Epicondylitis	Review	2022	Li et al. [107]
	Human	Tendinopathy	Review	2022	Barman et al. [108]
	Human	Tendinopathy	Review	2022	Cash et al. [103]
	Human	Tendinosis	Trial	2006	Mishra & Pavelko [109]
Ligaments	Porcine	ACL	Trial	2007	Murray et al. [110]
	Human	ACL	Review	2013	Braun et al. [111]
	Human	ACL/MCL	Review	2022	Kunze et al. [97]
IVD	Human	Degeneration	Review	2020	Chang et al. [112]
	Animal	Degeneration	Review	2017	Li et al. [113]
	Human	Degeneration	Review	2017	Basso et al. [114]
	Human	Low Back Pain	Trial	2022	Akeda et al. [115]
Menisci	Human	Sports	Review	2022	Herdea et al. [100]

OA = Osteoarthritis; ACL = Anterior cruciate ligament; MCL = Medial collateral ligament; IVD = intervertebral disc. Citations are representative of the field and many more exist in PubMed for some categories.

As an alternative to PRP as a source of growth factors and other anabolic molecules, some studies have used specific growth factors as supplements in an attempt to enhance the healing of injuries to these connective tissues [116–118]. Thus, some studies have used growth factors, such as angiogenic factors [119] and others, such as IGF-1 [118,120], in an attempt to enhance healing. One of the limitations of this approach is the short half-life of growth factors or the binding of growth factors to extracellular matrix (ECM) components, which makes them unavailable for interacting with cells.

Use of approaches that can lead to blockages of specific steps or molecules during healing has also been tried by many investigators, but with limited success as reviewed in [90,91,93,121]. Some of the limitations relate to the half-life of the molecules, the specificity of the interventions or a failure to disseminate throughout a dense scar tissue, even when early in the process [121].

It should be noted that most, if not all, of the previously described studies did not attempt to control any on-going inflammatory processes that were occurring in conjunction with the injury or any surgical intervention. The use of anti-inflammatory drugs immediately after induction of the injury and/or a surgical intervention is shown to inhibit sequelae to an injury or the inflammation in a joint. This is shown in both rabbits [49] and pig models [50]. However, one would have to be careful not to use concentrations of such anti-inflammatory molecules so as to not interfere with the normal healing process. Furthermore, high doses of drugs, such as high dose glucocorticoids, may have unintended consequences regarding mesenchymal stem cells in a joint that could participate in the healing process [70].

Thus, going forward, if attempts are made to improve the healing environment to achieve better healing outcomes, an approach using both anti-catabolic/inflammatory elements and a pro-anabolic aspect should be utilized in order to optimize their potential.

6. Use of Single Cell Preparations of Mesenchymal Stem/Signaling Cells (MSC) to Enhance the Repair/Regeneration of Damaged Mechanically Active Connective Tissues

Cells called mesenchymal stem cells, mesenchymal stromal cells or mesenchymal progenitor cells have been studied primarily since the early 1990s, reviewed in [122,123]. These cells express a subset of cell surface antigens and can be induced to differentiate in vitro into cells of various lineages, such as osteogenic, adipogenic and chondrogenic cell lineages. They can be isolated from bone marrow, adipose tissue, skin, brain and many tissues reviewed in [123]. Differences in the ability to differentiate towards the different lineages were noted between cells isolated from different locations [122–124]. However, MSC isolated from individual tissues demonstrates extensive heterogeneity, discussed in [123–125].

While the cells that were labeled mesenchymal stem cells ~30 years ago, attempts to use preparations of free MSC to repair damaged connective tissues by the injection of millions of cells into a local space, such as a joint, or systemically into the circulation, they have not yielded a reproducible effective repair of the damaged tissues, possibly due to a failure to home and be retained at the injury sites [108]. Some of this limitation may be overcome by engineering membrane expression of molecules to enhance localization, reviewed in [123]. While the injection of free MSC into osteoarthritic knees did not lead to overt repair of the damaged cartilage, it was noted that injection of such cells from various sources could lead to a lessening of the pain and inflammation of OA, reviewed in [126,127]. Some studies indicate that MSC were injected, but many reports used preparations that were not pure MSC and instead were a mixture of cells labeled Bone Marrow Aspirate Concentrate (BMAC) that may contain BM, MSC or BM stromal cells, but also other cells from the BM discussed in [87,128]. A recent report [129] indicated that BMAC was more efficacious than PRP, but it cannot be concluded that this was due to the MSC in the preparations.

This failure of free MSC to initiate effective repair of damaged tissues led Caplan to hypothesize that perhaps MSC should be relabeled Medicinal Signaling Cells (MSC) as they may function by secreting or releasing vesicles containing factors or mediators that enhance the ability of endogenous cells to initiate effective repair [130]. In this scenario, MSC would release factors that would interact with residual endogenous cells at a site of injury to then repair their own tissue. While this is an interesting possibility, such abilities may be compromised by an inflammatory environment at the site of tissue injury [131,132]. Thus, in the intra-articular environment is an inflammatory process which led to an alteration of the synovial fluid MSC that compromised their ability to aggregate, likely via the actions of the mediator MCP-1 [131]. Therefore, unless such inflammation is curtailed, the MSC might still be compromised in fulfilling a role as a signaling cell,

Furthermore, if indeed MSC are signaling cells, they should likely not be used as a treatment of "last resort" in situations like OA when the articular cartilage is in a severely degenerated state so there may be little template available to initiate effective repair. With regard to other injured tissues, such as tendons and ligaments, it remains to be determined whether the addition of MSC treatments will help "re-direct" early scar-forming cells towards a more normal tendon or ligament structure. Of note is that once injured, there is also an initial inflammatory environment, but also there is a loss of biomechanical integrity and thus the initial scar tissue is not biomechanically loaded in a real functional manner, and it takes some time for the collagen fibrils of a scar to realign to allow for function loading again [43]. How the MSC would interpret the direction of need in such circumstances remains to be determined.

While the concept of MSC with a signaling role may still have some potential limitations, there are recent lines of evidence that indicate that MSC can release extracellular vesicles (EV), sometimes labeled exosomes, that are membrane enclosed packets containing growth regulators, miRNAs, and other relevant molecules as reviewed in [133–136]. In a repair context, once MSC are localized they could release EV-containing molecules that could enhance the endogenous healing process, as these EV could then be taken up by endogenous cells leading to enhanced healing. Thus, the molecules contained in EV would be somewhat protected from degradation that might occur if they were secreted as individual molecules, particularly by proteinases or RNases in an inflammatory environment. However, while EV may enhance healing under controlled conditions, it remains to be determined how effective they may be when injected in vivo into inflammatory conditions.

In addition, it also remains to be determined whether the differentiation potential of MSC as progenitor or stem cells should be dismissed in favor of strictly a signaling role, discussed in [125]. Certainly, both roles may be useful in the enhanced repair of tissues when used in a tissue engineering approach to generate constructs that appear to enhance the healing of human cartilage defects when implanted [137]. Some applications of MSC for the treatment of MSK diseases and conditions for specific tissues are summarized in Table 2. As this is a large literature, the examples indicated are representative of this large and expanding field.

Table 2. Applications of Mesenchymal Stem/Progenitor Cells for Connective Tissue Repair and Regeneration.

Tissue	Species	Condition	Article Type	Year	Citation
Orthopedic Disease	Humans	Several	Review	2022	Malekpour et al. [138]
	Humans	Several	Review	2022	Ren et al. [139]
	Horses	Lameness	Original	2019	Longhini et al. [140]
Cartilage/OA	Human	General	Review	2021	Zha et al. [141]
	Human	General	Review	2021	Vahedi et al. [142]
	Human	Defects	Review	2021	Meng et al. [143]
	Human	Defects	Trial	2018	Shimomura et al. [137]
Tendons	Preclinical	General	Review	2016	Leong & Sun [144]
	General	Injury	Review	2021	Liu et al. [145]
	Human	Tendinopathy	Review	2021	Meeremans et al. [146]
Ligaments	Preclinical	ACL	Review	2015	Jang et al. [147]
	Human	ACL	Review	2015	Jang et al. [147]
Menisci	Preclinical	Injury	Review	2015	Yu et al. [148]
	Human	Injury	Review	2017	Chew et al. [149]
	All	Injury	Review	2021	Rhim et al. [150]
	All	Injury	Review	2022	Zhou et al. [151]
IVD	All	Degenerated	Review	2021	Croft et al. [152]
	All	Degenerated	Review	2022	Liang et al. [153]
	All	Degeneration	Review	2022	DiStefano et al. [154]
Muscle	Rat	Injury	Original	2021	Barbon et al. [155]

OA = Osteoarthritis; ACL = Anterior cruciate ligament; IVD = Intervertebral disc. Citations are representative of the field and many more exist in PubMed.

7. Use of MSC in Tissue Engineered Constructs to Enhance Repair of Injured or Diseased Tissues

While the use of free MSC has not yielded consistent success in repairing damaged tissues, using them in constructs generated in vitro has led to some successes for some tissues. MSC isolated from a variety of tissues have been isolated and then often incorporated into synthetic scaffolds, scaffolds with other ECM-like matrix components, an endogenous natural protein matrix or a hybrid synthetic/natural matrix, reviewed in [156–159]. Recent advances in bioprinting may offer more sophisticated and complex scaffold-cell constructs [160–162]. Such constructs are then implanted into defects in tissues or in an injury site. As many of the scaffolds used are biodegradable, it is then hoped that they will be replaced by a natural matrix over time. Interestingly, in nearly all reports of studies assessing the efficacy of such implants, there is no mention that the inflammatory environment generated by the implantation procedure has been controlled or addressed in any manner. In spite of this limitation, some successes are reported, particularly in the repair of articular cartilage which does not heal effectively without intervention.

While many studies have reported the use of MSC in biodegradable scaffolds, several studies using a Tissue-Engineered Construct (TEC) consisting of synovium-derived MSC in a matrix secreted by these cells in vitro for the repair and regeneration of difficult to repair tissues, such as articular cartilage [83,163], menisci [164–166] and intravertebral discs [167]. The initial cartilage studies were performed in a large animal model (pigs [83,163]), while more recent studies have resulted from implantation into patients with articular cartilage defects in a pilot "proof of principle" design [137,156]. The advantages of the autologous TEC approach are: (1) it does not require an artificial scaffold as the matrix generated by the in vitro culturing serves that purpose; (2) once released from the culture dish, it spontaneously aggregates into a construct with the cells and matrix intermixed; (3) when implanted into an injury site it adheres to the residual tissue and does not require fixation, possibly due to the fibronectin in the construct; (4) the cells in the TEC are not differentiated prior to implantation but then appear to differentiate in vivo and respond to local environmental factors including the mechanical loading conditions, or interact with endogenous cells to facilitate repair. In the case of articular cartilage repair, after implantation, the TEC leads to development of the layered structure of articular cartilage with a change in matrix molecules production appropriate for the in vivo conditions [84,165]. In the pilot studies in patients with articular cartilage defects, the implanted tissues have been assessed post-implantation [122,123]. While there is some variation in the structures resulting from the TEC implantation, all defects were filled, and some appeared to be regenerated [137,156].

It should be noted that the TEC implantation studies in both the preclinical models and the pilot human studies, the authors did not attempt to control any inflammation resulting from the implantation surgery or the initial event leading to the formation of the defects. In addition, they did not add any potential anabolic stimuli, such as PRP, to possibly negate the inflammation and provide further enhancement of healing. These latter points are interesting since all of the patients were skeletally mature (both males and females) and thus the post-implantation differentiation occurred in the absence of any factors present during early growth and maturation prior to puberty. As these studies were focused on limited defects in the articular cartilage, there was likely sufficient residual cartilage present to serve as a template and/or provide local factors required to maintain cartilage which were active when provided the right cell/matrix construct. It remains to be determined in detail how a local mechanical environment contributes to the development of articular cartilage, menisci, or intravertebral discs in conjunction with biological cues, discussed in [75].

While the studies in both preclinical models and in the pilot studies with patients with articular cartilage defects have exhibited very promising results, the outcomes are likely still in need of improvement. It was noted in the porcine studies that the repair cartilage exhibits hyaline-like characteristics but does not lead to regeneration of the surface layer [168] that has been called the lamina splendens [169]. As this superficial surface layer has been suggested to serve a barrier function for the hyaline cartilage [170] as well as a lubrication function [171], failure to regenerate this barrier may compromise long term survival of the repair tissue, reviewed in [172–174]. This conclusion is supported by the studies of Takada et al. [170] who reported that if the lamina splendens is disrupted, materials can gain entrance or exit from the cartilage, and it predisposes the development of OA. In rats, it appears that the lamina splendens arises during early post-natal life [170]; but whether it arises with a similar timeframe in humans could not be found. However, it has been found in other species discussed in [172]. Furthermore, what the stimulus is for the development of a lamina splendens is also not known, but it could relate to the biomechanical environment and the presence of unique growth and differentiation regulators in early post-natal life. However, the importance of the lamina splendens has led to some research effort to synthesize an artificial structure which could serve some of the functions of the lamina splendens [175]. This is an area for future research, given the lubrication and barrier function of the lamina splendens.

Additionally, regarding the presence of endogenous growth regulators, in the human TEC studies [137,156], the patients ranged in age from 28 to 46 years old and were therefore somewhat young. Preliminary preclinical studies in the porcine model demonstrate that the TEC approach is equally effective in skeletally immature and mature pigs [83], but studies were not performed with older pigs. Thus, it remains to be determined whether there is any age-dependent decline in the effectiveness of the TEC approach that may be attributed to an age-related decline in growth regulators in the tissues, or even after menopause in female patients. If such a decline is observed, it may be overcome with the use of autologous PRP [85,86], as long as the platelets in older individuals have not also been compromised.

With the preferred use of autologous MSC by both patients and some regulatory agencies, the use of MSC from young versus older/elderly patients is also a consideration as MSC numbers and function in some available depots also decline with age [176] and there is the potential for perhaps epigenetic alterations due to life experiences or exposure to chemicals that could potentially contribute to the compromised function [177]. To overcome this potential limitation of autologous MSC, some parents are having their children's cord blood MSC frozen in case they are needed, or the use of standardized allogeneic MSC from a source, such as an expanded cord blood MSC, has been proposed [178,179]. Thus, optimization of conditions to achieve the best success may require both the most appropriate MSC and the best in vivo conditions that can be obtained.

8. The Way Forward

While progress has been made toward using cell therapies to improve healing outcomes, there is still a need for further improvements. As discussed above, the attention has been focused on the implantation of cells, such as MSC, rather than trying to optimize the in vivo implantation environment. Likely, the way forward will require addressing more attention towards optimizing both what is implanted and the environment that it is implanted into. Clearly, attention to such variables will be complex and it is very likely that "one size does not fit all".

The adverse effect of inflammatory processes on the repair and regeneration potential of cellular therapies is of central concern, and such processes will have to be controlled if expectations of further success regarding cellular therapies are to be achieved. While the successes achieved thus far have provided support for further investment and research, some of the diseases or conditions that could benefit from cellular therapies will likely be more complex and challenging.

Using articular cartilage repair/regeneration as an important example due to the current successes and the need as cartilage-related conditions, such as osteoarthritis, affects tens of millions around the world, and with alternatives to cell therapies (i.e., drugs, exercise, injury prevention) of limited impact thus far. Osteoarthritis is both a disease of mechanics [180] and inflammation [181], and of the whole joint and not just the articular cartilage [182,183]. The term osteoarthritis is actually an umbrella term that encompasses several subtypes of OA including post-traumatic OA, metabolic OA associated with obesity reviewed in [20,42,181,184], post-menopausal onset OA discussed in [42], with idiopathic OA a large subpopulation of patients for which a link to a cause has not been clearly defined. Thus, OA, which can develop if a cartilage defect is not repaired, is heterogeneous and complex, and one cell therapy solution likely will not apply to all subtypes of the disease.

While the transition from repairing fresh cartilage defects using cell therapies may identify the need to develop several different lines of approach, there are likely some principles that the approaches should share. The first is that, unless the biomechanical environment is restored or the biomechanical compromise addressed, any cell therapy may not achieve long-term restoration of function. Second, as most OA patients are older and many will have co-morbidities, any cell therapy intervention will likely require a co-intervention to augment the need for an anabolic supplementation, such as PRP or EV. Thirdly, and relevant to point two, there will be a need to control inflammation, often in the context of diabetes, which can contribute to an altered inflammation [60]. In addition, rather than an acute inflammation as perhaps with a cartilage defect, those with OA may have converted to a chronic form which may require different strategies. Fourthly, the cell therapy cannot be considered the intervention of last resort when most of the articular cartilage is degraded, and the disease is considered end-stage. At this point there is little cartilage template remaining for a cellular therapy to enlist in the repair/regeneration effort. As the MSC in a TEC may not only contribute to the repair of the tissue damage by replacement, the undifferentiated MSC in the TEC can also release EV that could travel to remaining chondrocytes in the residual cartilage to contribute to the repair effort. Thus, repair via cell therapy interventions should likely be initiated early in the disease process rather than later.

While the above discussion has focused on repair of articular cartilage, some of the principles discussed can also be applied to the healing of other connective tissues of the MSK system, including menisci, intravertebral discs (IVD), as well as ligaments and tendons. In the case of tendons, tendons in different locations exhibit different properties [185], tendon properties can change with aging [74,186,187] and some tendons, such as the supraspinatus, can undergo age-related degeneration without overt symptoms [188–190]. Thus, cell therapy treatment could be envisioned to address tendinopathies rather than overt ruptures. Similar issues can likely also be applied to other tissues, such as menisci and IVD.

Another separate set of variables that could affect the efficacy of cellular therapies is genetics and epigenetics. Genetics and epigenetics could affect the MSCs, with age-related epigenetic changes potentially affecting the functionality of the MSC later in life when they are needed for cell therapies [191–193]. Genetics and epigenetics could also affect the target tissues of the cellular therapies. For example, some individuals with Marfan's Syndrome or the spectrum of Ehlers–Danlos Syndrome [194] may appear to have mutations in some of their ECM proteins that impair function and increase risk for injury or tissue failure reviewed in [193,195–197]. Thus, the outcome of the cellular therapy may not be optimal when using autologous cells and allogeneic cells may be preferred [198] or correcting the MSC via in vitro alterations [195]. While the examples presented are rare, there may also less overt variation in connective tissue molecules that predispose to injury or poor healing that do not present with symptoms, and these could also influence outcomes of cell therapies.

As the use of cellular therapies including the use of MSC continues to expand into more complex disease scenarios, it is clear that the use of multiple modalities in addition to the MSC will be needed. Some of the variables may be more readily assessed, but as continued improvement in genetic analysis and characterization of the epigenome become more common place, an element of precision medicine will be applied to the use of complex cellular therapies for MSK connective tissue repair and regeneration.

9. Conclusions

Attempts to enhance the repair and regeneration of injured connective tissues of the musculoskeletal system using cell therapies has been the subject of intense research over the past 30+ years. With the discovery of cells with the ability to differentiate into several relevant lineages (e.g., chondrocytes, bone cells, and others) reviewed in [122], this effort intensified. Using cells labeled mesenchymal stem cells (MSC), expectations ran high, but achieving success was more challenging.

Early after the discovery of adult "stem or progenitor" cells, there was considerable anticipation that they would rapidly be used to repair a variety of tissues damaged by injury or disease, particularly during the aging process. This hope rapidly became hype, and a number of what have been called rogue clinics and companies began selling stem cell-based cell therapy approaches directly to patients or consumers for a number of conditions [199–201]. Such rogue entities preyed on desperate patients, and many such clinics in North America have been recently curtailed by the FDA and Health Canada in the USA and Canada, respectively. Such rogue applications of stem cell therapies emphasize the need to continue to develop methods and interventions to use these cells more effectively and with a solid base of scientific and clinical justification. This will require building on past successes and failures to evaluate new directions and approaches.

Learning from these past scientifically and ethically approved research efforts, it is emerging that many of the relevant connective tissues that could benefit from stem cell interventions have complex structures, are designed to work in complex mechanical environments, and when injured this creates an inflammatory environment. Furthermore, when injured or subjected to a disease process, the situation arises as an adult or an elderly individual when the anabolic environment of youth (growth and maturation) is no longer evident. Thus, attention to the environment that cells, such as MSC, are placed in, either as individual cells or incorporated into constructs, needs to be addressed if the MSC are to achieve more of their potential to impact the return of functionality in these connective tissues. That is, control of an environment where a catabolic inflammatory process is needed, supplementation of the environment with appropriate anabolic mediators is also needed (either as molecules, PRP or extracellular vesicles), and for some circumstances using cellular therapy early in a disease process while the remaining endogenous tissue can serve a template function may additionally be critical. Finally, controlling the impact of co-morbidities (i.e., diabetes) may also be required. Thus, improving the environment into which the cells are placed may be critical for further success. Similarly, picking the right cells for the job may also be critical as MSC from different sources can exhibit different properties even though they can have a similar phenotype, as discussed in [123,124]. Thus, the right cells in the right environment at the right time are needed are discussed in [202–204], and there is likely a need for a more "precision medicine" approach as "one size does not fit all" [202].

While some progress is being made in the applications of cellular therapy, including MSC use in tissue engineered constructs as reviewed in [165,198], and many lessons have been learned as outlined above, several questions related to the issue of tissue regeneration still remain. The first relates to human heterogeneity and how such heterogeneity translates to variation in connective tissue structure and function. A second relates to the question of whether absolute regeneration is required to obtain optimal functioning in a specific mechanical environment? That is, would 80 or 90% regeneration at a structural level be sufficient for people in the 60–70 years old age range, but perhaps not acceptable for someone 30–40 years of age and wanting to maintain a very active lifestyle? Some of these philosophical questions may also need to be factored into the expectations of how success is defined going forward.

Author Contributions: D.A.H. drafted the initial version and then made subsequent changes. N.N. made additional revisions and edits to the drafts. All authors have read and agreed to the published version of the manuscript.

Funding: The preparation of this review was supported by funds from the Strategic Clinical Networks Program of Alberta Health Services.

Institutional Review Board Statement: N/A This review did not involve patients or patient data.

Informed Consent Statement: N/A This review did not involve patients or patient data.

Data Availability Statement: N/A No original data was presented in this article.

Acknowledgments: The authors thank the several colleagues for fruitful discussions on the topics presented over the past decade including Cyril B. Frank (deceased), Wataru Ando, Kazunori Shimomura, Ronald Zernicke, Nigel Shrive, Arin Sen and Roman Krawetz, as well as the contributions of many trainees over the past decades. We also apologize to all authors whose publications were not cited due to space considerations. This article is dedicated to the memory of Cyril B. Frank who was central to the initiation of many of the relevant studies discussed.

Conflicts of Interest: The authors declare no conflict of interest.

References

1. Lo, I.K.Y.; Marchuk, L.L.; Leatherbarrow, K.E.; Frank, C.B.; Hart, D.A. Collagen fibrillogenesis and mRNA levels in the maturing rabbit medial collateral ligament and patellar tendon. *Connect. Tis. Res.* **2004**, *45*, 11–22. [CrossRef] [PubMed]
2. Frank, C.B.; Matyas, J.; Hart, D. Aging of the medial collateral ligament. Interdisciplinary studies. In *Musculoskeletal Soft-Tissue Aging: Impact on Mobility*; Buckwalter, J.A., Goldberg, V.M., Woo, S.L.Y., Eds.; AAOS: Rosemont, IL, USA, 1993; pp. 305–326.
3. Eng, K.; Rangayyan, R.M.; Bray, R.C.; Frank, C.B.; Anscomb, J.; Veale, P. Quantitative analysis of the fine vascular anatomy of articular ligaments. *IEEE Trans. Biomed. Eng.* **1992**, *39*, 296–306. [CrossRef] [PubMed]
4. Bray, R.C.; Salo, P.T.; Lo, I.K.; Ackermann, P.; Rattner, J.B.; Hart, D.A. Normal ligament structure, physiology, and function. *Sports Med. Arthrosc. Rev.* **2005**, *13*, 127–135. [CrossRef]
5. Georgiev, G.P.; LLiev, A.; Kotov, G.; Kinov, P.; Slavcher, S.; Landzhov, B. Light and electron microscopic study of the medial collateral epiligament tissue in human knees. *World J. Orthop.* **2017**, *8*, 372–378. [CrossRef] [PubMed]
6. Hart, D.A.; Frank, C.B.; Bray, R. Inflammatory processes in repetitive motion and over-use syndromes: Potential role of neurogenic mechanisms in tendon and ligaments. In *Repetitive Motion Disorders of the Upper Extremity*; Gordon, S.L., Blair, S.J., Fine, L.J., Eds.; AAOS: Park Ridge, IL, USA, 1995; pp. 247–262.
7. Scott, A.; Lian, O.; Bahr, R.; Hart, D.A.; Duronio, V.; Khan, K.M. Increased mast cell numbers in human patellar tendinosis: Correlation with symptom duration and vascular hyperplasia. *Br. J. Sports Med.* **2008**, *42*, 753–757. [CrossRef]
8. Benjamin, M.; McGonagle, D. Entheses: Tendon and ligament attachment sites. *Scand. J. Med. Sci. Sports* **2009**, *19*, 520–527. [CrossRef]
9. Benjamin, M.; McGonagle, D. The enthesis organ concept and its relevance to the spondyloarthropathies. *Adv. Exp. Med. Biol.* **2009**, *649*, 57–70. [CrossRef]
10. Sevick, J.L.; Abusara, Z.; Andrews, S.H.; Xu, M.; Shurshid, S.; Chatta, J.; Hart, D.A.; Shrive, N.G. Fibril deformation underload of the rabbit Achilles tendon and medial collateral ligament femoral entheses. *J. Orthop. Res.* **2018**, *36*, 2506–2515. [CrossRef]
11. Sarmiento, P.; Little, D. Tendon and multiomics: Advantages, advances, and opportunities. *NPJ Regen. Med.* **2021**, *6*, 61. [CrossRef]
12. Huisman, E.S.; Andersson, G.; Scott, A.; Reno, C.R.; Hart, D.A.; Thornton, G.M. Regional molecular and cellular differences in the female rabbit Achilles tendon complex: Potential implications for understanding responses to loading. *J. Anat.* **2014**, *224*, 538. [CrossRef]
13. Arnoczky, S.P.; Cook, J.L.; Carter, T.; Turner, A.S. Translational models for studying meniscal repair and replacement: What they can and cannot tell us. *Tissue Eng. Part B Rev.* **2010**, *16*, 31–39. [CrossRef] [PubMed]
14. Andrews, S.H.; Rattner, J.B.; Abusara, Z.; Adesida, A.; Shrive, N.G.; Ronsky, J.L. Tie-fibre structure and organization in the knee menisci. *J. Anat.* **2012**, *224*, 531–537. [CrossRef] [PubMed]
15. Andrews, S.H.; Rattner, J.B.; Jamniczky, H.A.; Shrive, N.G.; Adesida, A. The structural and composition transition of the menisci roots into the fibrocartilage of the menisci. *J. Anat.* **2015**, *226*, 169–174. [CrossRef]
16. Andrews, S.H.J.; Adesida, A.; Abusara, Z.; Shrive, N.G. Current concepts on structure-function relationships in the menisci. *Connect. Tissue Res.* **2017**, *58*, 271–281. [CrossRef] [PubMed]
17. Frost, H.M. The mechanostat: A proposed pathogenic mechanism of osteoporosis and bone mass effects of mechanical and nonmechanical agents. *Bone Miner.* **1987**, *2*, 73–85.
18. Frost, H.M. Perspectives: A proposed general model of the "mechanostat" (suggestions from a new skeletal-biologic paradigm). *Anat. Rec.* **1996**, *244*, 139–147. [CrossRef]
19. Frost, H.M. Bone's mechanostat: A 2003 update. *Anat. Rec. A Discov. Mol. Cell. Evol. Biol.* **2003**, *275*, 1081–1101. [CrossRef]
20. Hart, D.A. Learning from human responses to deconditioning environments: Improved understanding of the "use it or lose it" principle. *Front. Sports Act. Living* **2021**, *3*, 685845. [CrossRef]
21. Hart, D.A.; Natsu-ume, T.; Sciore, P.; Tasevski, V.; Frank, C.B.; Shrive, N.G. Mechanobiology: Similarities and differences between in vivo and in vitro analysis at the functional and molecular levels. *Recent Res. Devel. Biophys. Biochem.* **2002**, *2*, 153–177.
22. Natsu-Ume, T.; Majima, T.; Reno, C.; Shrive, N.G.; Frank, C.B.; Hart, D.A. Menisci of the rabbit knee require mechanical loading to maintain homeostasis: Cyclic hydrostatic compression in vitro prevents derepression of catabolic genes. *J. Orthop. Sci.* **2005**, *10*, 396–405. [CrossRef]
23. Moretti, L.; Stalfort, J.; Barker, T.H.; Abebayehu, D. The interplay of fibroblasts, the extracellular matrix, and inflammation in scar formation. *J. Biol. Chem.* **2022**, *298*, 101530. [CrossRef] [PubMed]
24. Thornton, J.M.; Yin, K. Role of specialized pro-resolving mediators in modifying host defense and decreasing bacterial virulence. *Molecules* **2021**, *26*, 6970. [CrossRef] [PubMed]

25. Leuti, A.; Fava, M.; Pellegini, N.; Maccarrone, M. Role of specialized pro-resolving mediators in neuropathic pain. *Front. Pharmacol.* **2021**, *12*, 717993. [CrossRef]

26. Han, Y.H.; Lee, K.; Saha, A.; Han, J.; Choi, H.; Noh, M.; Lee, Y.H.; Lee, M.C. Specialized proresolving mediators for therapeutic interventions targeting metabolic and inflammatory disorders. *Biomol. Ther.* **2021**, *29*, 455–464. [CrossRef] [PubMed]

27. Blaudez, F.; Ivanovski, S.; Fournier, B.; Vaquette, C. The utilization of resolvins in medicine and tissue engineering. *Acta Biomater.* **2021**, *140*, 116–135. [CrossRef]

28. Bartczak, K.; Bialas, A.J.; Kotecki, M.J.; Gorski, P.; Piotrowski, W.J. More than a genetic code: Epigenetics of lung fibrosis. *Mol. Diagn. Ther.* **2020**, *24*, 665–681. [CrossRef]

29. Cai, Q.; Gan, C.; Tang, C.; Wu, H.; Gao, J. Mechanism and therapeutic opportunities of histone modifications in chronic liver disease. *Front. Pharmacol.* **2021**, *12*, 78491. [CrossRef]

30. Doody, K.M.; Bottini, N.; Firestein, G.S. Epigenetic alterations in rheumatoid arthritis fibroblast-like synoviocytes. *Epigenomics* **2017**, *9*, 479–492. [CrossRef]

31. Lin, Z.; Ding, Q.; Li, X.; Feng, Y.; He, H.; Huang, C.; Zhu, Y.Z. Targeting epigenetic mechanisms in vascular aging. *Front. Cardiovasc. Med.* **2022**, *8*, 806988. [CrossRef]

32. Guttzeit, S.; Backs, J. Post-translational modifications talk and crosstalk to class IIa histone deacetylases. *J. Mol. Cell Cardiol.* **2022**, *162*, 53–61. [CrossRef]

33. Kishi, K.; Okabe, K.; Shimizu, R.; Kubota, Y. Fetal skin possesses the ability to regenerate completely: Complete regeneration of skin. *Keio J. Med.* **2012**, *61*, 101–108. [CrossRef] [PubMed]

34. Pratsinis, H.; Mavrogonatou, E.; Kietsas, D. Scarless wound healing: From development to senescence. *Adv. Drug Deliv. Rev.* **2019**, *146*, 325–343. [CrossRef] [PubMed]

35. Yin, J.L.; Wu, Y.; Yuan, Z.W.; Gao, X.H.; Chen, H.D. Advances in scarless foetal wound healing and prospects for scar reduction in adults. *Cell Prolif.* **2020**, *53*, e12916. [CrossRef] [PubMed]

36. Gilliver, S.C. Sex steroids as inflammatory regulators. *J. Steroid Biochem. Mol. Biol.* **2010**, *120*, 105–115. [CrossRef]

37. Gilliver, S.C.; Ashworth, J.J.; Ashcroft, G.S. The hormonal regulation of cutaneous wound healing. *Clin. Dermatol.* **2007**, *25*, 56–62. [CrossRef]

38. Ashcroft, G.S.; Ashworth, J.J. Potential role of estrogens in wound healing. *Am. J. Clin. Dermatol.* **2003**, *4*, 737–743. [CrossRef]

39. Handman, M.J.; Ashcroft, G.S. Estrogen, not intrinsic aging, is the major regulator of delayed human wound healing in the elderly. *Genome Biol.* **2008**, *9*, R80. [CrossRef]

40. Arida, C.; Mastrokalos, D.S.; Panagopoulos, A.; Vlamis, J.; Triantafyllopoulos, I.K. A systematic approach for stronger documentation of anterior cruciate ligament graft choice. *Cureus* **2021**, *13*, e19017. [CrossRef]

41. Hayback, G.; Raas, C.; Rosenberger, R. Failure rates of common grafts used in ACL reconstructions: A systematic review of studies published in the last decade. *Arch. Orthop. Trauma Surg.* **2021**. [CrossRef]

42. Hart, D.A.; Werle, J.; Robert, J.; Kania-Richmond, A. Long wait times for knee and hip total joint replacement in Canada: An isolated health system problem, or a symptom of a larger problem? *Osteoarthr. Cartil. Open.* **2021**, *3*, 100141. [CrossRef]

43. Frank, C.; MacFarlane, B.; Edwards, P.; Rangayyan, R.; Liu, Z.Q.; Walsh, S.; Bray, R. A quantitative analysis of matrix alignment in ligament scars: A comparison of movement versus immobilization in an immature rabbit model. *J. Orthop. Res.* **1991**, *9*, 219–227. [CrossRef] [PubMed]

44. Achari, Y.; Chin, J.W.S.; Heard, B.J.; Rattner, J.B.; Shrive, N.G.; Frank, C.B.; Hart, D.A. Molecular events surrounding collagen fibril assembly in the early healing rabbit medial collateral ligament—Failure to recapitulate normal ligament development. *Connect. Tissue Res.* **2011**, *52*, 301–312. [CrossRef]

45. Marieswaran, M.; Jain, I.; Garg, B.; Sharma, V.; Kalyanasundaram, D. A review of biomechanics of anterior cruciate ligament and material for reconstruction. *Appl. Bionics Biomech.* **2018**, *2018*, 4657824. [CrossRef] [PubMed]

46. Blythe, A.; Tasker, T.; Zioupos, P. ACL fraft constructs: In-vitro fatigue testing highlights the occurrence of irrecoverable lengthening and the need for adequate (pre) conditioning to avert the reoccurance of knee instability. *Technol. Health Care.* **2006**, *14*, 335–347. [CrossRef]

47. Miller, R.M.; Rahnemai-Azar, A.A.; Suer, L.; Arilla, F.V.; Fu, F.H.; Debski, R.E.; Musahl, V. Tensile properties of a split quadriceps graft for ACL reconstruction. *Knee Surg. Sports Traumatol.* **2017**, *25*, 1249–1254. [CrossRef] [PubMed]

48. Bedi, A. *Editorial Commentary*: Buckle Up Surgeons: "Safety Belt" Reinforcement of Knee Anterior Cruciate Ligament Reconstruction Grafts. *Arthroscopy* **2018**, *34*, 500–501. [CrossRef]

49. Heard, B.J.; Barton, K.I.; Chung, M.; Achari, Y.; Shrive, N.G.; Frank, C.B.; Hart, D.A. Single intra-articular dexamethasone injection immediately post-surgery in a rabbit model mitigates early inflammatory response and post-traumatic osteoarthritis-like alterations. *J. Orthop. Res.* **2015**, *33*, 1826–1834. [CrossRef]

50. Sieker, J.T.; Ayturk, U.M.; Proffen, B.L.; Weissenberger, M.H.; Kiapour, A.M.; Murray, M.M. Immediate administration of intraarticular triamcinolone acetonide after joint injury modulates molecular outcomes associated with early synovitis. *Arthritis Rheum.* **2016**, *68*, 1637–1647. [CrossRef]

51. Bedrin, M.D.; Kartalias, K.; Yow, B.G.; Dickens, J.F. Degenerative joint disease after meniscectomy. *Sports Med. Arthrosc. Rev.* **2021**, *29*, e44–e50. [CrossRef]

52. Vinagre, G.; Cruz, F.; Alkhelaifi, K.; D'Hooghe, P. Isolated meniscus injuries in skeletally immature children and adolescents: State of the art. *J. ISAKOS* **2022**, *7*, 19–26. [CrossRef]

53. Kaul, G.; Cucchiarini, M.; Remberger, K.; Kohn, D.; Madry, H. Failed cartilage repair for early osteoarthritis defects: A biochemical, histological and immunohistochemical analysis of the repair tissue after treatment with marrow-stimulation techniques. *Knee Surg. Sports Traumatol. Arthrosc.* **2012**, *20*, 2315–2324. [CrossRef] [PubMed]

54. Erggelet, C.; Vavken, P. Microfracture for the treatment of cartilage defects in the knee joint—A golden standard? *J. Clin. Orthop. Trauma.* **2016**, *7*, 145–152. [CrossRef] [PubMed]

55. Gudeman, A.S.; Hinckel, B.B.; Oladeji, L.; Ray, T.E.; Gersoff, W.; Farr, J.; Sherman, S.L. Evaluation of commercially available knee cartilage restoration techniques stratified by FDA approval pathway. *Am. J. Sports Med.* **2021**, 1–7. [CrossRef] [PubMed]

56. Armoiry, X.; Cummins, E.; Connock, M.; Metcalfe, A.; Royle, P.; Johnston, R.; Rodrigues, J.; Waugh, N.; Mistry, H. Autologous chondrocyte implantation with chondrosphere for treating articular cartilage defects in the knee: An evidence review group perspective of a NICE single technology appraisal. *Pharmocoeconomics* **2019**, *37*, 879–886. [CrossRef] [PubMed]

57. Sharma, U.R.; Rathnakaran, A.N.; Raj, B.P.P.; Padinjakkara, G.; Das, A.; Vada, S.; Mudagal, M. The positive effect of pregnancy in rheumatoid arthritis and the use of medications for the management of rheumatoid arthritis during pregnancy. *Inflammopharmacology* **2021**, *29*, 987–1000. [CrossRef]

58. Gomez-Chavez, F.; Correa, D.; Navarrete-Meneses, P.; Cancino-Diaz, J.C.; Cancino-Diaz, M.E.; Rodriguez-Matinez, S. NF-kB and its regulators during pregnancy. *Front. Immunol.* **2021**, *12*, 679106. [CrossRef]

59. Hart, D.A.; Kydd, A.S.; Frank, C.B.; Hildebrand, K.A. Tissue repair in rheumatoid arthritis: Challenges and opportunities in the face of a systemic inflammatory disease. *Best Pract. Res. Clin. Rheumatol.* **2004**, *18*, 187–202. [CrossRef]

60. Ackermann, P.; Hart, D.A. Influence of comorbidities: Neuropathy, vasculopathy, and diabetes in healing response quality. *Adv. Wound Care* **2013**, *2*, 410–421. [CrossRef]

61. Ahmed, A.S.; Schizas, N.; Li, J.; Ahmed, M.; Ostenson, C.-G.; Salo, P.; Hewitt, C.; Hart, D.A.; Ackermann, P.W. Type 2 diabetes impairs tendon repair after injury in a rat model. *J. Appl. Physiol.* **2012**, *113*, 1784–1791. [CrossRef]

62. Ahmed, A.S.; Li, J.; Schizas, N.; Ahmed, M.; Ostenson, C.-G.; Salo, P.; Hewitt, C.; Hart, D.A.; Ackermann, P.W. Expressional changes in growth and inflammatory mediators during Achille's tendon repair in diabetic rats: New insights into a possible basis for compromised healing. *Cell Tissue Res.* **2014**, *357*, 109–117. [CrossRef]

63. Ahmed, A.S.; Li, J.; Abdul, A.M.D.; Ahmed, M.; Ostenson, C.-G.; Salo, P.T.; Hewitt, C.; Hart, D.A.; Ackermann, P.W. Compromised neurotrophic and angiogenic regenerative capability during tendon healing in a rat model of type-II diabetes. *PLoS ONE* **2017**, *12*, e0170748. [CrossRef] [PubMed]

64. Collins, K.H.; Herzog, W.; MacDonald, G.Z.; Reimer, R.A.; Rios, J.L.; Smith, I.C.; Zernicke, R.F.; Hart, D.A. Obesity, metabolic syndrome, and musculoskeletal disease: Common inflammatory pathways suggest a central role for loss of muscle integrity. *Front. Physiol.* **2018**, *9*, 112. [CrossRef] [PubMed]

65. Scott, A.; Zwerver, J.; Grewal, N.; de Sa, A.; Alktebi, T.; Granville, D.J.; Hart, D.A. Lipids, adiposity and tendinopathy: Is there a mechanistic link? Critical review. *Br. J. Sports Med.* **2015**, *49*, 984–988. [CrossRef] [PubMed]

66. De Sa, A.; Hart, D.A.; Khan, K.; Scott, A. Achilles tendon structure is negatively correlated with body mass index, but not influenced by statin use: A cross-sectional study using ultrasound tissue characterization. *PLoS ONE* **2018**, *13*, e0199645. [CrossRef]

67. Ghosh, N.; Kolade, O.O.; Shontz, E.; Rosenthal, Y.; Zuckerman, J.D.; Bosco, J.A., 3rd; Virk, M.S. Nonsteroidal anti-inflammatory drugs (NSAIDS) and their effect on musculoskeletal soft-tissue healing: A scoping review. *JBJS Rev.* **2019**, *7*, e4. [CrossRef]

68. White, A.E.; Henry, J.K.; Dziadosz, D. The effect of nonsteroidal anti-inflammatory drugs and selective COX-2 inhibitors on bone healing. *HSS J.* **2021**, *17*, 231–234. [CrossRef]

69. Kigera, J.W.M.; Gichangi, P.B.; Abdelmalek, A.K.M.; Ogeng'o, J.A. Age related effects of selective and non-selective COX-2 inhibitors on bone healing. *J. Clin. Orthop. Trauma* **2022**, *25*, 101763. [CrossRef]

70. Yasui, Y.; Hart, D.A.; Sugita, N.; Chijimatsu, R.; Koizumi, K.; Ando, W.; Moriguchi, Y.; Shimomura, K.; Myoui, A.; Yoshikawa, H.; et al. Time-dependent recovery of human synovial membrane mesenchymal stem cell function after high-dose steroid therapy: Case report and laboratory study. *Am. J. Sports Med.* **2018**, *46*, 695–701. [CrossRef]

71. Lee, S.; Kruger, B.T.; Ignatius, A.; Tuckerman, J. Distinct glucocorticoid receptor actions in bone homeostasis and bone diseases. *Front. Endocrinol.* **2022**, *12*, 815386. [CrossRef]

72. Roughley, P.J. Structural changes in the proteoglycans of human articular cartilage during aging. *J. Rheumatol.* **1987**, *14*, 14–15.

73. Chen, M.; Zhou, S.; Shi, H.; Gu, H.; Wen, Y.; Chen, L. Identification and validation of pivotal genes related to age-related meniscus degeneration based on gene exporession profiling analysis and in vivo and in vitro models detection. *BMC Med. Genom.* **2021**, *14*, 237. [CrossRef] [PubMed]

74. Jiang, X.; Wojtkiewicz, M.; Patwardhan, C.; Greer, S.; Kong, Y.; ZKuss, M.; Huang, X.; Liao, J.; Lu, Y.; Dudley, A.; et al. The effects of maturation and aging on the rotator cuff tendon-to-bone interface. *FASEB J.* **2021**, *35*, e22066. [CrossRef] [PubMed]

75. Hart, D.A.; Nakamura, N.; Shrive, N.G. Perspective: Challenges presented for regeneration of heterogeneous musculoskeletal tissues that normally develop in unique biomechanical environments. *Front. Bioeng. Biotechnol.* **2021**, *9*, 760273. [CrossRef] [PubMed]

76. Chen, C.H.; Li, D.L.; Chuang, A.D.C.; Dash, B.S.; Chen, J.P. Tension stimulation of tenocytes in aligned hyaluronic acid/platelet-rich plasma-polycaprolactone core-sheath nanofiber membrane scaffold for tendon tissue engineering. *Int. J. Mol. Sci.* **2021**, *22*, 11215. [CrossRef] [PubMed]

77. Szojka, A.R.; Li, D.X.; Sopcak, M.E.J.; Ma, Z.; Kunze, M.; Mulet-Sierra, A.; Adeeb, S.M.; Westover, L.; Jomha, N.M.; Adesida, A.B. Mechano-hypoxia conditioning of engineered human meniscus. *Front. Bioeng. Biotechnol.* **2021**, *9*, 739438. [CrossRef] [PubMed]

78. Hart, D.A.; Shrive, N.G.; Goulet, F. Tissue engineering of ACL replacements. *Sports Med. Arthrosc. Rev.* **2005**, *13*, 170–176. [CrossRef]

79. Kobayashi, M.; Spector, M. In vitro response of bone marrow-derived mesenchymal stem cells seeded in a type-1 collagen-glycosaminoglycan scaffold for skin wound repair under mechanical loading condition. *Mol. Cell Biomech.* **2009**, *6*, 217–227.

80. Yong, K.W.; Choi, J.R.; Choi, J.W.; Cowie, A.C. Recent advances in mechanically loaded human mesenchymal stem cells for bone tissue engineering. *Int. J. Mol. Sci.* **2020**, *21*, 5816. [CrossRef]

81. Pattappa, G.; Zellner, J.; Johnstone, B.; Docheva, D.; Angele, P. Cells under pressure- the relationship between hydrostatic pressure and mesenchymal stem cell chondrogenesis. *Eur. Cell Mater.* **2019**, *37*, 360–381. [CrossRef]

82. Zhang, S.; Yao, Y. The role of mechanical regulation in cartilage tissue engineering. *Curr. Stem Cell Res. Ther.* **2021**, *16*, 939–948. [CrossRef]

83. Shimomura, K.; Ando, W.; Tateishi, K.; Nansai, R.; Fujie, H.; Hart, D.A.; Kohda, H.; Kita, K.; Kanamoto, T.; Mae, T.; et al. The influence of skeletal maturity on allogenic synovial mesenchymal stem cell-based repair of cartilage in a large animal model. *Biomaterials* **2010**, *31*, 8004–8011. [CrossRef] [PubMed]

84. Fujie, H.; Nansai, R.; Ando, W.; Shimomura, K.; Moriguchi, Y.; Hart, D.A.; Nakamura, N. Zone-specific integrated cartilage repair using a scaffold-free tissue engineered construct derived from allogenic synovial mesenchymal stem cells: Biomechanical and histological assessments. *J. Biomech.* **2015**, *48*, 4101–4108. [CrossRef] [PubMed]

85. Andia, I.; Atilano, L.; Maffulli, N. Moving towards targeting the right phenotype with the right platelet-rich plasma (PRP) formulation for knee osteoarthritis. *Ther. Adv. Musculoskelet. Dis.* **2021**, *13*, 1759720X211004336. [CrossRef] [PubMed]

86. Kydd, A.S.R.; Hart, D.A. Efficacy and safety of platelet-rich plasma injections of osteoarthritis. *Curr. Treat. Options Rheum.* **2020**, *6*, 87–98. [CrossRef]

87. Burnham, R.; Smith, A.; Hart, D. The safety and efficacy of bone marrow concentrate injection for knee and hip osteoarthritis: A Canadian cohort. *Regen. Med.* **2021**, *16*, 619–628. [CrossRef]

88. Rodriguez-Merchan, E.C. Anterior cruciate ligament reconstruction: Is biological augmentation beneficial? *Int. J. Mol. Sci.* **2021**, *22*, 12566. [CrossRef]

89. Evans, C.H.; Huard, J. Gene therapy approaches to regenerating the musculoskeletal system. *Nat. Rev. Rheumatol.* **2015**, *11*, 234–242. [CrossRef]

90. Clutterbuck, A.L.; Asplin, K.E.; Harris, P.; Allaway, D.; Mobsdheri, A. Targeting matrix metalloproteinases in inflammatory conditions. *Curr. Drug Targets* **2009**, *10*, 1245–1254. [CrossRef]

91. Hildebrand, K.A.; Frank, C.B.; Hart, D.A. Gene intervention in ligament and tendon: Current status, challenges, future directions. *Gene Ther.* **2004**, *11*, 368–378. [CrossRef]

92. Hart, D.A.; Nakamura, N.; Marchuk, L.; Hiraoka, H.Y.; Boorman, R.; Kaneda, Y.; Shrive, N.G.; Frank, C.B. Complexity of determining cause and effect in vivo after antisense gene therapy. *Clin. Orthop. Relat. Res.* **2000**, *379*, S242–S251. [CrossRef]

93. Frank, C.; Shrive, N.; Hiraoka, H.; Nakamura, N.; Kaneda, Y.; Hart, D. Optimization of the biology of soft tissue repair. *J. Sci. Med. Sport.* **1999**, *2*, 190–210. [CrossRef]

94. Li, Z.J.; Yang, Q.Q.; Zhou, Y.L. Basic research on tendon repair: Strategies, evaluation, and development. *Front. Med.* **2021**, *8*, 664909. [CrossRef] [PubMed]

95. Padilla, S.; Sanchez, M.; Vaquerizo, V.; Malanga, G.A.; Fiz, N.; Azofra, J.; Rogers, C.J.; Samitier, G.; Sampson, S.; Seijas, R.; et al. Platelet-rich plasma applications for Achille's tendon repair: A bridge between biology and surgery. *Int. J. Mol. Sci.* **2021**, *22*, 824. [CrossRef] [PubMed]

96. Hutchinson, I.D.; Rodeo, S.A.; Perrone, G.S.; Murray, M.M. Can platelet-rich plasma enhance anterior cruciate ligament and meniscal repair? *J. Knee Surg.* **2015**, *28*, 19–28. [CrossRef]

97. Kunze, K.N.; Pakanati, J.J.; Vadhera, A.S.; Polce, E.M.; Williams, B.T.; Panvaresh, K.C.; Chahla, J. The efficacy of platelet-rich plasma for ligament injuries: A systematic review of basic science literature with protocol quality assessment. *Orthop. J. Sports Med.* **2022**, *10*, 23259671211066504. [CrossRef]

98. Collins, T.; Alexander, D.; Barkatali, B. Platelet-rich plasma: A narrative review. *EFORT Open Rev.* **2021**, *6*, 225–235. [CrossRef]

99. Belk, J.W.; Kraeulter, M.J.; Thon, S.G.; Littlefield, C.P.; Smith, J.H.; McCarty, E.C. Augmentation of meniscal repair with platelet-rich plasma: A systematic review of comparative studies. *Orthop. J. Sports Med.* **2020**, *8*, 2325967120926145. [CrossRef]

100. Herdea, A.; Struta, A.; Derihaci, R.P.; Ulici, A.; Costache, A.; Furtunescu, F.; Toma, A.; Charkaoui, A. Efficiency of platelet-rich plasma therapy for healing sports injuries in young athletes. *Exp. Ther. Med.* **2022**, *23*, 215. [CrossRef]

101. Sanchez, M.; Anitua, E.; Orive, G.; Mukika, I.; Andia, I. Platelet-rich therapies in the treatment of orthopaedic sport injuries. *ASports Med.* **2009**, *39*, 345–354. [CrossRef]

102. Sharun, K.; Jambagi, K.; Dhama, K.; Kumar, R.; Pawde, A.M. Amarpal, Therapeutic potential of platelet-rich plasma in canine medicine. *Arch. Razi Inst.* **2021**, *76*, 721–730. [CrossRef]

103. Cash, C.; Scott, L.; Walden, R.L.; Kuhn, A.; Bowman, E. Bibliometric analysis of the top 50 highly cited articles on platelet-rich plasma in osteoarthritis and tendinopathy. *Regen. Med.* **2022**, *17*, 491–506. [CrossRef] [PubMed]

104. Sax, O.C.; Chen, Z.; Mont, M.A.; Delanois, R.E. The efficacy of platelet-rich plasma for the treatment of knee osteoarthritis symptoms and structural changes: A systematic review and meta-analysis. *J. Arthroplast.* 2022, in press. [CrossRef] [PubMed]

105. Trams, E.; Malesa, K.; Pomianowski, S.; Kaminski, R. Role of platelets in osteoarthritis-updated systematic review and meta-analysis on the role of platelet-rich plasma in osteoarthritis. *Cells* **2022**, *11*, 1080. [CrossRef] [PubMed]
106. Venosa, M.; Calafiore, F.; Mazzoleni, M.; Romanini, E.; Cerciello, S.; Calvisi, V. Platelet-rich plasma and adipose-derived mesenchymal stem cells in association with arthroscopic microfracture of knee articular cartilage defects: A pilot randomized controlled trial. *Adv. Orthop.* **2022**, *2022*, 6048477. [CrossRef] [PubMed]
107. Li, S.; Yang, G.; Zhang, H.; Li, X.; Lu, Y. A systematic review on the efficacy of different types of platelet-rich plasma in the management of lateral epicondylitis. *J. Shoulder Elb. Surg.* **2022**, *13*, 1533–1544. [CrossRef]
108. Barman, A.; Sinha, M.K.; Sahoo, J.; Jena, D.; Patel, V.; Patel, S.; Bhattacharjee, S.; Baral, D. Platelet-rich plasma injection in the treatment of patellar tendinopathy: A systematic review and meta-analysis. *Knee Surg. Relat. Res.* **2022**, *34*, 22. [CrossRef]
109. Mishra, A.; Pavelko, T. Treatment of chronic elbow tendinosis with buffered platelet-rich plasma. *Am. J. Sports Med.* **2006**, *34*, 1774–1778. [CrossRef]
110. Murray, M.M.; Spindler, K.P.; Abreu, E.; Muller, J.A.; Nedder, A.; Kelly, M.; Frino, J.; Zurakowski, D.; Valenza, M.; Snyder, B.D.; et al. Collen-platelet ricj plasma hydrogel enhances primary repair of the porcine anterior cruciate ligament. *J. Orthop. Res.* **2007**, *25*, 81–91. [CrossRef]
111. Braun, H.J.; Wasterlain, A.S.; Dragoo, J.L. The use of PRP in ligament and meniscal healing. *Sports Med. Arthrosc. Rev.* **2013**, *21*, 206–212. [CrossRef]
112. Chang, Y.; Yang, M.; Ke, S.; Zhang, Y.; Xu, G.; Li, Z. Effect of platelet-rich plasma on intervertebral disc degeneration in vivo and in vitro: A critical review. *Oxid. Med. Cell Longev.* **2020**, *2020*, 8893819. [CrossRef]
113. Li, P.; Zhang, R.; Zhou, Q. Efficacy of platelet-rich plasma in retarding intervertebral disc degeneration: A meta-analysis of animal studies. *Biomed. Res. Int.* **2017**, *2017*, 7919201. [CrossRef] [PubMed]
114. Basso, M.; Cavagnaro, L.; Zanirato, A.; Divano, S.; Formica, C.; Formica, M.; Felli, L. What is the clinical evidence on regenerative medicine in intervertebral disc degeneration? *Musculoskelet. Surg.* **2017**, *101*, 93–104. [CrossRef] [PubMed]
115. Akeda, K.; Ohishi, K.; Takegami, N.; Sudo, T.; Yamada, J.; Fujiwara, T.; Niimi, R.; Matsumoto, T.; Nishimura, Y.; Ogura, T.; et al. Platelet-rich plasma releasate versu corticosteroid for the treatment of discogenic low back paIn A double-blind randomized controlled trial. *J. Clin. Med.* **2022**, *11*, 304. [CrossRef] [PubMed]
116. Williams, L.B.; Adesida, A.B. Angiogenic approaches to meniscal healing. *Injury* **2018**, *49*, 467–472. [CrossRef]
117. Anz, A.W.; Hackel, J.G.; Nilsson, E.C.; Andrews, J.R. Application of biologics in the treatment of the rotator cuff, meniscus, cartilage, and osteoarthritis. *J. Am. Acad. Orthop. Surg.* **2014**, *22*, 68–79. [CrossRef]
118. Halper, J. Advances in the use of growth factors for treatment of disorders of soft tissues. *Adv. Exp. Med. Biol.* **2014**, *802*, 59–76. [CrossRef]
119. Liu, X.; Zhu, B.; Li, Y.; Liu, X.; Guo, S.; Wang, C.; Li, S.; Wang, D. The role of vascular endothelial growth factor in tendon healing. *Front. Physiol.* **2021**, *12*, 766080. [CrossRef]
120. Molloy, T.; Wang, Y.; Murrell, G. The roles of growth factors in tendon and ligament healing. *Sports Med.* **2003**, *33*, 381–394. [CrossRef]
121. Nakamura, N.; Timmermann, S.A.; Hart, D.A.; Kaneda, Y.; Shrive, N.G.; Shino, K.; Ochi, T.; Frank, C.B. A comparison of in vivo gene delivery methods for antisense therapy in ligament healing. *Gene Ther.* **1998**, *5*, 1455–1561. [CrossRef]
122. Pittenger, M.F.; Dischert, D.E.; Peault, B.M.; Phinney, D.G.; Hare, J.M.; Caplan, A.I. Mesenchymal stem cell perspective: Cell biology to clinical progress. *npj Regen. Med.* **2019**, *4*, 22. [CrossRef]
123. Hart, D.A. What molecular recognition systems do mesenchymal stem cells/medicinal signaling cells (MSC) use to facilitate cell-cell and cell-matrix interactions? A review of evidence and options. *Int. J. Mol. Sci.* **2021**, *22*, 8637. [CrossRef] [PubMed]
124. Ando, W.; Kutcher, J.J.; Krawetz, R.; Sen, A.; Nakamura, N.; Frank, C.B.; Hart, D.A. Clonal analysis of synovial fluid stem cells to characterize and identify stable mesenchymal stromal cell/mesenchymal progenitor cell phenotypes in a porcine model: A cell source with enhancec commitment of the chondrogenic lineage. *Cytotherapy* **2014**, *16*, 776–788. [CrossRef] [PubMed]
125. Hart, D.A. Perspective: Is it time to rename MSC (mesenchymal stem cells/medicinal signaling cells) with a name that reflects their combined in vivo functions and their in vitro abilities?—Possibly "pluripotent mesenchymal regulatory cells (PMRC)". *J. Biomed. Sci. Eng.* **2021**, *14*, 317–324. [CrossRef]
126. Motejunas, M.W.; Bonneval, L.; Carter, C.; Reed, D.; Ehrhardt, K. Biologic therapy in chronic pain management: A review of the clinical data and future investigations. *Curr. Pain Headache Rep.* **2021**, *25*, 30. [CrossRef] [PubMed]
127. Hernandez, P.A.; de la Mata Lloyd, J. Expanded mesenchymal stromal cells in knee osteoarthritis: A systematic literature review. *Rheumatol. Clin.* **2022**, *18*, 49–55. [CrossRef]
128. Hernigou, J.; Vertongen, P.; Rasschaert, J.; Hernigou, P. Role of scaffolds, subchondral, intra-articular injections of fresh autologous bone marrow concentrate regenerative cells in treating human knee cartilage lesion: Different approaches and different results. *Int. J. Mol. Sci.* **2021**, *22*, 3844. [CrossRef]
129. El-Kadiry, A.E.; Lumbao, C.; Salame, N.; Rafei, M.; Shammaa, R. Bone marrow aspirate concentrate versus platelet-rich plasma for treating knee osteoarthritis: A one-year non-randomized retrospective comparative study. *BMC Musculoskelet. Disord.* **2022**, *23*, 23. [CrossRef]
130. Caplan, A.I. Mesenchymal stem cells: Time to change the name! *Stem Cells Transl. Med.* **2017**, *6*, 1445–1451. [CrossRef]
131. Harris, Q.; Seto, J.; O'Brien, K.; Lee, P.S.; Kondo, C.; Heard, B.J.; Hart, D.A.; Krawetz, R.J. Monocyte chemotactic protein-1 inhibits chondrogenesis of synovial mesenchymal progenitor cells: An in vitro study. *Stem Cells* **2013**, *31*, 2253–2265. [CrossRef]

132. Krawetz, R.J.; Wu, Y.E.; Martin, L.; Rattner, J.B.; Matyas, J.R.; Hart, D.A. Synovial fluid progenitors expressing CD90+ from normal but not osteoarthritic joints undergo chondrogenic differentiation without micro-mass culture. *PLoS ONE* **2012**, *7*, e43616. [CrossRef]

133. Cabral, J.; Ryan, A.E.; Griffin, M.D.; Ritter, T. Extracellular vesicles as modulators of wound healing. *Adv. Drug Deliv. Rev.* **2018**, *129*, 394–406. [CrossRef] [PubMed]

134. Harrell, C.R.; Jovicic, N.; Djonov, V.; Arsenijevic, N.; Volarevic, V. Mesenchymal stem cell-derived exosomes and other extracellular vesicles as new remedies in the therapy of inflammatory diseases. *Cells* **2019**, *8*, 1605. [CrossRef] [PubMed]

135. Casado-Diaz, A.; Quesada-Gomez, J.M.; Dorada, G. Extracellular vesicles derived from mesenchymal stem cells (MSC) in regenerative medicine: Applications in skin wound healing. *Front. Bioeng. Biotechnol.* **2020**, *8*, 146. [CrossRef] [PubMed]

136. Roefs, M.T.; Sluijter, J.P.G.; Vader, P. Extracellular vesicle-associated proteins in tissue repair. *Trends Cell Biol.* **2020**, *30*, 990–1013. [CrossRef]

137. Shimomura, K.; Uasui, Y.; Koisumi, K.; Chijimatsu, R.; Hart, D.A.; Yonetani, Y.; Ando, W.; Nishi, T.; Kanamoto, T.; Horibe, S.; et al. First-in-human pilot study of implantation of a scaffold-free tissue-engineered construct generated from autologous synovial mesenchymal stem cells for the repair of knee chondral lesions. *Am. J. Sports Med.* **2018**, *46*, 2384–2393. [CrossRef]

138. Malekpour, K.; Hazrati, A.; Zahar, M.; Markov, A.; Zekiy, A.O.; Navashenaq, J.G.; Roshangar, L.; Ahmadi, M. The potential use of mesenchymal stem cells and their derived exosomes for orthopedic diseases treatment. *Stem Cell Rev. Rep.* **2022**, *18*, 933–951. [CrossRef]

139. Ren, S.; Wang, C.; Guo, S. Review of the role of mesenchymal stem cells and exosomes derived from mesenchymal stem cells in the treatment of orthopedic disease. *Med. Sci. Monit.* **2022**, *28*, e935937. [CrossRef]

140. Longhini, A.L.; Salazar, T.E.; Vieira, C.; Trinh, T.; Duan, Y.; Pay, L.M.; Calzi, S.L.; Losh, M.; Johnston, N.A.; Xie, H. Peripheral blood-derived mesenchymal stem cells demonstrate immunomodulatory potential for therapeutic use in horses. *PLoS ONE* **2019**, *14*, e0212642. [CrossRef]

141. Zha, K.; Li, X.; Yang, Z.; Tian, G.; Sun, Z.; Sui, X.; Dai, Y.; Liu, S.; Guo, Q. Heterogeneity of mesenchymal stem cells in cartilage regeneration: From characterization to application. *NPJ Regen. Med.* **2021**, *6*, 14. [CrossRef]

142. Vahedi, P.; Moghaddamshahabi, R.; Webster, T.J.; Koyuncu, A.C.C.; Ahmadian, E.; Khan, W.S.; Mohamed, A.J.; Eftekhari, A. The use of infrapatellar fat pad-derived mesenchymal stem cells in articular cartilage regeneration: A review. *Int. J. Mol. Sci.* **2021**, *22*, 9215. [CrossRef]

143. Meng, H.Y.H.; Lu, V.; Khan, W. Adipose tissue-derived mesenchymal stem cells as a potential restorative treatment for cartilage defects: A PRISMA review and meta-analysis. *Pharmaceuticals* **2021**, *14*, 1280. [CrossRef] [PubMed]

144. Leong, D.J.; Sun, H.B. Mesenchymal stem cells in tendon repair and regeneration: Basic understanding and translational challenges. *Ann. N. Y. Acad. Sci.* **2016**, *1383*, 88–96. [CrossRef] [PubMed]

145. Liu, P.P.Y. Mesenchymal stem cell-derived extracellular vesicles for the promotion of tendon repair-an update of literature. *Stem Cell Rev. Rep.* **2021**, *17*, 379–389. [CrossRef]

146. Meeremans, M.; Van de Walle, G.R.; Van Vlierberghe, S.; De Schauwer, C. The lack of a representative tendinopathy model hampers fundamental mesenchymal stem cell research. *Front Cell Dev. Biol.* **2021**, *9*, 651164. [CrossRef] [PubMed]

147. Jang, K.M.; Lim, H.C.; Bae, J.H. Mesenchymal stem cells for enhancing biologic healing after anterior cruciate ligament injuries. *Curr. Stem Cell Res.* **2015**, *10*, 535. [CrossRef]

148. Yu, H.; Adesida, A.; Jomha, N.M. Meniscus repair using mesenchymal stem cells—A comprehensive review. *Stem Cell Res. Ther.* **2015**, *6*, 86. [CrossRef]

149. Chew, E.; Prakash, R.; Khan, W. Mesenchymal stem cells in human meniscal regeneration: A systematic review. *Ann. Med. Surg.* **2017**, *24*, 3–7. [CrossRef]

150. Rhim, H.C.; Jeon, O.H.; Han, S.B.; Bae, J.H.; Suh, D.W.; Jang, K.M. Mesenchymal stem cells for enhancing biological healing after meniscal injuries. *World J. Stem Cells* **2021**, *13*, 1005–1029. [CrossRef]

151. Zhou, Y.F.; Zhang, D.; Yan, W.T.; Lian, K.; Zhang, Z.Z. Meniscus regeneration with multipotent stromal cell therapies. *Front. Bioeng. Biotechnol.* **2022**, *10*, 796408. [CrossRef]

152. Croft, A.S.W.; Illien-Junger, S.; Grad, S.; Guerrero, J.; Wangler, S.; Gantenbein, B. The application of mesenchymal stromal cells and their homing capabilities to regenerate intervertebral disc. *Int. J. Mol. Sci.* **2021**, *22*, 3519. [CrossRef]

153. Laing, W.; Han, B.; Hai, Y.; Sun, D.; Yin, P. Mechanism of action of mesenchymal stem cell-derived exosomes in the intervertebral disc degeneration treatment and bone repair and regeneration. *Front. Cell Dev. Biol.* **2022**, *9*, 833840. [CrossRef] [PubMed]

154. DiStefano, T.J.; Vaso, K.; Danias, G.; Chionuma, H.N.; Weiser, J.R.; Iatridis, J.C. Extracellular vesicles as an emerging treatment option for intervertebral disc degeneration: Therapeutic potential, translational pathways, and regulatory considerations. *Adv. Healthc. Mater.* **2022**, *11*, e100596. [CrossRef] [PubMed]

155. Barbon, S.; Rajendran, S.; Bertalot, T.; Piccione, M.; Gasparella, M.; Parnigotto, P.P.; Di Liddo, R.; Concorni, M.T. Growth and differentiation of circulating stem cells after extensive ex vivo expansion. *Tissue Eng. Regen. Med.* **2021**, *18*, 411–427. [CrossRef] [PubMed]

156. Shimomura, K.; Hamada, H.; Hart, D.a.; Ando, W.; Nishi, T.; Trattnig, S.; Nehrer, S.; Nakamura, N. Histological analysis of cartilage defects repaired with an autologous human stem cell construct 48 weeks postimplantation reveals structural details not detected by T2-mapping MRI. *Cartilage* **2021**, *13* (Suppl. S1), 694S–706S. [CrossRef]

157. Sart, S.; Jeske, R.; Chen, X.; Ma, T.; Li, Y. Engineering stem cell-derived extracellular matrices: Decellularization, characterization, and biological function. *Tissue Eng. Part B Rev.* **2020**, *26*, 402–422. [CrossRef]

158. Venkataiah, V.S.; Yahata, Y.; Kitagawa, A.; Inagaki, M.; Kakiuchi, Y.; Nakano, M.; Suzuki, S.; Handa, K.; Saito, M. Clinical applications of cell-scaffold constructs for bone regeneration therapy. *Cells* **2021**, *10*, 2687. [CrossRef]

159. Ahmed, E.; Saleh, T.; Xu, M. Recellularization of native tissue derived acellular scaffold with mesenchymal stem cells. *Cells* **2021**, *10*, 1787. [CrossRef]

160. Luo, T.; Tan, B.; Zhu, L.; Wang, Y.; Liao, J. A review on the design of hydrogels with different stiffness and their effects on tissue repair. *Front. Bioeng. Biotechnol.* **2022**, *10*, 817391. [CrossRef]

161. Messaoudi, O.; Henrionnet, C.; Bourge, K.; Loeuille, D.; Gillet, P.; Pinzano, A. Stem cells and extrusion 3D printing for hyaline cartilage engineering. *Cells* **2020**, *10*, 2. [CrossRef]

162. Tharakan, S.; Khondkar, S.; Ilyas, A. Bioprinting of stem cells in multimaterial scaffolds and their applications in bone tissue engineering. *Sensors* **2021**, *21*, 7477. [CrossRef]

163. Ando, W.; Tateishi, K.; Hart, D.A.; Katakai, D.; Tanaka, Y.; Nakata, K.; Hashimoto, J.; Fujie, H.; Shino, K.; Yoshikawa, H.; et al. Cartilage repair using an in vitro generated scaffold-free tissue-engineered construct derived from porcine synovial mesenchymal stem cells. *Biomaterials* **2007**, *28*, 5462–5470. [CrossRef] [PubMed]

164. Moriguchi, Y.; Tateishi, K.; Ando, W.; Shimomura, K.; Yonetani, Y.; Kita, K.; Hart, D.A.; Gobbi, A.; Shino, K.; Yoshikawa, H.; et al. Repair of meniscal lesions using a scaffold-free tissue-engineered construct derived from allogenic synovial MSCs in a miniature swine model. *Biomaterials* **2013**, *34*, 2185–2193. [CrossRef] [PubMed]

165. Shimomura, K.; Ando, W.; Fujie, H.; Hart, D.A.; Yoshikawa, H.; Nakamura, N. Scaffold-free tissue engineering for injured joint surface restoration. *J. Exp. Orthop.* **2018**, *5*, 2. [CrossRef] [PubMed]

166. Shimomura, K.; Rothrauff, S.; Hart, D.A.; Hamamoto, S.; Kobayashi, M.; Yoshikawa, H.; Tuan, R.S.; Nakamura, N. Enhanced repair of meniscal hoop structure injuries using an aligned electrospun nanofibrous scaffold combined with a mesenchymal stem cell-derived tissue engineered construct. *Biomaterials* **2019**, *192*, 346–354. [CrossRef]

167. Ishiguro, H.; Kaito, T.; Yarimitsu, S.; Hashimoto, K.; Okada, R.; Kushioka, J.; Chijimatsu, R.; Takenka, S.; Makino, T.; Saka, Y.; et al. Interbertebral disc regeneration with an adipose mesenchymal stem cell-derived tissue-engineered construct in a rat nucleotomy model. *Acta Biomater.* **2019**, *87*, 118–129. [CrossRef]

168. Ando, W.; Fujie, H.; Moriguchi, Y.; Nansai, R.; Shimomura, K.; Hart, D.A.; Yoshikawa, H.; Nakamura, N. Detection of abnormalities in the superficial zone of cartilage repaired using a tissue engineered construct derived from synovial stem cells. *Eur. Cell Mater.* **2012**, *24*, 292. [CrossRef]

169. MacConaill, M.A. The movements of bones and joints. IV. The mechanical structure of articulating cartilage. *J. Bone Jt. Surg.* **1951**, *B33*, 251–257. [CrossRef]

170. Takada, N.; Wada, I.; Sugimura, I.; Maruyama, H.; Matsui, N. A possible barrier function of the articular surface. *Kaibogaku Zasshi.* **1999**, *74*, 631–637.

171. Kumar, P.; Oka, M.; Toguchida, J.; Kobayashi, M.; Uchida, E.; Nakamura, T.; Tanaka, K. Role of uppermost superficial layer of articular cartilage in the lubrication mechanism of joints. *J. Anat.* **2001**, *1999*, 241–250. [CrossRef]

172. Hollander, A.P.; Dickinson, S.C.; Kafienah, W. Stem cells and cartilage development: Complexities of a simple tissue. *Stem Cells* **2010**, *28*, 1992–1996. [CrossRef]

173. Wu, J.P.; Kirk, T.B.; Zheng, M.H. Study of the collagen structure in the superficial zone and physiological state of articular cartilage using a 3D confocal imaging technique. *J. Orthop. Surg. Res.* **2008**, *3*, 29. [CrossRef] [PubMed]

174. Bhatnagar, R.; Christian, R.G.; Nakano, T.; Aherne, F.X.; Thompson, J.R. Age related changes and osteochondrosis in swine articular and epiphyseal cartilage: Light and electron microscopy. *Can. J. Comp. Med.* **1981**, *45*, 188–195. [PubMed]

175. Wan, H.; Ren, K.; Kaoer, H.J.; Sharma, P.K. Cartilage lamina splendens inspired nanostructured coating for biomaterial lubrication. *J. Colloid Interface Sci.* **2021**, *594*, 435–445. [CrossRef] [PubMed]

176. Choudhery, M.S. Strategies to improve regenerative potential of mesenchymal stem cells. *World J. Stem Cells* **2021**, *13*, 1845. [CrossRef] [PubMed]

177. Wang, R.; Wang, Y.; Zhu, L.; Liu, Y.; Li, W. Epigenetic regulation in mesenchymal stem cell aging and differentiation and osteoporosis. *Stem Cells Int.* **2020**, *2020*, 8836258. [CrossRef] [PubMed]

178. Sadlik, B.; Jaroslawski, G.; Gladusz, D.; Puszkarz, M.; Markowska, M.; Paweler, K.; Bonuczkowski, D.; Oldak, T. Knee cartilage regeneration with umbilical cord mesenchymal stem cells embedded in collagen scaffold using dry arthroscopy technique. *Adv. Exp. Med. Biol.* **2017**, *1020*, 113–122. [CrossRef] [PubMed]

179. Lee, N.H.; Na, S.M.; Ahn, H.W.; Kang, J.K.; Seon, J.K.; Song, E.K. Allogenic human umbilical cord blood-derived mesenchymal stem cells are more effective than bone marrow aspiration concentrate for cartilage regeneration after high tibial osteotomy in medial unicompartment osteoarthritis of the knee. *Arthroscopy* **2021**, *37*, 2521–2530. [CrossRef]

180. Felson, D.T. Osteoarthritis as a disease of mechanics. *Osteoarthr. Cartil.* **2013**, *21*, 10–15. [CrossRef]

181. Berenbaum, F.; Walker, C. Osteoarthritis and inflammation: A serious disease with overlapping phenotypic patterns. *Postgrad. Med.* **2020**, *132*, 377–384. [CrossRef]

182. Radin, E.L.; Burr, D.B.; Caterson, B.; Fyhrie, D.; Brown, T.D.; Boyd, R.D. Mechanical determinants of osteoarthrosis. *Semin. Arthritis Rheum.* **1991**, *21* (Suppl. S2), 12–21. [CrossRef]

183. Frank, C.B.; Shrive, N.G.; Boorman, R.S.; Lo, I.K.Y.; Hart, D.A. New perspectives on bioengineering of joint tissues: Joint adaptation creates a moving target for engineering replacement tissues. *Ann. Biomed. Eng.* **2004**, *32*, 458. [CrossRef] [PubMed]

184. Courties, A.; Berenbaum, F.; Sellam, J. The phenotypic approach to osteoarthritis: A look at metabolic syndrome-associated osteoarthritis. *Jt. Bone Spine* **2019**, *86*, 725–730. [CrossRef] [PubMed]

185. Thornton, G.M.; Shao, X.; Chung, M.; Sciore, P.; Boorman, R.S.; Hart, D.A.; Lo, I.K.Y. Changes in mechanical loading lead to tendon-specific alterations in MMP and TIMP expression: Influence of stress deprivation and intermittent cyclic hydrostatic compression on rat supraspinatus and Achilles tendon. *Br. J. Sports Med.* **2010**, *44*, 698–703. [CrossRef] [PubMed]

186. Sarbacher, C.A.; Halger, J.T. Connective tissue and age-related diseases. *Subcell. Biochem.* **2019**, *91*, 281–310. [CrossRef]

187. Siadat, S.M.; Zamboulis, D.E.; Thorpe, C.T.; Ruberti, J.W.; Connizzo, B.K. Tendon extracellular matrix assembly, maintenance and dysregulation throughout life. *Adv. Exp. Med. Biol.* **2021**, *1348*, 45–103. [CrossRef]

188. Gumina, S.; Villani, C.; Arcen, V.; Fagnani, C.; Nistico, L.; Venditto, T.; Castagna, A.; Candela, V. Rotator cuff degeneration: The role of genetics. *J. Bone Jt. Surg. Am.* **2019**, *101*, 600–605. [CrossRef]

189. Keener, J.D.; Patterson, B.M.; Orvets, N.; Chamberlain, A.M. Degenerative rotator cuff tears: Refining surgical indications based on natural history data. *J. Am. Acad. Orthop. Surg.* **2019**, *27*, 156. [CrossRef]

190. Long, Z.; Nakagawa, K.; Wang, Z.; Amadio, P.C.; Zhao, C.; Gingery, A. Age-related cellular and microstructural changes to the rotator cuff enthesis. *J. Orthop. Res.* 2021; *in press.* [CrossRef]

191. Klett, C.C. Hereditary disorders of connective tissue: A review. *Wound Repair Regen.* **1997**, *5*, 3–11. [CrossRef]

192. Trudgian, J.; Trotman, S. Ehlers-Danlos syndrome and wound healing: Injury in a collagen disorder. *Br. J. Nurs.* **2011**, *20*, S10. [CrossRef]

193. Handa, K.; Abe, S.; Suresh, V.V.; Fujieda, Y.; Oshikawa, M.; Orimoto, A.; Kobayashi, Y.; Yamada, S.; Yamaba, S.; Murakami, S.; et al. Fibrillin-1 insufficiency alters periodontal wound healing failure in a mouse model of Marfan syndrome. *Arch. Oral Biol.* **2018**, *90*, 53–60. [CrossRef] [PubMed]

194. Malfait, F.; Castori, M.; Francomano, C.A.; Giunta, C.; Kosho, T.; Byers, P.H. The Ehlers-Danlos syndromes. *Nat. Rev. Dis. Primers* **2020**, *6*, 64. [CrossRef] [PubMed]

195. Islam, M.; Chang, C.; Gershwin, M.E. Ehlers-Danlos syndrome: Immunologic contrasts and connective tissue comparisons. *J. Transl. Autoimmun.* **2020**, *4*, 100077. [CrossRef] [PubMed]

196. Park, J.W.; Yan, L.; Stoddard, C.; Wang, X.; Yue, Z.; Crandall, L.; Robinson, T.; Chang, Y.; Denton, K.; Li, E.; et al. Recapitulating and correcting Marfan syndrome in a cellular model. *Int. J. Biol. Sci.* **2017**, *13*, 588. [CrossRef]

197. Prentice, D.A.; Pearson, W.A.; Fogarty, J. Vascular Ehlers-Danlos syndrome: Treatment of a complex abdominal wound with vitamin C and mesenchymal stromal cells. *Adv. Skin Wound Care* **2021**, *34*, 1–6. [CrossRef]

198. Shimomura, K.; Hamamoto, S.; Hart, D.A.; Yoshikawa, H.; Nakamura, N. Meniscal repair and regeneration: Current strategies and future perspectives. *J. Clin. Orthop. Trauma* **2018**, *9*, 247–253. [CrossRef]

199. Murray, I.R.; Chahla, J.; Frank, R.M.; Piuzzi, N.S.; Mandelbaum, B.R.; Dragoo, J.L.; Members of the Biologics Association. Rogue stem cell clinics. *Bone Jt. J.* **2020**, *102-B*, 148–154. [CrossRef]

200. Liska, M.G.; Crowley, M.G.; Borlongan, C.V. Regulated and unregulated clinical trials of stem cell therapies for stroke. *Transl. Stroke Res.* **2017**, *8*, 93–103. [CrossRef]

201. Yeo-The, N.S.L.; Tang, B.L. Moral obligations in conducting stem cell-based therapy trials for autism spectrum disorder. *J. Med. Ethics* **2022**, *48*, 343–348. [CrossRef]

202. Eliasberg, C.L.; Rodeo, S.A. Editorial Commentary: Cell-based therapies for articular cartilage repair require precise progenitor cell characterization and determination of mechanism of action. *Arthroscopy* **2021**, *37*, 3357–3359. [CrossRef]

203. Knapik, D.M.; Evuarherhe, A., Jr.; Frank, R.M.; Steinwaches, M.; Rodeo, S.; Mumme, M.; Cole, B.J. Nonoperative and operative soft-tissue and cartilage regeneration and orthopaedic biologics of the knee: An orthoregeneration network (ON) foundation review. *Arthroscopy* **2021**, *37*, 2704–2721. [CrossRef] [PubMed]

204. Rodeo, S. Cell therapy in orthopaedics: Where are we in 2019? *Bone Jt. J.* **2019**, *101-B*, 361–364. [CrossRef] [PubMed]

biomedicines

In Situ Gene Expression in Native Cryofixed Bone Tissue

Krisztina Nikovics [1,*,†], Cédric Castellarin [1,†], Xavier Holy [2], Marjorie Durand [3], Halima Morin [4], Abdelhafid Bendahmane [4] and Anne-Laure Favier [1]

[1] Imagery Unit, Department of Platforms and Technology Research, French Armed Forces Biomedical Research Institute, 91223 Brétigny-sur-Orge, France; cedric.castellarin@intradef.gouv.fr (C.C.); anne-laure.favier@intradef.gouv.fr (A.-L.F.)

[2] Department of Platforms and Technology Research, French Armed Forces Biomedical Research Institute, 91223 Brétigny-sur-Orge, France; xavier.holy@intradef.gouv.fr

[3] Osteo-Articulary Biotherapy Unit, Department of Medical and Surgical Assistance to the Armed Forces, French Armed Forces Biomedical Research Institute, 91223 Brétigny-sur-Orge, France; marjorie1.durand@intradef.gouv.fr

[4] National Research Institute for Agriculture, Food and the Environment (INRAE), Institute of Plant Sciences Paris-Saclay (IPS2), University Paris-Saclay, 91400 Orsay, France; halima.morin@inrae.fr (H.M.); abdelhafid.bendahmane@inrae.fr (A.B.)

* Correspondence: krisztina.nikovics@def.gouv.fr; Tel.: +33-(0)-1-78-65-13-33

† These authors contributed equally to this work.

Abstract: Bone is a very complex tissue that is constantly changing throughout the lifespan. The precise mechanism of bone regeneration remains poorly understood. Large bone defects can be caused by gunshot injury, trauma, accidents, congenital anomalies and tissue resection due to cancer. Therefore, understanding bone homeostasis and regeneration has considerable clinical and scientific importance in the development of bone therapy. Macrophages are well known innate immune cells secreting different combinations of cytokines and their role in bone regeneration during bone healing is essential. Here, we present a method to identify mRNA transcripts in cryosections of non-decalcified rat bone using in situ hybridization and hybridization chain reaction to explore gene expression in situ for better understanding the gene expression of the bone tissues.

Keywords: cryofixation; bone; in situ hybridization; hybridization chain reaction (HCR); macrophage

Citation: Nikovics, K.; Castellarin, C.; Holy, X.; Durand, M.; Morin, H.; Bendahmane, A.; Favier, A.-L. In Situ Gene Expression in Native Cryofixed Bone Tissue. *Biomedicines* **2022**, *10*, 484. https://doi.org/10.3390/biomedicines10020484

Academic Editor: Mike Barbeck

Received: 27 January 2022
Accepted: 15 February 2022
Published: 18 February 2022

Publisher's Note: MDPI stays neutral with regard to jurisdictional claims in published maps and institutional affiliations.

1. Introduction

Vertebrate bone is a dynamically changing tissue that constantly adapts throughout life. For successful bone healing, coordinated cross talk is needed between inflammatory and bone-forming cells [1–5]. Nowadays, the exact mechanisms of bone regeneration remains to be elucidated [6–9]. Bones and bone marrow contain different types of macrophages: (i) erythroid island macrophages; (ii) hematopoietic stem cell macrophages; and (iii) osteoclasts [10]. Macrophages play an important role both in osteoblast-mediated bone formation [9] and in osteoclast development [10]. In addition, the newly-discovered osteal macrophages, so called "osteomacs", have a fundamental role during bone regeneration [6,9,11,12]. The exact role of these cells is still under study. The cytokines and other soluble factors secreted by macrophages can induce the bone formation in vitro [2,5,9,13]. Cytokines are the critical actors in coordinating an efficient repair of damaged bone tissue [14]. Examination of cytokine expressions during bone regeneration is essential for establishing new diagnostic and therapeutic approaches for bone tissue repair. Protein expression analysis of bone cells mostly uses immunofluorescence technique methods to detect proteins in situ [15]. In spite of immunolabeling being very convenient and well reproducible, there are some disadvantages, such as non-specific labeling with certain antibodies [16,17]. Additional difficulty is that cytokines are usually secreted, so the identification of cells producing this peptide or protein is problematic [16].

In contrast, in situ hybridization (ISH) is one of the most suitable methods to investigate localization of gene expression in situ and based on the detection of mRNA product of genes involved in protein translation [16,18–21]. The basis of this technique is that complementary RNA and DNA sequences form hybrids with one another by hydrogen bonding. ISH is a very powerful technique; however, the probe design is complex, and the different steps are fastidious, needing a high level of optimization. An additional problem is that the conservation of the RNA as mRNA is very sensitive to degradation and RNase enzymes can be found everywhere [20].

The digoxigenin technique (in situ-DIG) is the most commonly used non-radioactive IHS method [22,23]. It is highly sensitive, but unfortunately only allows the analysis of a single gene in a single sample. In the 1980s, the fluorescence in situ hybridization (FISH) technique was published [24]. This method has the advantage of allowing the simultaneous analysis of several genes in the same sample, but is only suitable for the analysis of highly expressed genes [25]. Recently, a new method has been published in which in situ hybridization is coupled with the detection by hybridization chain reaction (in situ-HCR) [21,26–28]. One of the main advantages of this technique is its high sensitivity that makes it suitable for testing low-expression genes. In addition, several genes can be analyzed simultaneously. However, in situ-HCR is less efficient than the in situ-DIG method [25].

We found that fixation in buffered formalin, decalcification in EDTA and embedding in paraffin is a good compromise, as it provides not only a good morphology and excellent conditions for immunohistochemistry but also allows DNA- and RNA-based molecular studies [15,29,30] (Table 1).

Table 1. Comparison of the different approaches.

	Resin (R)/Paraffin (P)-Section	**Cryo-Section**
Section preparation	Long method	Short method
Toxic substances	Long period	Short period
Size of the section	5 µm	5 µm
Quality of morphology	Good	Medium
Application	Histological staining (R, P) Immunolabeling (P) In situ hybridization (P, medium sensitivity)	Histological staining Immunolabeling In situ hybridization (strong sensitivity) Laser microdissection

Cryosectioning of hard tissue has been introduced several decades ago [31] and was optimized by the tape technique described by Kawamoto et al. [32–34]. The main advantage of this system is that there is no fixation and embedding before section preparation, so it is much faster than the conventional method and more useful for in situ hybridization (Table 1). As mRNA is very sensitive to degradation, the challenge resided in the ability to cut cryofixed bone tissue and to preserve mRNA for analysis. To overcome these limitations, cryosectioning of the bone of rat was combined with ISH on the entire femur together with the muscle.

2. Materials and Methods

2.1. Rat Animal Model

All experiments were approved by the IRBA Institutional Animal Care and Use Committee (protocol 65 DEF_IGSSA_SP). Interventions were carried out in an accredited animal facility. 8-week-old (200 g average weight) male Sprague Dawley rats (Charles River Laboratories, Freiburg, Germany) were housed individually in cages, in a temperature- and light-controlled environment, with food and water ad libitum. Before collecting the femurs with muscles, animals were euthanized at 12 weeks old with an overdose of sodium pentobarbital (150 mg/kg) administrated intraperitoneally.

2.2. Slides Coating

Slides were manually coated in two steps using coating and pretreatment solutions (Leica Microsystems, Richmond, IL, USA). Slides were first pretreated with A solution (5 mL of A buffer concentrate (39475270, Leica, Wetzlar, Germany), 1.25 mL of 0.1 M acetic acid and 25 mL of acetone in 500 mL distilled water) (Figure 1A). After 30 min in dark, the A solution was completed with acetone (q.s.p. 500 mL), filtered and stored until 6 months at 4 °C in a dark bottle. Slides were immersed 3 times in A solution and pulled out diagonally to avoid streaks. Slides were kept overnight at RT or warmed at 90 °C for 5 min before the coating process. A total of 15 µL of B solution (39475271, Leica) was coated on pretreated slides to obtain a thin and homogenous coating (Figure 1B,C). The surface of the coating was adapted to match with the cryosection surface (Figure 1D).

2.3. Embedding and Cryosectioning of Entire Femur of the Rat

RNase-free instruments, materials and buffers were used to collect bone samples. After euthanasia of the rat, the whole femur was cleaned rapidly and a part of the muscles around the bone was kept. The femur was placed at the bottom of the embedding mold and covered with cryomount medium (CM) (00890-EX, HistoLab, Askim, Norvege). Samples were snap-frozen with 2-methylbutan cooled in liquid nitrogen to obtain a block (Figure 1E). When entirely frozen, the sample was transferred on dry ice to −80 °C, wrapped in foil aluminum. Then, it was stored at −80 °C until further processing. A cryostat (Cryostat FSE Shandon, Thermo Electron Corporation, San Diego, CA, USA) with a Leica CryoJane (9194701, Leica) system was used for cryosectioning. The block was fixed with cryomount medium (Figure S1A–D). Tools were precooled within the cryostat to avoid warming up the sample during block trimming (Figure 1F). The surface of the adhesive film (39475214, Leica) was adapted to the surface of the block (Figure 1G) and CryoJane tape transfer system was applied (39475205, Leica) to obtain high-quality sections.

The following points are crucial to the success of the experiment:

Briefly: (i) Positioning the adhesive film on the surface of the block (Figure S1E). (ii) Applying the roller while exerting a certain force to improve the adhesion of the film (Figure S1F). (iii) Cutting the sample slowly and uninterruptedly (Figure S1G). Obtaining 5 µm tissue sections on the adhesive film (Figures 1H and S1H). (iv) Transferring the section from the film to a standard histological slide, manually pretreated with Leica's A and B solutions (Figure S1I). (v) Fixation of the section on the coated side of the slide by CryoJane UV flash system (Figure S1J–L). (vi) Removal of the adhesive film (Figure S1M) and optimization the transfer of the cryo-section (Figure S2A–F). Four points are decisive to ensure good quality of the cryosections: (1) The CM block is very important for a good section (the tissue alone does not adhere well to the film). The film must be in contact with the CM and with the tissue. (2) The entire surface of the adhesive film must be in contact with the CM block. The film must not be wider than Sections 3 and 4. Very important is to hold the bottom of the film with a pair of pliers when cutting—this avoids heating the film which must remain cold and is safer (Figure 1M).

2.4. Histological Staining

Hematoxylin and phloxin (HP) staining was performed as followed: the sections were incubated in several successive baths: 40 s in hemalum (11487, Merck, Darmstadt, Germany) buffer (0.2 g hemalum, 5 g aluminum potassium sulfate in 100 mL distilled water), 3 min in water, 30 s in phloxin (15926, Merck) buffer (0.5 g phloxin in 100 mL distilled water), 1 min in water, 2 min in 70% ethanol, 30 s in 95% ethanol, 1 min in 100% ethanol, 1 min in 100% ethanol. At the end, nuclei were colored in blue and cytoplasm in pink (Figure 1I–K).

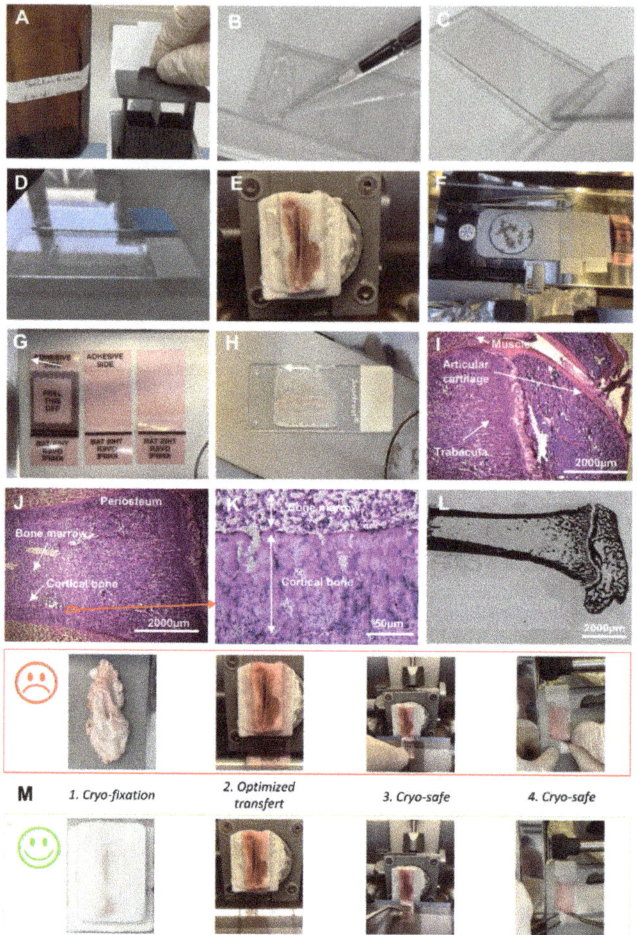

Figure 1. Cryo-fixation and cryo-section steps. Slides were manually coated with (**A**) A and (**B**) B buffer. (**C**) Thin B buffer layer on the slide surface. (**D**) Adjusted coating of the manually prepared slide compared with the manufactured one. (**E**) Entire cryo-embedded and trimmed femur of rat. (**F**) Cryo-bar to precool slides, adhesive film and tweezers. (**G**) Several surface size of adhesive film. (**H**) Thin (5 μm) cryo-section of femur deposed on a pre-coated slide. (**I,J**) Histological coloration of femur stained with Hemalin–phloxine–saffron. (**K**) Expanded view: high magnification image of the area within the red rectangle in image (**J**). (**L**) Von Kossa staining of the mouse and rat bone. (**M**) Upper panel: technical practices to avoid; down panel: to improve the result in comparison with the appropriate practices.

Von Kossa staining was performed as followed: Sections were rinsed with distilled water and incubated for 30 min in the dark with silver nitrate solution (1 g silver nitrate in 100 mL distilled water) at room temperature. After washes with distilled water, the sections were incubated under UV for one hour (sections should be covered with distilled water). Finally, the short passage in 95% and 100% ethanol, followed by xylene, were carried out. For the good conservation of the cryo-section, Eukitt mounting solution was used (Figure 1L).

2.5. Immunofluorescence

Sections of rat femur were fixed in paraformaldehyde (PFA (P6148, Sigma, Lezennes, France); 4% (*w/v*) in PBS (Phosphate-Buffered Salin without Ca and Mg, GAUPBS0001, Eurobio, Les Ulis, France)). After three washes in PBS, the sections were permeabilized for 15 min with PBS containing 0.5% Triton X100 (*v/v*). The non-specific binding sites were blocked with Emerald Antibody Diluent (Sigma 936B-08) for 1 h. The sections were incubated overnight at 4 °C with the primary rabbit anti-CD68 (ab125212, Abcam, Amsterdam, The Netherlands) antibody at 1:1000 dilutions. Then they were washed in PBS and incubated with the secondary anti-rabbit Alexa Fluor 488 (A-21206, Thermo Scientific, Villebon sur Yvette, France) antibody at 1:500 dilution for 2 h at room temperature. Finally, sections were washed in PBS for 20 min and mounted using a Fluoroshield mounting medium with DAPI (Abcam, ab104139). Fluorescence was detected using an epifluorescence microscope DM6000 (Leica, Germany) equipped with monochrome and color digital cameras.

2.6. In Situ-DIG Hybridization

For in situ-DIG hybridization, a couple of primers were chosen to get an approximately 1000 bp (β-actin (944 bp, CD68 1059 bp) PCR amplicon (Table S1A). β-actin and CD68 DNA primers were designed using ApE software (Table S1B). Two other primers, including 62 bp of the PCR product, were designed to recognize the T3 and T7 promoter sequences to perform the in vitro transcription (Figure 2). DNA oligos were synthesized by Eurogentec and were dissolved in ddH2O and stored at −20 °C.

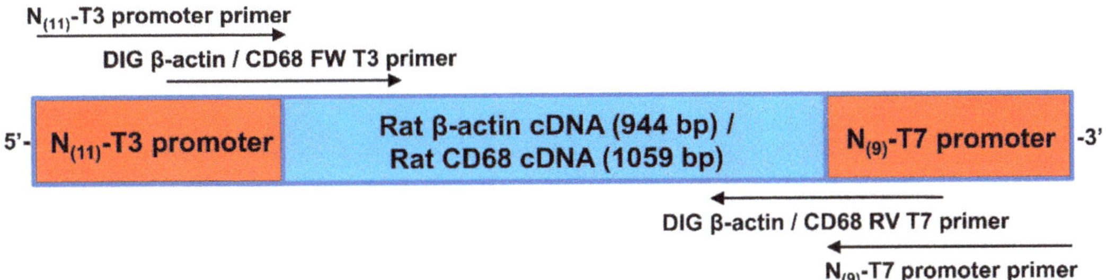

Figure 2. Schematic representation of the PCR product (together with primer localization and orientation) used for in vitro transcription.

2.6.1. cRNA Probes for In Situ-DIG Hybridization

Frozen femur samples were homogenized in liquid nitrogen. RNA was isolated using RNeasy Fibrous Tissue mini kit (HB-0485, Qiagen, Courtaboeuf, France) according to the manufacturer's recommendations. RNA extracts were eluted with 20 µL RNase-free water. Reverse transcription (RT) was performed with oligo(dT) primers following the instructions of the Sensiscript transcription kit (205211, Qiagen). Reactions were carried out using 50 ng of RNA with 10 µM oligo(dT) primers, RNase inhibitor (2 IU) and Sensiscript reverse transcriptase. cDNA was synthesized at 37 °C during 60 min. This cDNA library was used for the amplification of the β-actin templates for in vitro transcription. These templates contained the T3 and T7 promoters. The cRNA labeling was generated by in vitro transcription with T3 and T7 RNA polymerases, both in antisense and sense direction. Sense probes were used as controls. The PCR fragments were purified by agarose gel electrophoresis and specific bands were isolated using PCR Clean-Up kit (740,609.10, Macherey-Nagel, Hoerdt, France).

Next, RNA was labeled using an in vitro transcription Kit (P1450, Promega, Charbonnières les Bains, France) according to manufacturer's recommendations. RNA was labeled with digoxigenin, by addition of a modified nucleotide, DIG-11-UTP. Transcription was carried out in buffer containing dNTPs, DIG-11-UTP, RNase inhibitor and RNA polymerase at 37 °C for 1 h. DNAse was then added and the mix further incubated for 30 min at 37 °C to degrade DNA.

A total of 10 µL (10 mg mL^{-1}) transfer RNAs (1010945001, Roche, Boulogne-Billancourt, France) were added to the mix and the probes were precipitated with 10 M ammonium acetate (A1542, Sigma) and 100% cold ethanol overnight at −20 °C. After centrifugation at 15,000× g at 4 °C for 30 min, the pellet was rinsed with 70% cold ethanol. To improve probe penetration for probes longer than 500 nucleotides, hydrolysis of the probe is needed. For hydrolysis, the probe was suspended in 50 µL carbonate buffer (120 mM Na$_2$CO$_3$, 80 mM NaHCO$_3$, pH 10.2) and incubated at 60 °C for 55 min. The reaction was stopped with a buffer containing 10 µL of 10% acetic acid, 12 µL 3 M sodium acetate (pH 4.8) and 312 µL 100% cold ethanol and incubated at −20 °C for at least 30 min. After centrifugation, the pellet was washed with 70% cold ethanol and suspended in 50% RNase-free formamide.

The effectiveness of labeling was analyzed by dot-blots. cRNA were dot-blotted on a nitrocellulose membrane (88018, Thermo Fisher Sci., Illkirch, France), and detected with an anti-digoxigenin (11093274910, Roche) DIG-specific antibody. From each serial dilution of the probes, 1 µL was spotted on a nitrocellulose membrane, dried and UV-crosslinked for 1 min. To prevent nonspecific antibody binding, the membrane was blocked with 1% bovine serum albumin (BSA; GAUBSA01, Eurobio, Les Ulis, France) in 100 mM Tris pH 7.5 and 150 mM NaCl (Tris-NaCl) buffer for 15 min. Afterwards the membrane was incubated for 30 min with anti-DIG antibody at 1:2000 dilutions in BSA/Tris-NaCl and then washed three times for 5 min in BSA/Tris-NaCl and once with Tris-NaCl. Finally, the membrane was stained with an NBT kit (NBT/BCIP; S3771, Roche) according to the manufacturer's recommendations.

2.6.2. Fixation and Pretreatment of Sections for In Situ-DIG Hybridization

All the following steps were performed under a laminar flow cabinet and under RNase-free conditions. The tissues were fixed in 4% (w/v) in PBS/paraformaldehyde (PBS, w/o Ca and Mg, GAUPBS0001, Eurobio; PFA, P6148, Sigma) for 30 min, and then treated with 100% methanol for 15 min and air-dried. Sections were incubated in 0.125 mg mL^{-1} Proteinase K (P2308, Sigma) in 200 mL 100 mM Tris pH 7.5 and 50 mM EDTA buffer for 10 min at 37 °C to degrade the proteins and to improve the probes' access to the target mRNA. Proteinase K reaction was stopped with 0.2% glycine (G7126, Sigma) in 1X PBS. Then sections were treated with 0.5% acetic anhydride (A6404, Sigma) in triethanolamine solution (0.1 M pH 8) to avoid non-specific hybridization. The sections were then washed twice for 2 min with PBS and dehydrated with successive baths of saline solution and ethanol: 30 s in 30% ethanol, 0.85% NaCl buffer; 30 s in 50% ethanol, 0.85% NaCl buffer; 30 s in 75% ethanol, 0.85% NaCl buffer; 30 s in 85% ethanol, 0.42% NaCl buffer; 30 s in 96% ethanol; 30 s in 96% ethanol; 1 min in 100% ethanol. Slides were then stored at −20 °C.

2.6.3. Prehybridization and Hybridization for In Situ-DIG

The sections were pre-hybridized for 2 h at 45 °C in a pre-hybridization buffer (50% formamide (GHYFOR0402, Eurobio), 0.5× sodium chloride citrate (SSC) (GHYSSC007, Eurobio) buffer, 50 µg mL^{-1} heparin (H3393, Sigma), 100 µg mL^{-1} transfer RNA and 0.1% (v/v) Tween 20 (822184, Merck). Finally, the sections were incubated overnight at 45 °C with the RNA probes (2 µL probe in 200 µL hybridization buffer (50% formamide, 100 µg mL^{-1} transfer RNA, 7.5% (v/v) Tween 20, 8.5% NaCl, 20% dextran sulfate (Eurobio GHYDEX000T) and 2.5× Denhardt's Solution (50× stock, D2532, Sigma)), which were previously denaturized for 2 min at 80 °C in the hybridization buffer.

Non-specific hybrids were dissociated with following washes: 30 min in 0.1× SSC + 0.5% SDS at 45 °C, 2 h in 2× SSC + 50% formamide at 45 °C, 5 min in NTE (0.5 M NaCl, 10 mM Tris pH 8, 1 mM EDTA) at 45 °C, 30 min in NTE + 10 mg ml-1 Rnase A (10109169001, Roche) at 37 °C, 1 h in 2× SSC + 50% formamide at 45 °C, 2 min in 0.1× SSC at 45 °C and finally 15 min in PBS at RT.

2.6.4. Detection for In Situ-DIG

Immunodetection of the DIG-labeled probes was performed using an anti-DIG antibody coupled to alkaline phosphatase, as described by the manufacturer (11093274910, Roche). For the immunological detection step, the sections were incubated in a first buffer (0.5% Blocking reagent (1110961176001, Roche) in 100 mM Tris pH 7.5 and 150 mM NaCl) for 1 h and in a second one (1% BSA in 0.5% (v/v) Triton X100, 100 mM Tris pH 7.5 and 150 mM NaCl) for 1 h to block unspecific sites. Sections were incubated with anti-digoxygenin antibody 1:1250 for 1 h, and then washed three times for 20 min with 1% BSA in 0.5% (v/v) Triton X100, 100 mM Tris pH 7.5 and 150 mM NaCl solution and next incubated for 15 min in the same solution without BSA. Finally, this solution was replaced with the last buffer (100 mM Tris pH 9.5, 100 mM NaCl and 50 mM MgCl$_2$) for 15 min.

Staining was initiated at alkali pH. The sections were incubated for 1–2 days in a buffer containing 337 µL BCIP (5-bromo-4-chloro-3-indolyl-phosphate) and 225 µL NBT (Nitroblue tetrazolium chloride) in 50 mL solution (100 mM Tris pH 9.5, 100 mM NaCl and 50 mM MgCl$_2$) until a blue precipitate adhering to the sections was formed. The reaction was stopped by adding a stop solution (10 mM Tris pH 7.5 and 5 mM EDTA) for 10 min. The DIG sections were observed with an epifluorescence microscope DM6000 (Leica, Germany) equipped with monochrome and color digital cameras while the HCR sections were observed with a confocal microscope (LSM700, Zeiss, Dresden, Germany).

2.7. In Situ-HCR Hybridization

The HCR protocol of Choi and colleagues (2014, 2016) was performed with some modifications as described below to enhance mRNA localization in the femur of the rat [21,26].

2.7.1. β-Actin Probe Design for In Situ-HCR Hybridization

The probes were designed using ApE software. For each gene, five probes were designed. The entire gene sequence was used to localize the introns, and probes were designed exactly at the boundaries between two exons. This approach increased the capacity of probes to hybridize with the mRNA and not with the genomic DNA. A specific additional sequence was included to interact with the hairpin coupled with the fluorophore [21]. DNA oligos were synthesized by Eurogentec. Details of probe sequences are described in Table S1B. All oligos were dissolved in ddH2O and stored at −20 °C.

2.7.2. Fixation and Pretreatment of Sections for In Situ-HCR Hybridization

This process was common for both DIG and HCR in situ hybridization methods.

2.7.3. Prehybridization and Hybridization for In Situ-HCR

The sections were pre-hybridized for 10 min at RT in a hybridization buffer (50% formamide, 5× SSC, 9 mM citric acid pH 6, 50 µg mL^{-1} heparin, 1× Denhardt's Solution, 0.1% (v/v) Tween 20 and 10% dextran-sulfate). Previously, the hybridization probes (2 pmol per slide) were denatured for 2 min at 80 °C. Finally, the sections were incubated in a hybridization buffer together with probes overnight at 45 °C. Nonspecific hybrids were dissociated with the following washes: 30 min in 0.1× SSC + 0.5% SDS at 45 °C, followed by 2 h in 2× SSC + 50% formamide at 45 °C, and then 2 min in 0.1× SSC at 45 °C and finally 15 min in PBS at RT.

2.7.4. Detection for In Situ-HCR

Sections were first incubated for 2 h at RT with an amplification buffer (5× SSC, 0.1% (v/v) Tween 20, 10% dextran-sulfate and 100 µg mL^{-1} salmon sperm ADN) and subsequently for 12 to 16 h with the DNA hairpins marked with a fluorophore (Alexa Fluor488) (diluted in amplification buffer, as described previously). The hairpins were previously heated at 95 °C for 90 s and cooled to RT for 30 min. The sections were then washed 2× 30 min in 5× SSCT (5× SSC and 1% (*v/v*) Tween 20) and 5 min with 5× SSC without Tween at RT.

2.8. Microspectrofluorimetry

Emission fluorescence spectra was measured between 460 and 650 nm (5 nm bandwidth) with a Leica TCS SP8 Confocal Microscope. Cyan fluorescence was excited at 488 nm. Washes: 30 min in 0.1× SSC + 0.5% SDS at 45 °C, followed by 2 h in 2× SSC + 50% formamide at 45 °C, and then 2 min in 0.1× SSC at 45 °C and finally 15 min in PBS at RT.

3. Results and Discussion

The particular mechanism of bone regeneration is under active examination. However, preparing a histology section from bone is quite difficult due to the mineralization of this tissue. Mineralized tissues should be decalcified from 1 to 2 weeks before embedding and sectioning. Decalcification can alter the antigenicity of certain proteins and can cause degradation of RNA molecules. Therefore, immunohistochemistry and ISH are less possible on the sections [35–37]. The success of the ISH technique realization extremely depends on the quick preparation of good-quality bone tissue sections. The aim of our work was to develop a new approach for obtaining high-quality undecalcified bone sections applicable to various ISH analyses.

Because ISH is based on mRNA analysis, it is essential to develop a procedure for maintaining an RNase-free lab (RNase-free instruments and materials, wear gloves and work quickly to avoid storing samples at RT) so as to conserve mRNAs of sufficient quality and quantity for subsequent analyses. Rapid techniques without any prolonged aqueous phase steps are crucial to prevent RNA degradation. Manual coating of slides greatly helped to optimize sample attachment to the slide during in situ hybridization (Figure 1A–D), with the area of coating to be adjusted to the sample area (Figure 1D).

First, the entire femur together with muscle was embedded in a cryo-embedding medium, frozen and trimmed (Figure 1E). All tools were maintained cold into the cryo-bar (Figure 1F). Several shapes of adhesive film were prepared to fix equivalent surfaces of cryosections (Figure 1G). Thin (5 μm) cryosections were cut (Figures 1H and S1A–M), stained first with HP (Figure 1I–K) and then with von Kossa medium (Figure 1L) to confirm the capacity of the cryosection technique to retain morphological structures and mineralization of the bone. To optimize the cryosectioning and transfer, four technical points were described in detail (Figure 1M).

Subsequently, the bone sections were analyzed using a conventional immunofluorescence technique (with anti-CD68 antibody) to identify the macrophages (Figure 3). One of the most widely used markers for the analysis of monocytes/macrophages is the Cluster of Differentiation (CD) CD68 protein [38,39]. Although weakly expressed in other mononuclear phagocyte cells, this glycoprotein is highly expressed in macrophages. Very weak expression can be detected in other non-hematopoietic cells (mesenchymal stem cells, fibroblast, endothelial and tumor cells) [40,41]. Monocytes/macrophages were detected in the bone marrow (Figure 3A), and in the interface between periosteum and cortical bone (Figure 3B). Osteoclast and macrophages have similar origins and both produce CD68 protein [10,42,43]. We identified the presence of osteoclasts as multinucleated cell expressing CD68 protein (Figure 3B). No expression was identified in the negative control (Figure 3C,D).

Next, ISH with a digoxigenin-labeled probe (in situ-DIG) was performed. Specific probes were designed and synthetized to target the mRNAs of β-actin in bone, a ubiquitous protein with a strong expression in almost every cell [44–46], then a 944 bp complementary RNA (cRNA) was generated (Figure 2, Table S1A). An mRNA probe was chosen because the RNA–RNA interaction is more efficient than the RNA–DNA interaction [20].

The cellular localization of the β-actin was analyzed by the in situ-DIG method (Figure 4). When the RNA probe used for hybridization was the same sense (not complementary) as the mRNA (negative control), no labeling was observed (Figure 4A,C), whereas the expression of β-actin was very intense in the bone marrow (Figure 4B) and in the periosteum (Figure 4D).

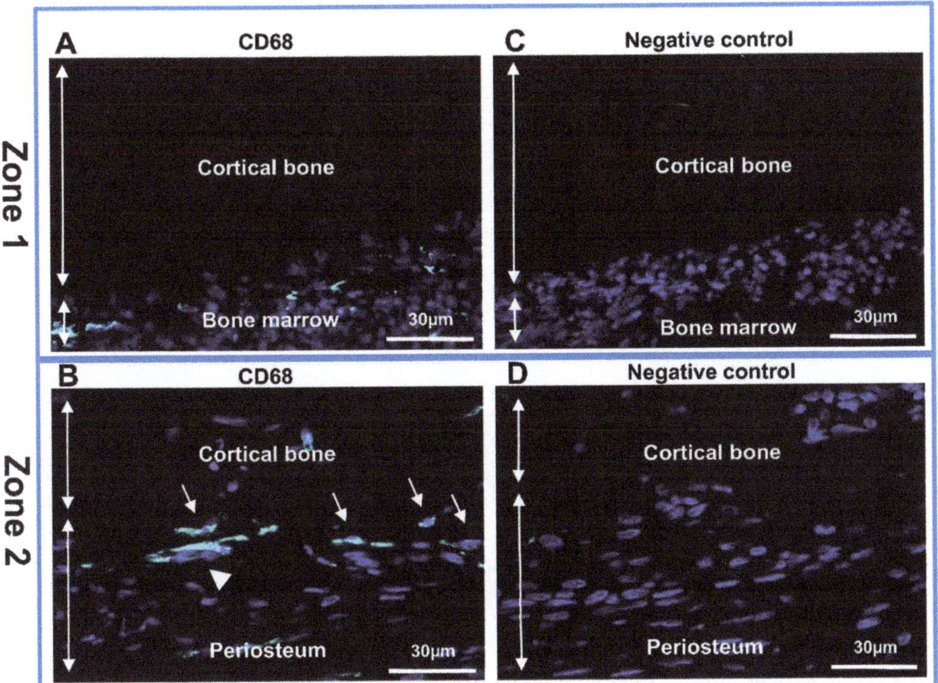

Figure 3. Identification of macrophages and osteoclast in rat femurs. Immunolabeling with anti-CD68 antibody in sections of two representative zones of the sample. (**A,C**) Zone 1-cortical bone and bone marrow; (**B,D**) Zone 2-periosteum and cortical bone. (**A,B**) Anti-CD68 (Alexa488, turquoise fluorescence), labeling the macrophages and osteoclasts. (**C,D**) Negative control. Nuclear staining with DAPI (blue fluorescence). Thin arrow: macrophages; arrowhead: osteoclast.

In situ hybridization coupled with hybridization chain reaction detection (in situ-HCR) was chosen because this approach is more sensitive than fluorescence in situ hybridization (FISH) [18,47,48] and allows identification of several genes at the same time [16,21,28,49]. Indeed, β-actin expression was analyzed in a rat bone animal model, using a DNA probe linked to a fluorophore instead of an enzyme. In the absence of a probe (negative control), no labeling was observed (Figure 4E,G), whereas a strong labeling was detected both in bone marrow (Figure 4F) and in the periosteum (Figure 4H). DAPI was used as nuclear counterstain.

To go further with our investigation, the identification of macrophages we used both the in situ-DIG and HCR techniques. Macrophages were identified based on the expression of CD68 mRNA (Figure 5). CD68 mRNA expression was strong in the bone marrow (Figure 5A,C) and in the interface between the periosteum and cortical bone (Figure 5B,D) with both in situ techniques.

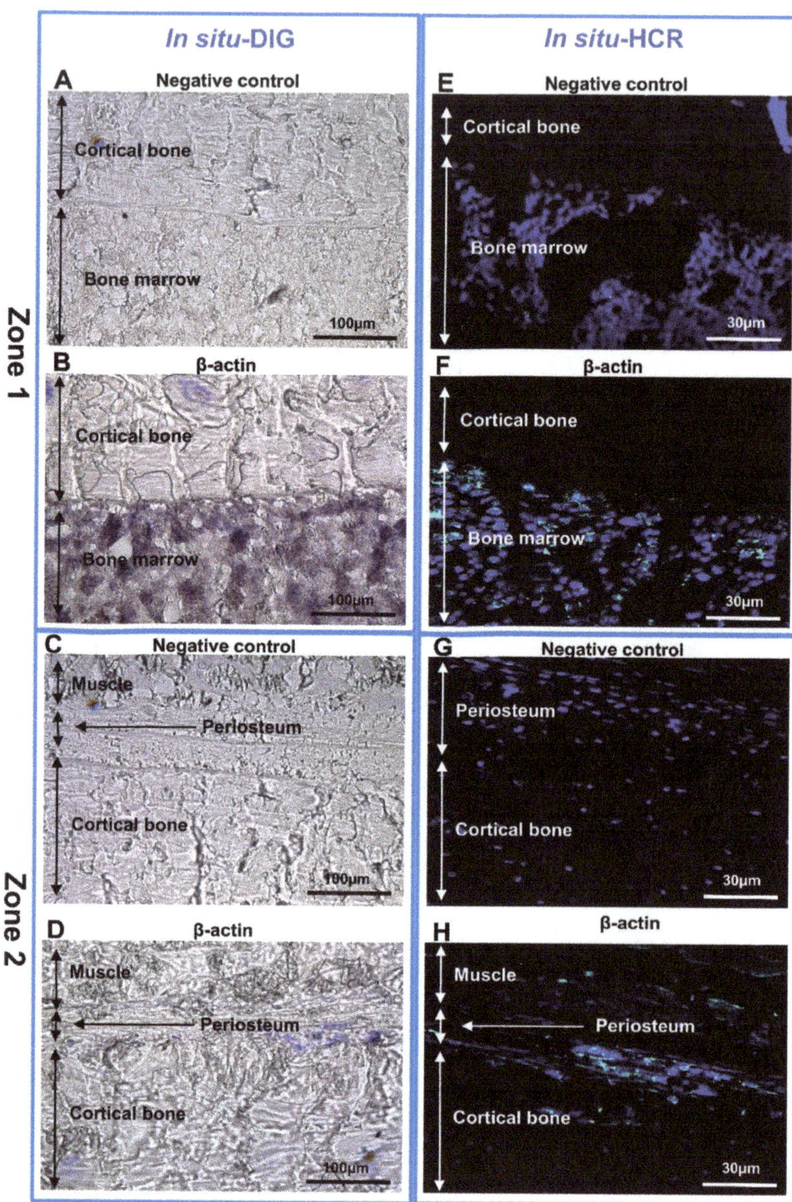

Figure 4. In situ hybridization analysis of non-decalcified rat bone section. (**A–H**) β-actin expression in sections of two representative zones of the sample. (**A,B,E,F**) Zone 1-cortical bone and bone marrow; (**C,D,G,H**) Zone 2-muscle, periosteum and cortical bone. (**A–D**) In situ-DIG; (**E–H**) in situ-HCR. (**A,C,E,G**) Negative control. (**B,D,F,H**) Expression of β-actin mRNA in non-decalcified rat bone.

Autofluorescence of bone tissue was previously described [50–52]. To confirm that the detected fluorescence resulted from an in situ-HCR signal and did not derive from nonspecific binding of an Alexa Fluor 488 molecule or from autofluorescence, hybridization was evaluated by microspectrofluorimetry (Figure 6). The emission spectra of the fluorescence peaked at 520 nm, corresponding to Alexa Fluor 488 fluorophore spectra.

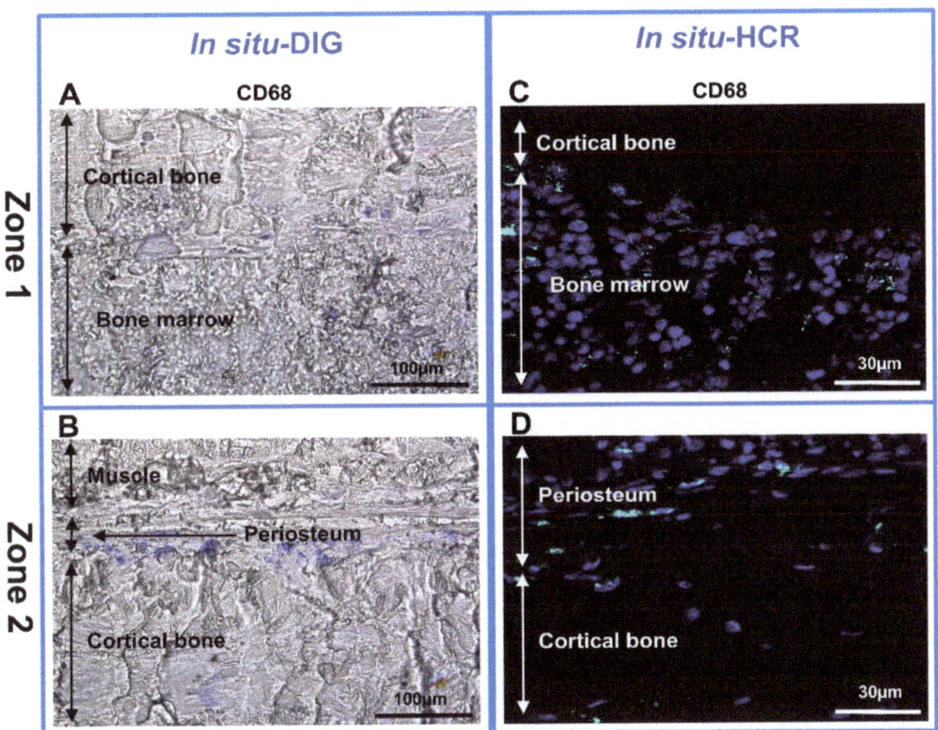

Figure 5. In situ hybridization analysis in non-decalcified rat bone section. (**A–D**) CD68 mRNA expression in section of two representative zones of the sample. (**A,C**) Zone 1-cortical bone and bone marrow. (**B,D**) Zone 2-muscle, periosteum and cortical bone. (**A,B**) In situ-DIG; (**C,D**) in situ-HCR.

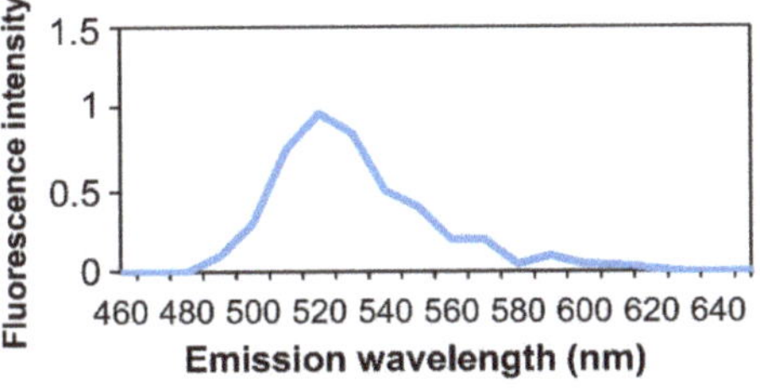

Figure 6. Microspectroscopy analysis of non-decalcified rat bone section. In situ microspectroscopy analysis of the Alexa Fluor 488 fluorescence.

ISH analysis is an important technique used in order to understand the molecular mechanism of bone tissue regeneration. Our aim was to analyze the section of the entire rat femur. Paraffin embedding was not appropriate in our case, because the muscle slowed down the penetration of the decalcifying solution (EDTA) into the bone. Thus, the decalcifying treatment was very long (around 5 weeks), which prevented the ISH on these samples (data not shown). We developed an improved version of the CryoJane tape transfer system to prepare cryosections of undecalcified rat femurs with good bone tissue morphology and applicable for ISH analyses. We demonstrated that in situ-HCR is a promising new technique for visualizing macrophages and studying the expression of different genes in bone tissue.

4. Conclusions

Bone healing is very complex process. For proper regeneration, dynamic interplay between external and internal signals is essential. Cytokines of the macrophages and other cellular factors act at diverse times, and have indispensable functions during repair. In order to make progress in bone regeneration understanding, the development of new tools is essential. These tools will provide opportunities to explore in situ the spatial actors involved in inflammation and bone tissue regeneration. Indeed, the application of distinct ISH approaches can bring new comprehension to bone gene expression and tissue regeneration. Combination of cryofixation with ISH techniques is a relevant approach to study the molecular and spatial biological mechanisms of bone regeneration and provides advanced perspectives in the field of regenerative medicine to induce bone regeneration for developing new treatments.

Supplementary Materials: The following supporting information can be downloaded at: https://www.mdpi.com/article/10.3390/biomedicines10020484/s1. Figure S1: Workflow of cryo-sample preparation. (A) Covering mandrel with cryomount solution. (B) Positioning the block on the mandrel. (C) Hardening of the cryomount and fixation of the sample on the mandrel. (D) Trimming of the sample. (E) Positioning of the adhesive film on the surface of the block. (F) Application of the roller to improve the adhesion of the film. (G) Cutting the sample slowly and uninterruptedly by holding the lower part of the adhesive film. (H) Obtaining of 5 μm tissue section on adhesive film. (I) Positioning the adhesive film on the precoated slide. (J) Application of the roller to improve the adhesion to the slide. (K) Positioning the slide in CryoJane flash unit. (L) Triggering two flashes at 30 s intervals. (M) Removing the film with cold tool. Figure S2: Workflow to optimize the transfer of the cryo-section. (A) Block trimming to adjust to the sample surface. (B) Positioning the block on the mandrel. (C) Estimation of the adhesive film surface. (D) Cutting of the adhesive film. (E) Positioning of the adhesive film on the surface of the block. (F) Adhesion of the film on the surface of the sample. Table S1: (A) The cDNA sequence of the β-actin used for in vitro transcription. The T3 and T7 promoter sequences labeled with red. (B) Oligos used for in situ hybridization experiments. T3 (T3 promoter), T7 (T7 promoter), FW (forward primer), RV (reverse primer).

Author Contributions: C.C. performed cryofixation, cryosectionning of bone; K.N. and H.M. performed ISH and HCR experimentations, designed probes and produced cDNA; M.D. performed rat animal model experimentation; C.C. and K.N. collected the data; X.H., K.N. and A.-L.F. designed the work; K.N., A.-L.F. and X.H. performed data analysis, interpretation; K.N. and A.-L.F. drafted the paper; C.C., K.N., X.H., M.D., H.M., A.B. and A.-L.F. approved the final version to be published. All authors have read and agreed to the published version of the manuscript.

Funding: Work was supported by the Délégation Générale de l'Armement (DGA) (PDH2-NRBC-4-NR-4306).

Institutional Review Board Statement: This study was approved by the French Army Animal Ethics Committee (N°2011/22.1). All rats were treated in compliance with the European legislation (dir 2010/63/EU) implemented into French law (decree 2013-118) regulating animal experimentation.

Acknowledgments: We are very grateful to Martine Miquel for her helpful advice and critical reviewing of our manuscript. We are thankful to Zsolt Kelemen for his invaluable help on microspectrofluorimetry. We thank the Imagery platform of INRA-Versailles for technical assistance on confocal microscopy. This work was supported by the Service de Santé des Armées and a grant (NBC-4-NR-4306) from Direction Générale de l'Armement (DGA, Paris, France).

Conflicts of Interest: Authors declare no conflict of interest.

References

1. Kaur, S.; Raggatt, L.J.; Batoon, L.; Hume, D.A.; Levesque, J.-P.; Pettit, A.R. Role of Bone Marrow Macrophages in Controlling Homeostasis and Repair in Bone and Bone Marrow Niches. *Semin. Cell Dev. Biol.* **2017**, *61*, 12–21. [CrossRef] [PubMed]
2. Horwood, N.J. Macrophage Polarization and Bone Formation: A Review. *Clin. Rev. Allergy Immunol.* **2016**, *51*, 79–86. [CrossRef] [PubMed]
3. Tan, B.; Tang, Q.; Zhong, Y.; Wei, Y.; He, L.; Wu, Y.; Wu, J.; Liao, J. Biomaterial-Based Strategies for Maxillofacial Tumour Therapy and Bone Defect Regeneration. *Int. J. Oral Sci.* **2021**, *13*, 9. [CrossRef] [PubMed]

4. Muñoz, J.; Akhavan, N.S.; Mullins, A.P.; Arjmandi, B.H. Macrophage Polarization and Osteoporosis: A Review. *Nutrients* **2020**, *12*, 2999. [CrossRef]
5. Weitzmann, M.N. Bone and the Immune System. *Toxicol. Pathol.* **2017**, *45*, 911–924. [CrossRef]
6. Batoon, L.; Millard, S.M.; Raggatt, L.J.; Pettit, A.R. Osteomacs and Bone Regeneration. *Curr. Osteoporos. Rep.* **2017**, *15*, 385–395. [CrossRef]
7. Lebaudy, E.; Fournel, S.; Lavalle, P.; Vrana, N.E.; Gribova, V. Recent Advances in Antiinflammatory Material Design. *Adv. Healthc. Mater.* **2021**, *10*, e2001373. [CrossRef]
8. Eggold, J.T.; Rankin, E.B. Erythropoiesis, EPO, Macrophages, and Bone. *Bone* **2019**, *119*, 36–41. [CrossRef]
9. Pajarinen, J.; Lin, T.; Gibon, E.; Kohno, Y.; Maruyama, M.; Nathan, K.; Lu, L.; Yao, Z.; Goodman, S.B. Mesenchymal Stem Cell-Macrophage Crosstalk and Bone Healing. *Biomaterials* **2019**, *196*, 80–89. [CrossRef]
10. Sun, Y.; Li, J.; Xie, X.; Gu, F.; Sui, Z.; Zhang, K.; Yu, T. Macrophage-Osteoclast Associations: Origin, Polarization, and Subgroups. *Front. Immunol.* **2021**, *12*, 778078. [CrossRef]
11. Chang, M.K.; Raggatt, L.-J.; Alexander, K.A.; Kuliwaba, J.S.; Fazzalari, N.L.; Schroder, K.; Maylin, E.R.; Ripoll, V.M.; Hume, D.A.; Pettit, A.R. Osteal Tissue Macrophages Are Intercalated throughout Human and Mouse Bone Lining Tissues and Regulate Osteoblast Function in Vitro and in Vivo. *J. Immunol.* **2008**, *181*, 1232–1244. [CrossRef] [PubMed]
12. Miron, R.J.; Bosshardt, D.D. OsteoMacs: Key Players around Bone Biomaterials. *Biomaterials* **2016**, *82*, 1–19. [CrossRef] [PubMed]
13. Batoon, L.; Millard, S.M.; Wullschleger, M.E.; Preda, C.; Wu, A.C.-K.; Kaur, S.; Tseng, H.-W.; Hume, D.A.; Levesque, J.-P.; Raggatt, L.J.; et al. CD169($^+$) Macrophages Are Critical for Osteoblast Maintenance and Promote Intramembranous and Endochondral Ossification during Bone Repair. *Biomaterials* **2019**, *196*, 51–66. [CrossRef]
14. Altan-Bonnet, G.; Mukherjee, R. Cytokine-Mediated Communication: A Quantitative Appraisal of Immune Complexity. *Nat. Rev. Immunol.* **2019**, *19*, 205–217. [CrossRef] [PubMed]
15. Liu, H.; Zhu, R.; Liu, C.; Ma, R.; Wang, L.; Chen, B.; Li, L.; Niu, J.; Zhao, D.; Mo, F.; et al. Evaluation of Decalcification Techniques for Rat Femurs Using HE and Immunohistochemical Staining. *Biomed. Res. Int.* **2017**, *2017*, 9050754. [CrossRef] [PubMed]
16. Nikovics, K.; Morin, H.; Riccobono, D.; Bendahmane, A.; Favier, A. Hybridization-chain-reaction Is a Relevant Method for in Situ Detection of M2d-like Macrophages in a Mini-pig Model. *FASEB J.* **2020**, *34*, 15675–15686. [CrossRef] [PubMed]
17. Sicherre, E.; Favier, A.-L.; Riccobono, D.; Nikovics, K. Non-Specific Binding, a Limitation of the Immunofluorescence Method to Study Macrophages In Situ. *Genes* **2021**, *12*, 649. [CrossRef]
18. Gupta, R.; Cooper, W.A.; Selinger, C.; Mahar, A.; Anderson, L.; Buckland, M.E.; O'Toole, S.A. Fluorescent In Situ Hybridization in Surgical Pathology Practice. *Adv. Anat. Pathol.* **2018**, *25*, 223–237. [CrossRef]
19. Chu, Y.-H.; Hardin, H.; Zhang, R.; Guo, Z.; Lloyd, R.V. In Situ Hybridization: Introduction to Techniques, Applications and Pitfalls in the Performance and Interpretation of Assays. *Semin. Diagn. Pathol.* **2019**, *36*, 336–341. [CrossRef]
20. Jensen, E. Technical Review: In Situ Hybridization. *Anat Rec.* **2014**, *297*, 1349–1353. [CrossRef]
21. Choi, H.M.T.; Beck, V.A.; Pierce, N.A. Next-Generation in Situ Hybridization Chain Reaction: Higher Gain, Lower Cost, Greater Durability. *ACS Nano* **2014**, *8*, 4284–4294. [CrossRef] [PubMed]
22. Looi, L.M.; Cheah, P.L. In Situ Hybridisation: Principles and Applications. *Malays. J. Pathol.* **1992**, *14*, 69–76.
23. Veselinyová, D.; Mašlanková, J.; Kalinová, K.; Mičková, H.; Mareková, M.; Rabajdová, M. Selected In Situ Hybridization Methods: Principles and Application. *Molecules* **2021**, *26*, 3874. [CrossRef] [PubMed]
24. Schramm, A.; De Beer, D.; Wagner, M.; Amann, R. Identification and Activities in Situ of Nitrosospira and Nitrospira Spp. as Dominant Populations in a Nitrifying Fluidized Bed Reactor. *Appl. Environ. Microbiol.* **1998**, *64*, 3480–3485. [CrossRef] [PubMed]
25. Nikovics, K.; Favier, A.-L. Macrophage Identification In Situ. *Biomedicines* **2021**, *9*, 1393. [CrossRef] [PubMed]
26. Choi, H.M.T.; Calvert, C.R.; Husain, N.; Huss, D.; Barsi, J.C.; Deverman, B.E.; Hunter, R.C.; Kato, M.; Lee, S.M.; Abelin, A.C.T.; et al. Mapping a Multiplexed Zoo of MRNA Expression. *Development* **2016**, *143*, 3632–3637. [CrossRef] [PubMed]
27. Choi, H.M.T.; Schwarzkopf, M.; Fornace, M.E.; Acharya, A.; Artavanis, G.; Stegmaier, J.; Cunha, A.; Pierce, N.A. Third-Generation in Situ Hybridization Chain Reaction: Multiplexed, Quantitative, Sensitive, Versatile, Robust. *Development* **2018**, *145*, dev165753. [CrossRef]
28. Choi, H.M.T.; Schwarzkopf, M.; Pierce, N.A. Multiplexed Quantitative In Situ Hybridization with Subcellular or Single-Molecule Resolution within Whole-Mount Vertebrate Embryos: QHCR and DHCR Imaging (v3.0). *Methods Mol. Biol.* **2020**, *2148*, 159–178. [CrossRef]
29. Lei, Z.; van Mil, A.; Xiao, J.; Metz, C.H.G.; van Eeuwijk, E.C.M.; Doevendans, P.A.; Sluijter, J.P.G. MMISH: Multicolor MicroRNA in Situ Hybridization for Paraffin Embedded Samples. *Biotechnol. Rep.* **2018**, *18*, e00255. [CrossRef]
30. Kremer, M.; Quintanilla-Martínez, L.; Nährig, J.; von Schilling, C.; Fend, F. Immunohistochemistry in Bone Marrow Pathology: A Useful Adjunct for Morphologic Diagnosis. *Virchows Arch.* **2005**, *447*, 920–937. [CrossRef]
31. Rijntjes, N.V.; Van de Putte, L.B.; Van der Pol, M.; Guelen, P.J. Cryosectioning of Undecalcified Tissues for Immunofluorescence. *J. Immunol. Methods* **1979**, *30*, 263–268. [CrossRef]
32. Kawamoto, T. Light Microscopic Autoradiography for Study of Early Changes in the Distribution of Water-Soluble Materials. *J. Histochem. Cytochem.* **1990**, *38*, 1805–1814. [CrossRef] [PubMed]
33. Kawamoto, T. Use of a New Adhesive Film for the Preparation of Multi-Purpose Fresh-Frozen Sections from Hard Tissues, Whole-Animals, Insects and Plants. *Arch. Histol. Cytol.* **2003**, *66*, 123–143. [CrossRef] [PubMed]

34. Kawamoto, T.; Kawamoto, K. Preparation of Thin Frozen Sections from Nonfixed and Undecalcified Hard Tissues Using Kawamoto's Film Method (2020). *Methods Mol. Biol.* **2021**, *2230*, 259–281. [CrossRef]
35. Salie, R.; Li, H.; Jiang, X.; Rowe, D.W.; Kalajzic, I.; Susa, M. A Rapid, Nonradioactive in Situ Hybridization Technique for Use on Cryosectioned Adult Mouse Bone. *Calcif. Tissue Int.* **2008**, *83*, 212–221. [CrossRef]
36. Kramer, I.; Salie, R.; Susa, M.; Kneissel, M. Studying Gene Expression in Bone by in Situ Hybridization. *Methods Mol. Biol.* **2012**, *816*, 305–320. [CrossRef]
37. Yang, Y.; Liu, Q.; Zhang, L.; Fu, X.; Chen, J.; Hong, D. A Modified Tape Transfer Approach for Rapidly Preparing High-Quality Cryosections of Undecalcified Adult Rodent Bones. *J. Orthop. Translat.* **2021**, *26*, 92–100. [CrossRef]
38. Kosmac, K.; Peck, B.D.; Walton, R.G.; Mula, J.; Kern, P.A.; Bamman, M.M.; Dennis, R.A.; Jacobs, C.A.; Lattermann, C.; Johnson, D.L.; et al. Immunohistochemical Identification of Human Skeletal Muscle Macrophages. *Bio-Protocol* **2018**, *8*, e2883. [CrossRef]
39. Iqbal, A.J.; McNeill, E.; Kapellos, T.S.; Regan-Komito, D.; Norman, S.; Burd, S.; Smart, N.; Machemer, D.E.W.; Stylianou, E.; McShane, H.; et al. Human CD68 Promoter GFP Transgenic Mice Allow Analysis of Monocyte to Macrophage Differentiation in Vivo. *Blood* **2014**, *124*, e33–e44. [CrossRef]
40. Chistiakov, D.A.; Killingsworth, M.C.; Myasoedova, V.A.; Orekhov, A.N.; Bobryshev, Y.V. CD68/Macrosialin: Not Just a Histochemical Marker. *Lab. Investig.* **2017**, *97*, 4–13. [CrossRef]
41. Chistiakov, D.A.; Bobryshev, Y.V.; Orekhov, A.N. Changes in Transcriptome of Macrophages in Atherosclerosis. *J. Cell Mol. Med.* **2015**, *19*, 1163–1173. [CrossRef] [PubMed]
42. Udagawa, N.; Takahashi, N.; Akatsu, T.; Tanaka, H.; Sasaki, T.; Nishihara, T.; Koga, T.; Martin, T.J.; Suda, T. Origin of Osteoclasts: Mature Monocytes and Macrophages Are Capable of Differentiating into Osteoclasts under a Suitable Microenvironment Prepared by Bone Marrow-Derived Stromal Cells. *Proc. Natl. Acad. Sci. USA* **1990**, *87*, 7260–7264. [CrossRef] [PubMed]
43. Razafimahefa, J.; Gosset, C.; Mongiat-Artus, P.; Andriamampionona, T.F.; Verine, J. Stromal Osseous Metaplasia in Urothelial Carcinoma of the Bladder: A Rare Case Report and Literature Review. *Diagn. Pathol.* **2019**, *14*, 75. [CrossRef]
44. Gan, F.; Luk, G.D.; Gesell, M.S. Nonradioactive in Situ Hybridization Techniques for Routinely Prepared Pathology Specimens and Cultured Cells. *Null* **1994**, *17*, 313–319. [CrossRef]
45. Hoock, T.C.; Newcomb, P.M.; Herman, I.M. Beta Actin and Its MRNA Are Localized at the Plasma Membrane and the Regions of Moving Cytoplasm during the Cellular Response to Injury. *J. Cell Biol.* **1991**, *112*, 653–664. [CrossRef] [PubMed]
46. Taneja, K.L.; Singer, R.H. Detection and Localization of Actin MRNA Isoforms in Chicken Muscle Cells by in Situ Hybridization Using Biotinated Oligonucleotide Probes. *J. Cell Biochem.* **1990**, *44*, 241–252. [CrossRef]
47. Bertram, S.; Schildhaus, H.-U. Fluorescence in situ hybridization for the diagnosis of soft-tissue and bone tumors. *Pathologe* **2020**, *41*, 589–605. [CrossRef]
48. Frickmann, H.; Zautner, A.E.; Moter, A.; Kikhney, J.; Hagen, R.M.; Stender, H.; Poppert, S. Fluorescence in Situ Hybridization (FISH) in the Microbiological Diagnostic Routine Laboratory: A Review. *Crit. Rev. Microbiol.* **2017**, *43*, 263–293. [CrossRef]
49. Tsuneoka, Y.; Funato, H. Modified in Situ Hybridization Chain Reaction Using Short Hairpin DNAs. *Front. Mol. Neurosci.* **2020**, *13*, 75. [CrossRef]
50. Fauch, L.; Palander, A.; Dekker, H.; Schulten, E.A.; Koistinen, A.; Kullaa, A.; Keinänen, M. Narrowband-Autofluorescence Imaging for Bone Analysis. *Biomed. Opt. Express* **2019**, *10*, 2367–2382. [CrossRef]
51. Su, W.; Yang, L.; Luo, X.; Chen, M.; Liu, J. Elimination of Autofluorescence in Archival Formaldehyde-Fixed, Paraffin-Embedded Bone Marrow Biopsies. *Arch. Pathol. Lab. Med.* **2019**, *143*, 362–369. [CrossRef] [PubMed]
52. Capasso, L.; D'Anastasio, R.; Guarnieri, S.; Viciano, J.; Mariggiò, M. Bone Natural Autofluorescence and Confocal Laser Scanning Microscopy: Preliminary Results of a Novel Useful Tool to Distinguish between Forensic and Ancient Human Skeletal Remains. *Forensic Sci. Int.* **2017**, *272*, 87–96. [CrossRef] [PubMed]

 *biomedicines*

Technical Note

Macrophages Characterization in an Injured Bone Tissue

Krisztina Nikovics [1,*], **Marjorie Durand** [2], **Cédric Castellarin** [1], **Julien Burger** [3], **Emma Sicherre** [1], **Jean-Marc Collombet** [2], **Myriam Oger** [1], **Xavier Holy** [4] **and Anne-Laure Favier** [1]

1   Imagery Unit, Department of Platforms and Technology Research, French Armed Forces Biomedical Research Institute, 91223 Brétigny-sur-Orge, France; cedric.castellarin@intradef.gouv.fr (C.C.); emma.sicherre@supbiotech.fr (E.S.); myriam.oger@intradef.gouv.fr (M.O.); anne-laure.favier@intradef.gouv.fr (A.-L.F.)
2   Osteo-Articulary Biotherapy Unit, Department of Medical and Surgical Assistance to the Armed Forces, French Armed Forces Biomedical Research Institute, 91223 Brétigny-sur-Orge, France; marjorie1.durand@intradef.gouv.fr (M.D.); jean-marc.collombet@intradef.gouv.fr (J.-M.C.)
3   Microbiology and Infectious Diseases Department, French Armed Forces Biomedical Research Institute, 91223 Brétigny-sur-Orge, France; julien.burger@intradef.gouv.fr
4   Department of Platforms and Technology Research, French Armed Forces Biomedical Research Institute, 91223 Brétigny-sur-Orge, France; xavier.holy@intradef.gouv.fr
*   Correspondence: krisztina.nikovics@def.gouv.fr or krisztina.nikovics@intradef.gouv.fr; Tel.: +33-(0)-1-78-65-13-331

Abstract: Biomaterial use is a promising approach to facilitate wound healing of the bone tissue. Biomaterials induce the formation of membrane capsules and the recruitment of different types of macrophages. Macrophages are immune cells that produce diverse combinations of cytokines playing an important role in bone healing and regeneration, but the exact mechanism remains to be studied. Our work aimed to identify in vivo macrophages in the Masquelet induced membrane in a rat model. Most of the macrophages in the damaged area were M2-like, with smaller numbers of M1-like macrophages. In addition, high expression of IL-1β and IL-6 cytokines were detected in the membrane region by RT-qPCR. Using an innovative combination of two hybridization techniques (in situ hybridization and in situ hybridization chain reaction (in situ HCR)), M2b-like macrophages were identified for the first time in cryosections of non-decalcified bone. Our work has also demonstrated that microspectroscopical analysis is essential for macrophage characterization, as it allows the discrimination of fluorescence and autofluorescence. Finally, this work has revealed the limitations of immunolabelling and the potential of in situ HCR to provide valuable information for in vivo characterization of macrophages.

Keywords: macrophages; hybridization chain reaction (HCR); cryosection; bone; cytokines; Masquelet induced membrane

Citation: Nikovics, K.; Durand, M.; Castellarin, C.; Burger, J.; Sicherre, E.; Collombet, J.-M.; Oger, M.; Holy, X.; Favier, A.-L. Macrophages Characterization in an Injured Bone Tissue. *Biomedicines* **2022**, *10*, 1385. https://doi.org/10.3390/biomedicines10061385

Academic Editor: Chikafumi Chiba

Received: 25 May 2022
Accepted: 8 June 2022
Published: 11 June 2022

Publisher's Note: MDPI stays neutral with regard to jurisdictional claims in published maps and institutional affiliations.

1. Introduction

Macrophages have an essential role both in osteoblast-mediated bone formation [1] and in osteoclast development [2,3], but the detailed function of these cells is not yet fully understood. In addition, the cytokines and other soluble factors secreted by macrophages can induce bone formation in vitro and in vivo [1,4–8].

For a long time, in vitro cultures were used to study the phenotypic characterization of macrophages as a model to control the extracellular environment [9–11]. In vitro macrophages can be classified into two families: (i) M1 macrophages and (ii) M2 macrophages. The M1 family expresses pro-inflammatory cytokines, such as tumor necrosis factor (TNF), interleukin-1 beta (IL-1β), interleukin-6 (IL-6), interleukin-12 (IL-12), interleukin-15 (IL-15), interleukin-18 (IL-18), interleukin-23 (IL-23), and interleukin-28 (IL-28) mediating inflammation. The M2 family expresses anti-inflammatory cytokines, such as interleukin-10 (IL-10), interleukin-1β receptor antagonist (IL-1RA), transforming growth factor-beta (TGF-β) and proangiogenic cytokines (vascular endothelial growth factor (VEGF) that resolve inflammation and also modulate the extracellular matrix (ECM). In vitro studies have shown that the M2 family characterization is more complex and can be subdivided into four subtypes: M2a, M2b, M2c, and M2d [12–16]. Each subtype expresses a distinctive panel of cytokines and plays a different role in tissue regeneration (Figure 1) [17–23]. Surprisingly, M2b and M2d macrophages also express pro-inflammatory cytokines including TNF, IL-1β, IL-6, or IL-12 [16,24]. This division is not perfect because specific (nontypical) macrophages do not belong to either group but play an important immunoregulatory role. An in vitro approach does not allow the study of other immune cells and their respective secreted cytokines normally present at the site of the regeneration. Indeed, the host response in vivo is more complex highlighting the difficulty to deduce in vivo results from in vitro observations. The knowledge of macrophage phenotypes under in vivo conditions is still poorly understood and further investigations are essential, especially in our case, to study macrophage involvement during bone regeneration. Accordingly, in vivo macrophages are named M1-like and M2-like macrophages or resolving macrophages (Figure 1) [23,25–30].

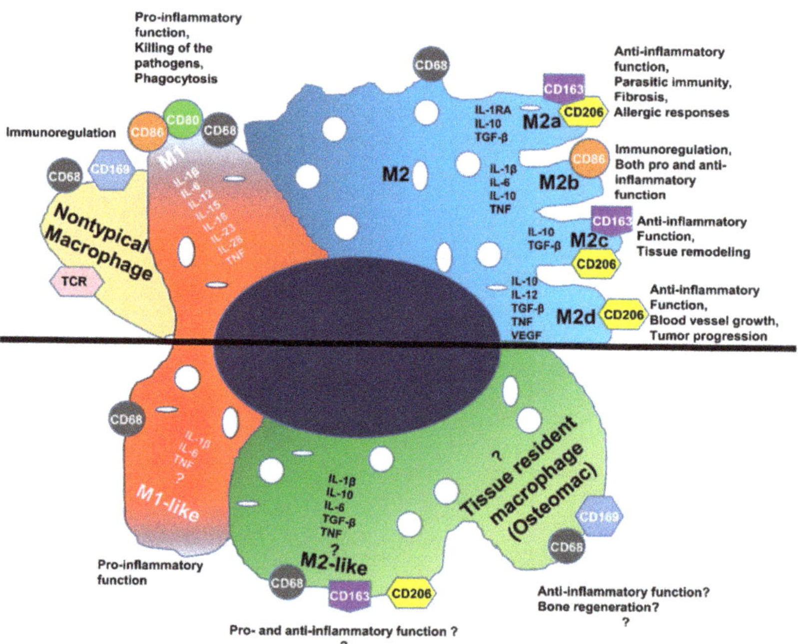

Figure 1. Macrophage polarization subtypes; different cytokine expressions and functions of the macrophage populations in vitro and in vivo.

An additional category of macrophages called "Tissue-resident macrophages" is thought to participate in bone repair. Contained in almost all tissues, they are called 'osteomacs' when localized in the bone [1,2,31,32].

Biomaterial-based therapy is a useful method to improve bone regeneration; however, its underlying repair mechanism is not yet elucidated [33–40]. About 30 years ago, the French surgeon Alain-Charles Masquelet developed a new technique to repair bone defects called the Masquelet induced membrane [41]. The surgeon places a fixator along the bone and completes its missing parts with a biomaterial, an inert polymer that has been engineered for interacting with biological systems, usually polymethylmethacrylate (PMMA). Later on, a membrane called the Masquelet induced membrane will be generated all around the biomaterial as an immune reaction against a foreign body. It is essential for further bone regeneration [41–43]. In the literature, biomaterials are described to induce the appearance of the macrophages at the bone injury site [1,7,31,32,44–46].

One approach to characterize macrophages is the identification of their expressed cytokines. Different techniques, such as Northern blot, qPCR, microarray, flow cytometry analysis, and next-generation deep sequencing methods can be used. These methods only provide a result from a mixture of different cell types. In addition, measuring the expression level of the cytokines is not sufficient; the localization of the cytokine-expressing cells in the tissue should also be determined [47].

To localize the cytokine-expressing cells in the tissue, different methods are available: (i) expression of reporter constructs, but the limitation of this technique is the requirement of a transgenic animal; (ii) classical immunostaining techniques using antibodies against markers to detect specific proteins of the macrophages. However, since cytokines are generally secreted, it is difficult to determine exactly which cells produce this protein or peptide [48]. This technique can only distinguish between M1-like and M2-like macrophages. However, it is not appropriate for identifying M2-like subtypes, as there is currently no cell surface marker available to distinguish between the different subtypes; and (iii) in situ hybridization is one of the most convincing methods to identify the cytokine-expressing cells because it is based on messenger RNA (mRNA) detection of the targeted genes [47].

In the present study, the cryosections of non-decalcified rat femur surrounded by muscle were investigated by immunostaining, in situ hybridization, and in situ HCR to identify phenotypes of macrophages involved in bone regeneration. In these challenging conditions, expression of CD68, CD163, IL-1β, IL-6 and β-actin genes was successfully detected resulting for the first time in the identification and localization of M2b-like macrophages in vivo in the bone of rats during bone regeneration.

2. Materials and Methods

2.1. Rat Animal Model

All experiments were approved by the IRBA Institutional Animal Care and Use Committee (protocol 65 DEF_IGSSA_SP). Surgeries were carried out in an accredited animal facility. Eight-week-old (200 g average weight) male Sprague Dawley rats (Charles River Laboratories, Freiburg, Germany) were housed individually in cages in a temperature and light-controlled environment with food and provided water ad libitum. Before collecting femurs with muscles, animals were euthanized at 12 weeks old with an overdose of sodium pentobarbital (150 mg/kg) administrated intraperitoneally.

2.2. Embedding and Cryosectioning of the Entire Femur (Bone and Muscle Together) of the Rat

Embedding and cryosectioning methods were performed as described in [49]. RNase-free instruments, materials, and buffers were used to collect bone samples. After euthanasia of the rat, the whole femur was cleaned rapidly, and a part of the muscles around the bone was kept. The femur was placed at the bottom of the embedding mold and covered with cryomount medium (CM) (00890-EX, HistoLab, Askim, Norway). Samples were snap-frozen with 2-methyl butane cooled in liquid nitrogen to obtain a block.

2.3. Histological Staining

Histological staining was performed as described in [49]. Hematoxylin and phloxin (HP) staining was performed as follows: the sections were incubated in several successive baths: 40 s in a hemalum (11,487, Merck, Darmstadt, Germany) buffer (0.2 g hemalum, 5 g aluminum potassium sulfate in 100 mL distilled water), 3 min in water, 30 s in a phloxin (15,926, Merck, Darmstadt, Germany) buffer (0.5 g phloxin in 100 mL distilled water), 1 min in water, 2 min in 70% ethanol, 30 s in 95% ethanol, 1 min in 100% ethanol, and 1 min in 100% ethanol. In the end, nuclei were colored in blue and cytoplasm in pink.

2.4. Immunofluorescence

Sections of the rat femur were fixed in 4% (w/v) paraformaldehyde (PFA (P6148, Sigma, Lezennes, France) in PBS (phosphate-buffered saline without Ca and Mg, GAUPBS0001, Eurobio, Les Ulis, France). After three washes in PBS, the sections were permeabilized for 15 min with 0.5% (v/v) Triton X100 buffered with PBS. The non-specific binding sites were blocked with Emerald Antibody Diluent (936B-08, Sigma, Lezennes, France) for 1 h at room temperature. Then, sections were incubated overnight at 4 °C with the primary rabbit anti-CD68 (ab125212, Abcam, Amsterdam, The Netherlands) antibody at 1:1000 dilution; the primary goat anti-CD206 (C20) (sc-34577, Santa Cruz Bio., Heidelberg, Germany) antibody at 1:1000 dilution; the primary rabbit anti-CD163 (ab182422, Abcam, Amsterdam, The Netherlands) antibody at 1:500 dilution; and the primary rabbit anti-Iba1 (ab178846, Abcam, Amsterdam, The Netherlands) antibody at 1:200 dilution. Then, the sections were washed in PBS and incubated with the secondary anti-rabbit Alexa Fluor 488 (A-21206, Thermo Scientific, Waltham, MA, USA) antibody at 1:500 dilution and the secondary anti-goat Alexa Fluor 568 (ab175704, Abcam, Amsterdam, The Netherlands) antibody at 1:500 dilution for 2 h at room temperature. Finally, sections were washed in PBS for 20 min and mounted using a Fluoroshield mounting medium with DAPI (ab104139, Abcam, Amsterdam, The Netherlands). The fluorescence was detected using an epifluorescence microscope DM6000 (Leica, Schönwaldeglien, Germany) equipped with monochrome and color digital cameras.

2.5. RT-qPCR

Frozen femur samples were homogenized in liquid nitrogen. Total RNA was then isolated using an RNeasy Fibrous Tissue mini kit (HB-0485, Qiagen, Courtaboeuf, France) according to the m anufacturer's recommendations. RNA extracts were recovered with 20 µL RNase-free water. Reverse transcription (RT) was performed with oligo-dT primers following the instructions of the Sensiscript transcription kit (205211, Qiagen, Courtaboeuf, France); cDNA synthesis was carried out for 1 h at 37 °C using 50 ng of RNA with 10 µM oligod(T) primers, RNase inhibitor (2 IU), and Sensiscript reverse transcriptase.

Real-time qPCR was carried out in a 20 µL final volume using LC480 SybrGreen I Mastermix (Roche Applied Science, Mannheim, Germany) using 0.25 µL of cDNA. Oligos designed for RT-qPCR experiments are listed in Table S1.

2.6. In Situ DIG Hybridization

In situ hybridization methods were performed as described in [49]. Oligos designed for in situ hybridization experiments are listed in Table S2.

Digoxigenin (DIG)-labeled cRNA probes were used for in situ hybridization. Briefly, the tissues were fixed in 4% (w/v) in PBS/paraformaldehyde (PBS, w/o Ca and Mg, GAUPBS0001, Eurobio, Les Ulis, France; PFA, P6148, Sigma, Lezennes, France) for 30 min and then treated with 100% methanol for 15 min and air-dried. The sections were pre-hybridized for 2 h at 45 °C in a pre-hybridization buffer (50% formamide (GHYFOR0402, Eurobio, Les Ulis, France), 0.5× sodium ch loride citrate (SSC) (GHYSSC007, Eurobio, Les Ulis, France) buffer, 50 µg mL^{-1} heparin (H3393, Sigma, Lezennes, France), 100 µg mL^{-1} transfer RNA, and 0.1% (v/v) Tween 20 (822184, Merck, Darmstadt, Germany). Finally, the sections were incubated overnight at 45 °C with the RNA probes (2 µL probe in 200 µL hybridization buffer (50% formamide, 100 µg mL^{-1} transfer RNA, 7.5% (v/v) Tween 20, 8.5% NaCl, 20% dextran sulfate (GHYDEX000T, Eurobio, Les Ulis, France), and 2.5× Denhardt's Solution (50× stock, D2532, Sigma, Lezennes, France)), which were previously denaturized for 2 min at 80 °C in the hybridization buffer. Non-specific hybrids were dissociated with the following washes: 30 min in 0.1× SSC + 0.5% SDS at 45 °C, 2 h in 2× SSC + 50% formamide at 45 °C, 5 min in NTE (0.5 M NaCl, 10 mM Tris pH 8, 1 mM EDTA) at 45 °C,

30 min in NTE + 10 mg mL^{-1} Rnase A (10109169001, Roche, Boulogne-Billancourt, France) at 37 °C, 1 h in 2× SSC + 50% formamide at 45 °C, 2 min in 0.1× SSC at 45 °C, and finally 15 min in PBS at RT.

Immunodetection of the DIG-labeled probes was performed using an anti-DIG antibody coupled to alkaline phosphatase as described by the manufacturer (11093274910, Roche, Boulogne-Billancourt, France). Afterward, the sections were incubated for 1–2 days in a buffer containing 337 μL BCIP (5-Bromo-4-chloro-3-indolyl-phosphate) and 225 μL NBT (Nitroblue tetrazolium chloride) in a 50 mL solution (100 mM Tris pH 9.5, 100 mM NaCl and 50 mM MgCl$_2$) until a blue precipitate adhering to the sections were formed. The DIG sections were observed with an epifluorescence microscope DM6000 (Leica, Schönwaldeglien, Germany) equipped with monochrome and color digital cameras, while the HCR sections were observed with a confocal microscope (LSM700, Zeiss, Dresden, Germany).

2.7. In Situ HCR Hybridization

In situ hybridization methods were performed as described in [49]. Oligos designed for in situ hybridization experiments are listed in Table S2.

The HCR protocol of Choi and colleagues (2014, 2016, 2018, 2020) was performed with some modifications as described below to enhance mRNA localization in the femur of the rat [50–53].

The sections were pre-hybridized for 10 min at RT in a hybridization buffer (50% formamide, 5× SSC, 9 mM citric acid pH 6, 50 μg mL^{-1} heparin, 1× Denhardt's Solution, 0.1% (*v/v*) Tween 20 and 10% dextran-sulfate). Previously, the hybridization probes (2 pmol per slide) were denatured for 2 min at 80 °C. Finally, the sections were incubated in a hybridization buffer together with probes overnight at 45 °C. Nonspecific hybrids were dissociated with the following washes: 30 min in 0.1× SSC + 0.5% SDS at 45 °C, followed by 2 h in 2× SSC + 50% formamide at 45 °C, then 2 min in 0.1× SSC at 45 °C, and finally 15 min in PBS at RT.

Sections were first incubated for 2 h at RT with an amplification buffer (5× SSC, 0.1% (*v/v*) Tween 20, 10% dextran-sulfate, and 100 μg mL^{-1} salmon sperm ADN) and subsequently for 12 to 16 h with the DNA hairpins marked with a fluorophore (Alexa Fluor488) (diluted in amplification buffer as previously described). The hairpins were previously heated at 95 °C for 90 s and cooled to RT for 30 min. The sections were then washed 2× 30 min in 5× SSCT (5× SSC and 1% (*v/v*) Tween 20) and 5 min with 5× SSC without Tween at RT.

2.8. Microspectroscopical Analysis

For spectroscopic analysis, an LSM 780 (Carl Zeiss, Jena, Germany) confocal microscope was used to acquire a lambda stack of 10 nm wavebands between 425 nm and 625 nm. Then, Zen Black software (Carl Zeiss, Jena, Germany) built-in plugin was used to perform linear unmixing using the automatic component extraction algorithm. It was possible to extract Alexa 488 spectrum, and the second spectrum was considered the sum of all other sources of endogenous fluorescence (autofluorescence).

2.9. Quantification of HIS by ImageJ

To quantify the percentage of the marked region on in situ hybridization slides, each slide was acquired with the Hamamatsu Nanozoomer S60 at 20x (Tokyo, Japan). On the virtual slide, three regions of interest (ROI) were drawn in the immediate neighborhood of the biomaterial; three others were done away from it (internal negative control). Each zone was analyzed using the Fiji software. The ratio between the surface of the marked region and the surface of the entire ROI was computed after color deconvolution and thresholding (using the MaxEntropy algorithm).

Statistical analyses were performed with a *t*-test.

3. Results

3.1. Detection of Immune Cell Accumulation around Biomaterial

It was shown that biomaterial triggered the recruitment of macrophages at the site of the bone wound and enhanced wound healing [1,7,32,54]. Rat femur was operated on with biomaterial (with the Masquelet induced membrane technique), and three weeks later the whole bone together with femur and biomaterial was harvested. The non-operated femur from the second leg of this rat was used as a control (called non-operated femur). The Masquelet induced membrane technique consists of two different operative phases. In the first operation, a fixator was placed around the bone, and the missing bone fragments were filled with biomaterial. Three weeks after surgery, a Masquelet induced membrane was formed around the biomaterial. Three weeks later, during the second operation, the biomaterial was removed, but the Masquelet induced membrane remained, and the biomaterial was replaced with a bone graft. To examine the architecture and the regeneration region of the bone, cryosections of femurs were analyzed by Hematoxylin and phloxin (HP) histological coloration (Figure 2). The control femur was used as a reference to observe bone and muscle in their native structure (Figure 2A). In the operated femur, the Masquelet induced membrane, the regenerated region close to the Masquelet induced membrane, and the biomaterial were observed (Figure 2B). Some immune cells were identified by their structure at higher microscope magnification in the regeneration region (Figure 2C), hypothesized as macrophages.

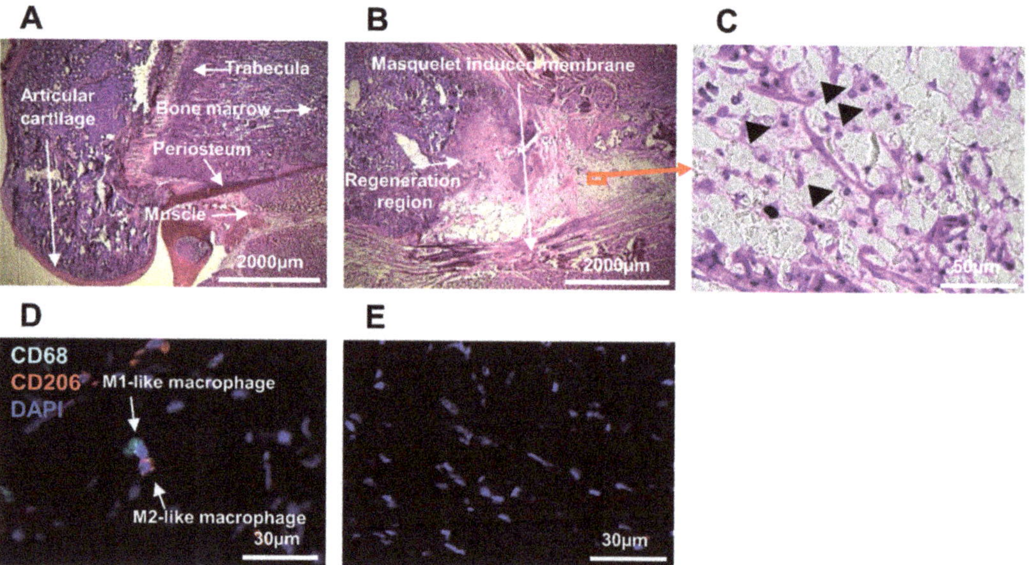

Figure 2. Histologic and microscopic analysis of rat femur: (**A–C**) histologic analysis of the femurs; histopathological image of (**A**) non-operated; and (**B**) operated femurs of rats stained with hematoxylin and phloxin; (**C**) expanded view: high magnification image of the area within the red rectangle in image (**B**); (**D**,**E**) identification of macrophages M1 and M2 in the operated femur; anti-CD68 (Alexa488, turquoise fluorescence), labeling the M1 and M2 macrophages; anti-CD206 (Alexa568, red fluorescence) labeling the M2 macrophages and satellite cells; nuclear staining with DAPI (blue fluorescence).

In situ visualization of macrophages is quite problematic. The cluster of differentiation (CD) CD68 protein is one of the most common monocytes/macrophages marker proteins [55], but in other mononuclear phagocyte cells and non-hematopoietic cells (mesenchymal stem cells, fibroblast, endothelial, and tumor cells) weak expression can be detected [56]. Other markers such as CD206 and CD163 mostly recognize M2 macrophages, but CD206 protein is also expressed in satellite and CD163 protein in dendritic cells [56,57]. In situ identification of human and mouse M2 macrophages can be performed by double immunolabeling with CD206 or CD163 together with CD68 antibodies (Table 1) [47,58].

Table 1. Cell phenotypes.

	CD68	CD206	CD163	Iba1
M1-like macrophages	+	−	−	+
M2-like macrophages	+	+	+	+
Satellite cells	−	+	−	−

The same region was examined by the traditional immunostaining method (with anti-CD68 and anti-CD206 antibodies) to detect if M1-like or M2-like macrophages were localized in the regeneration region (Table 1). We observed M1-like macrophages expressing CD68 protein only and M2-like macrophages expressing both CD68 and CD206 proteins (Figure 2D). In the absence of a primary antibody (negative control), no expression of these markers was detected (Figure 2E).

To go further in our investigation, immunostaining was performed on both the non-operated femur and the operated femur (Figure 3) in large sections (2 mm × 1.2 mm). Using bright field microscopy, the non-operated femur showed a representative architecture of the bone (Figure 3A) and an absence of fluorescence signal (Figure 3B). In the operated femur, the three expected regions (the regeneration region, the Masquelet induced membrane, and surrounding muscle) were observed (Figure 3C). Using CD68 and CD206 immunostaining, both M1-like and M2-like macrophages were detected (Figure 3D). However, the M1-like and M2-like macrophage repartition differed among the three zones (Figure 3E–G). A similar amount of M1-like and M2-like macrophages were observed in the regenerating region (Figure 3E) compared to the interface (between the regeneration region and Masquelet induced membrane) region, where mostly M1-like macrophages (Figure 3F) and a predominantly M2-like macrophage population were detected (Figure 3G).

Another widely accepted macrophage marker is the ionized calcium-binding adaptor molecule 1 (Iba1) (Table 1) [59–61]. Labeling with the anti-Iba1 antibody together with the anti-CD206 antibody showed very similar results (Figure S1). The only difference that has been detected is that fewer M1-like macrophages have been detected by Iba1 labeling in the interface region than with CD68 labeling. This result suggests that single CD68 positive cells include other cells than macrophages that have not yet been identified.

In summary, both M1-like and M2-like macrophages were identified in the injured region of the operated femur. After quantification of 1000 cells, 16.55% (CD68/CD206) and 17.5% (Iba1/CD206) were identified as M2-like macrophages compared to 6.17% (CD68/CD206) and 4.1% (Iba1/CD206) of M1-like macrophages (Table 2). To confirm this result, immunostaining with an anti-CD163 antibody was performed because this protein is also expressed by M2-like macrophages [16]. Next, the CD163 and CD206 co-labeling was performed to identify the M2-like macrophages and satellite cells in the same pictures (Figure S2). In this condition, 21.6% of M2-like macrophages were identified (Figure 4A and Table 1), while no expression of the protein was detected in the negative control (Figure 4B).

Table 2. Quantification of M1-like and M2-like macrophages in the regeneration region of the operated femur (quantitative analysis, based on random examination of 1000 cells in each of the conditions).

	M1-like Macrophages	M2-like Macrophages
CD68/CD206	6.2%	16.6%
Iba1/CD206	4.1%	17.5%
CD163	/	21.6%

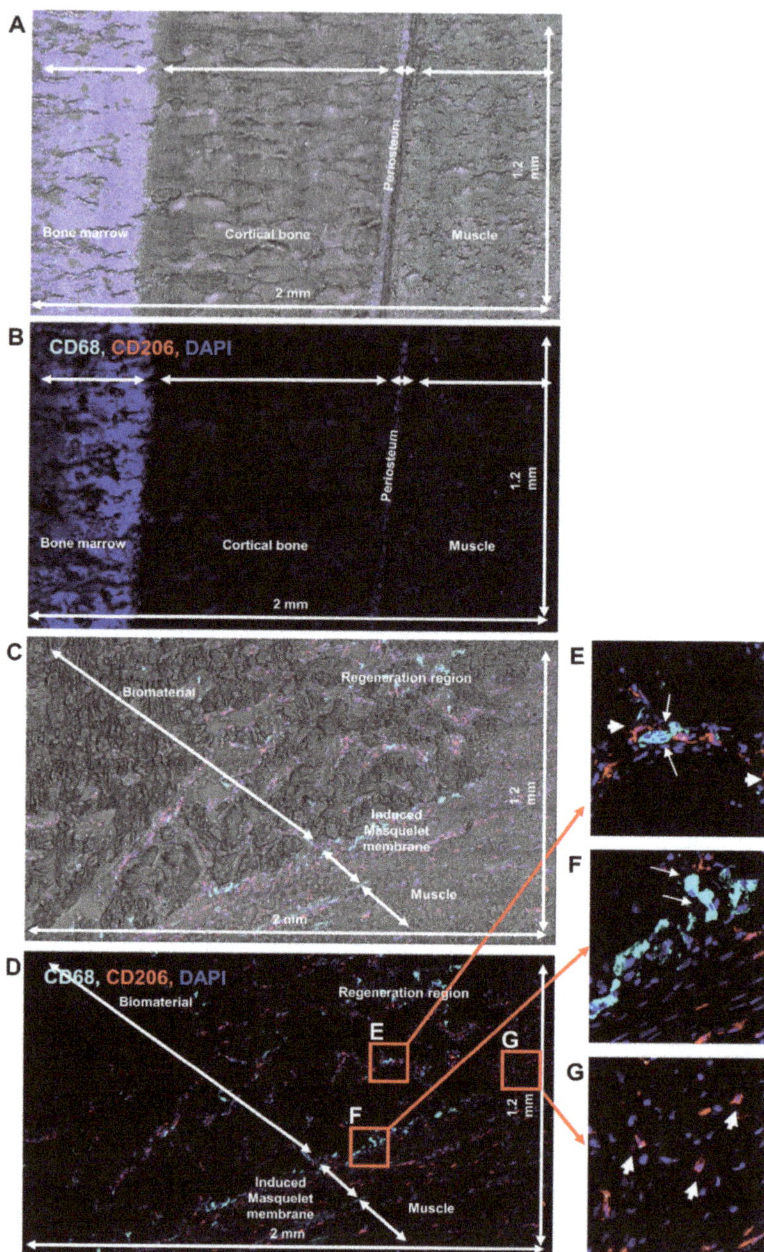

Figure 3. Identification of M1-like and M2-like macrophages in the rat femurs: (**A,B**) immuno-labeling with anti-CD68 and CD206 antibodies of the non-operated; and (**C–G**) operated femurs; bright-field image together with immunolabeling of the non-operated (**A**) and operated (**C**) femurs; (**E–G**) expanded view: high magnification image of the area within the red rectangle in image C; anti-CD68 (Alexa488, turquoise fluorescence), labeling the M1 and M2 macrophages; anti-CD206 (Alexa568, red fluorescence) labeling the M2 macrophages and satellite cells; nuclear staining with DAPI (blue fluorescence); thin arrow: M1-like macrophages; thick arrow: M2-like macrophages.

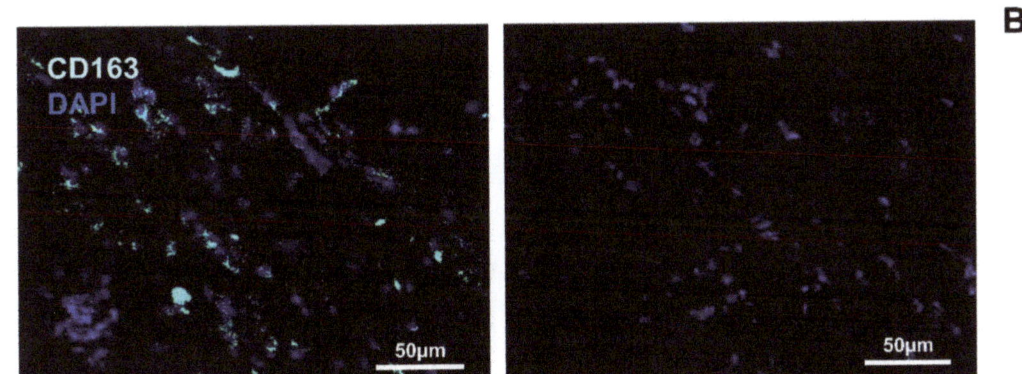

Figure 4. Identification of M2-like macrophages in the operated femur: (**A**) immunolabeling of the operated femur with anti-CD163; (**B**) negative control of the immunolabeling; anti-CD163 (Alexa488, turquoise fluorescence), labeling the M2 macrophages; nuclear staining with DAPI (blue fluorescence).

3.2. Detection of Macrophages in Other Regions

Previously we showed the capacity to detect macrophages using immunostaining in a non-decalcified cryo-fixed bone in the regeneration region (region 1). We further examined whether the different regions of the femur far from the injury region contained macrophages with phenotypic differences. In this context, three other regions were investigated: the muscle far from the bone or the wound (region 2), the periosteum (region 3), and the bone (region 4) (Figures 5 and S3). No difference between operated and non-operated femur tissues were observed (Figure 5). In region 2, no macrophages were present; only satellite cells could be detected (Figure 5B,C). In region 3, both M1-like and M2-like macrophages were detected close to osteoclasts, as it is multinucleated cells expressing CD68 (Figure 5D,E). In region 4, several cells expressing CD68 protein were detected. This result seemed to be coherent as macrophages were also localized in the bone marrow. Surprisingly, cells expressing only the CD206 protein were also detected (Figure 5F,G). This result requires further examination. No expression of both CD68 and CD206 markers was detected in the negative control (Figure S3).

3.3. Detection of Macrophages Based on Their Autofluorescence Feature

It is well known that bone tissue has strong autofluorescence [62,63]. This makes it rather difficult to analyze the operated tissue with immunolabeling (using fluorophore-labeled antibody) because there was an abundant autofluorescence in the regenerating region (Figure 6). Autofluorescence may appear from structural proteins, such as collagen and elastin, but other endogenous fluorophores, such as Flavin-type molecules, are also often localized in cells [64,65]. In particular, immune cells, such as macrophages and granulocytes contain a large amount of phagosome/phagolysosome in their cytosol [66]. Sections of the operated femur tissues had relatively intense (excitation with 488 nm (Figure 6A–D,F) and excitation with 568 nm (Figure 6B–D,F)) autofluorescences exclusively in the cells with endosome-like structure (Figure 6E,G). The sections of the operated femur were subjected to different treatments, TrueVIEW Autofluorescence Quenching Kit, Blue Evans, or Black Soudan, to get rid of endogenous fluorescence (data not shown). Unfortunately, the endogen fluorescence was resistant to all treatments used so far.

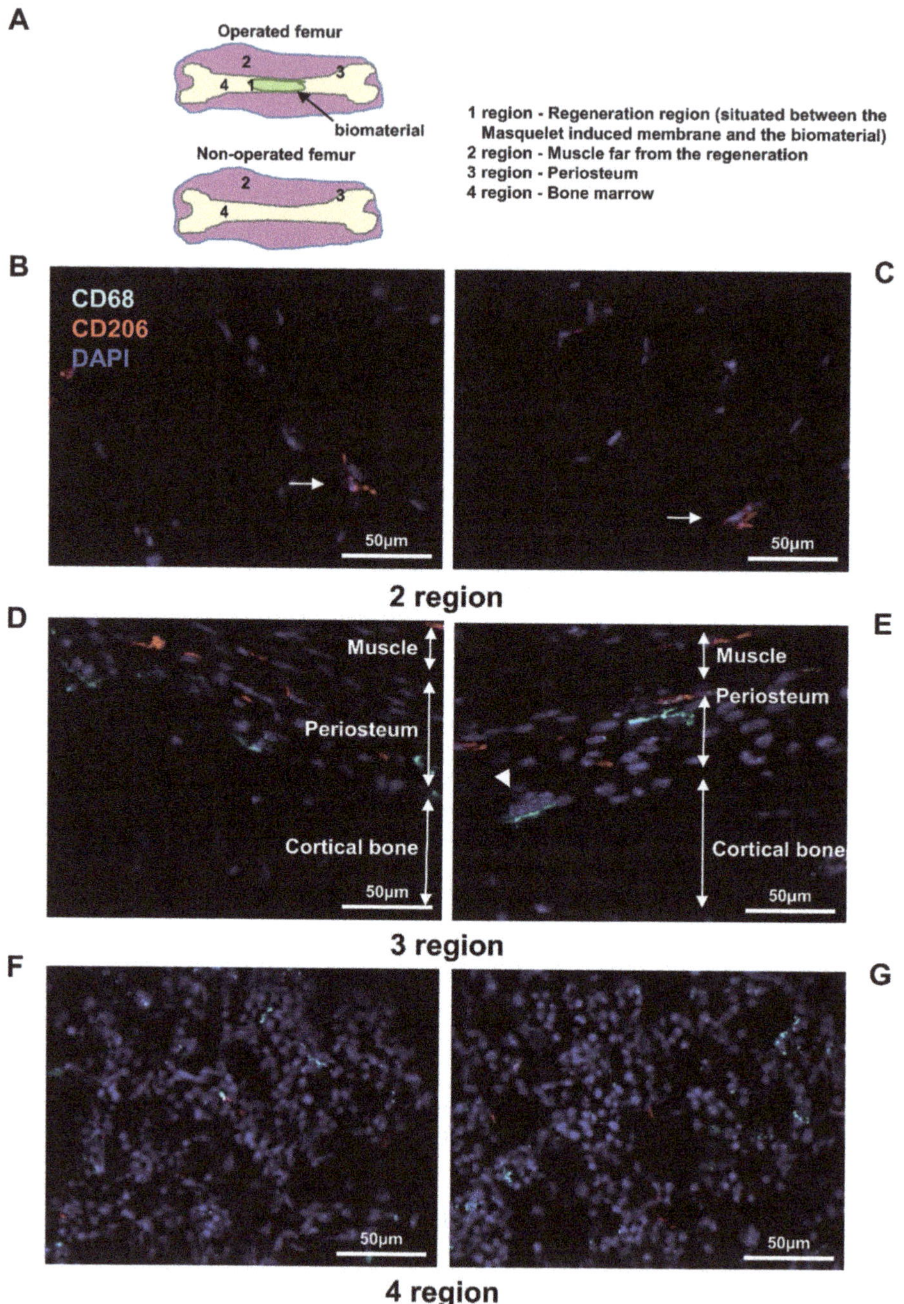

Figure 5. Identification of M1-like and M2-like macrophages in the rat femurs: (**A**) presentation of the different regions of the femurs; (**B,D,F**) operated; and (**C,E,G**) non-operated femurs labeled with anti-CD68 and anti-CD206 antibodies; anti-CD68 (Alexa488, turquoise fluorescence), labeling the M1 and M2 macrophages; anti-CD206 (Alexa568, red fluorescence) labeling the M2 macrophages and satellite cells; nuclear staining with DAPI (blue fluorescence); thin arrow: satellite cells; arrowhead: osteoclast.

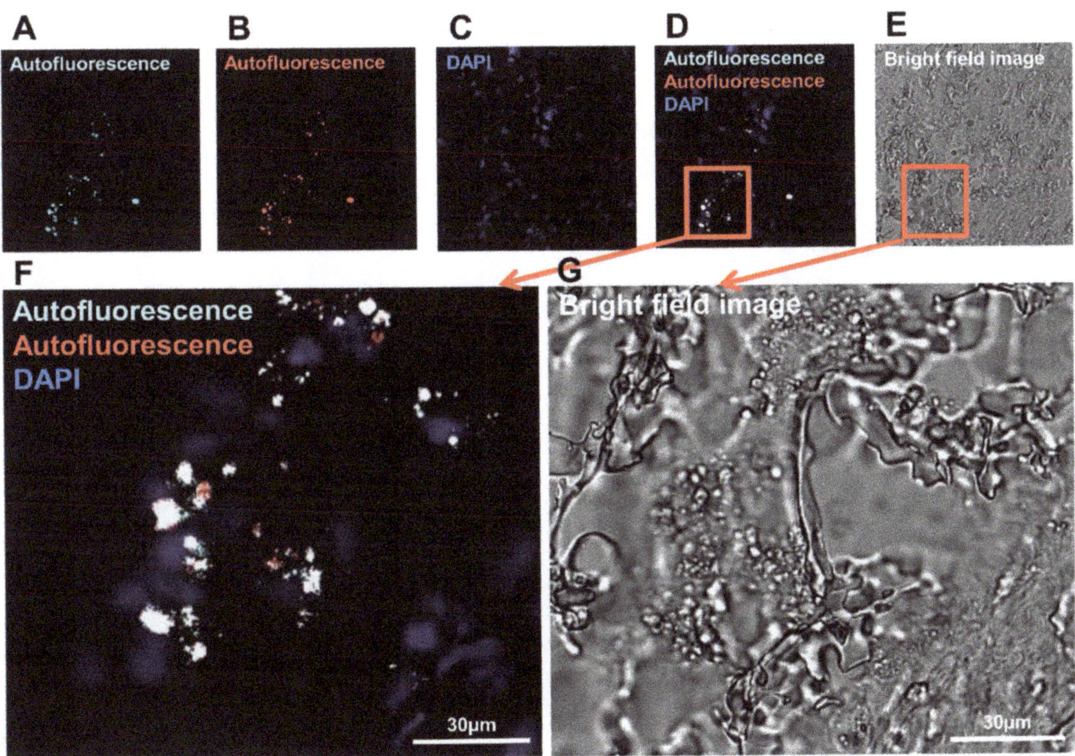

Figure 6. Observation of the autofluorescence in the operated rat femur: (**A**) detected autofluorescence after excitation with light (488 wavelengths, turquoise fluorescence); (**B**) detected autofluorescence after excitation with light (568 wavelengths, red fluorescence); (**C**) nuclear staining with DAPI (blue fluorescence); (**D**) merged image; (**E**) bright-field image; (**F**) expanded view: high magnification image of the area within the red rectangle in image (**D**); (**G**) expanded view: high magnification image of the area within the red rectangle in image (**E**).

3.4. Investigation of Macrophage Fluorescence by Microspectroscopic Analysis

As the endogen fluorescence was very strong in the bone tissue, further investigation was needed for macrophage characterization. Turquoise fluorescence was detected in macrophages, but the autofluorescence prevented any conclusion as both specific fluorescence (coming from the immune signal) and autofluorescence were detected in a mixed signal (Figure 7A). Microspectroscopical analysis in situ facilitates the differentiation between specific fluorescence and autofluorescence. The ratio between the intensity of specific fluorescence and autofluorescence should be higher than three to conclude that the autofluorescence is not disturbing. This technique was applied to excited signals at 488 nm to separate them from their characteristic emission spectra. The emission spectrum of specific fluorescence reached a maximum at 520 nm, whereas the emission spectrum of autofluorescence peaked at approximately 550 nm (Figure 7B). The intensity of the specific CD68 fluorescence was approximately the same as the intensity of the autofluorescence (Figure 7B). Using this technique, two subtracted images were extracted from Figure 7C: the autofluorescence signal (Figure 7D) and the specific CD68 fluorescence of the macrophages (Figure 7E). In the resulting image, we can distinguish autofluorescence from the CD68 specific signal in cells (Figure 7E).

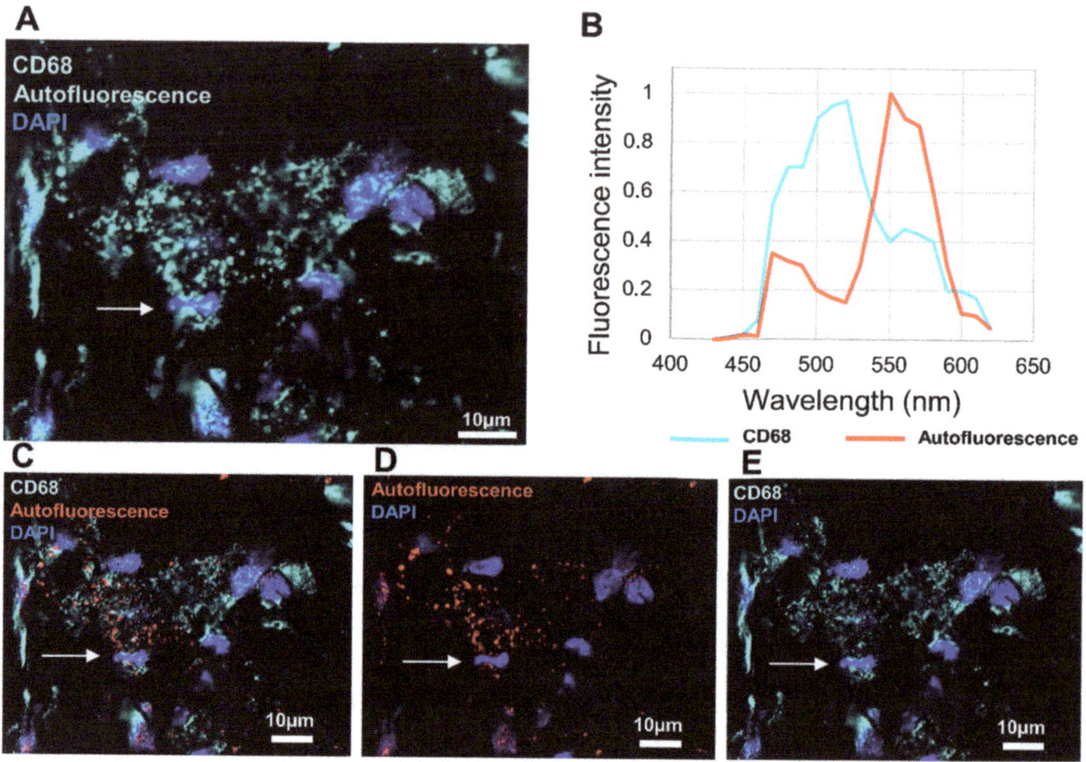

Figure 7. Identification of the in situ autofluorescence by microspectroscopy analysis. In the operated femur, macrophages were labeled with anti-CD68 antibody. Tissues were excited at 488 nm. (**A**) Emission of the autofluorescence and the antigen-specific fluorescence (Alexa Fluor 488) with turquoise fluorescence. (**B**) Microspectroscopycal analysis of Alexa Fluor 488 emission (turquoise line) and autofluorescence (red line). (**C**) Separation of autofluorescence emission (red fluorescence) and antigen-specific fluorescence (Alexa Fluor 488) (turquoise fluorescence). (**D**) Emission of the autofluorescence (red fluorescence). (**E**) Emission of the antigen-specific fluorescence (Alexa Fluor 488) (turquoise fluorescence). Nuclear staining with DAPI (blue fluorescence). Fine arrow: macrophages.

3.5. Detection of IL-1β and IL-6 Cytokines in the Operated Femur

The presence of M1-like and mostly M2-like macrophages was shown earlier in the wounded bone tissue with immunofluorescence techniques. The next step was to determine the M2-like macrophages based on their cytokine expressions. Each subtype of M2-like macrophages (M2a, M2b, M2c, or M2d) secretes a different panel of cytokines. The next step was to identify the types of cytokines produced in the operated femur using RT-qPCR. All three markers showed higher expression in the operated femur (2.2 times more CD68, 1.61 times more CD163, and 2.29 times more CD206 mRNA) (Figure 8). Among the tested genes, two cytokines showed much higher expression, IL-1β (6.89 times more) and IL-6 (4.44 times more), in the operated femur compared to the non-operated femur (Figure 8). The only M2 macrophage that produces these two cytokines is the M2b macrophage (Figure 1). The in situ DIG technique is a commonly used, non-radioactive IHS method because it is very sensitive, but it has the limitation that it only allows analysis of a single gene in a sample.

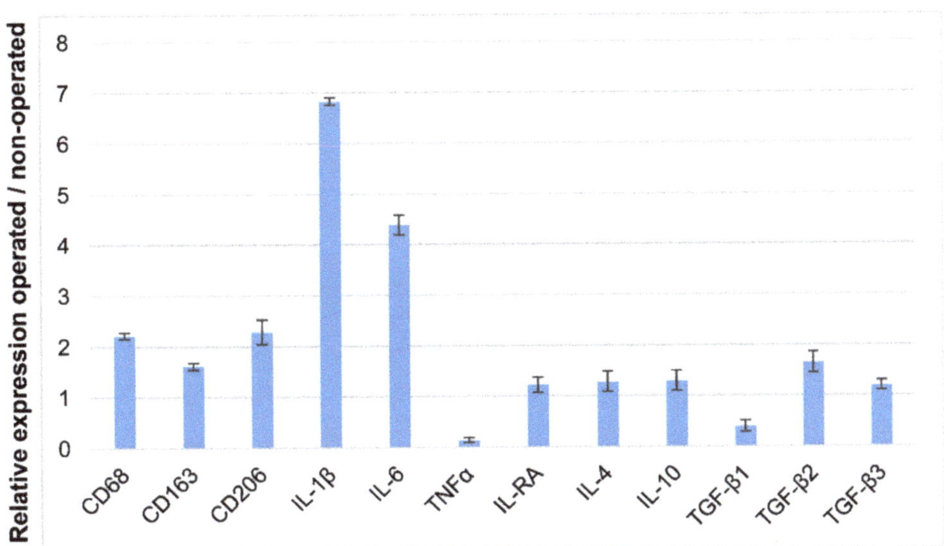

Figure 8. Expression of the different marker genes and cytokines in the operated femur; qRT-PCR analysis of the different gene expressions. This result is the average of three measurements.

3.6. Identification of Macrophage Phenotype Using Both I DIG and In Situ HCR Techniques

In situ HCR was chosen to further characterize the macrophages. The advantage of this technique is that it is more sensitive than FISH, and unlike in situ DIG, it allows the identification of several genes at once. Our next question was whether we could detect M2b-like macrophages in situ using this technique in the operated rat bone. Since cytokines are usually secreted [48,67], immunostaining is not suitable to distinguish between different subtypes. Indeed, it is quite challenging to identify which cells produce the secreted protein. For this reason, an in situ hybridization technique was used to reveal the subtypes of M2-like macrophages as it allows the detection of the mRNA coding for those cytokines.

For the first time, in situ hybridization with a digoxigenin-labeled cRNA probe (in situ DIG) was used because the labeling remains stable, and it is an advantage to examine the labeling architecture. This technique is the most widely used non-radioactive ISH method because it is very sensitive, unfortunately, only one mRNA per section can be detected [68]. The expression of β-actin, IL-1β, IL-6, CD68, and CD163 genes was identified (Figure 9). The β-actin was used as a positive control (Figure 9A). In the absence of a specific probe, no expression was observed (Figure 9B). The mRNA expression of CD68, CD163, IL-1β, and IL-6 genes (Figure 9C–F) was detected around the biomaterial. Both CD68 (Figure 9C) and IL-1β mRNA expression (Figure 9D) was very intense. The CD163 (Figure 9E) and IL-6 (Figure 9F) mRNA expressions were much weaker, but a higher magnification showed a clear labeling (Figure 9G,H). In addition, β-actin, CD68, CD163, IL-1β, and IL-6 signals were quantified in the operated femur. Upregulation of these genes was detected in the Masquelet induced membrane region.

The same labeling was performed on the non-operated femur sections. In the non-operated section, abundant expression of the β-actin mRNA was detected (Figure S4A), and in the negative control (Figure S4B), expression was not observed. The CD68 (Figure S4C) and IL-1β mRNA (Figure S4D) expressions were intense in the bone marrow, but no expression in the other part of the femur was detected. No expression was found with a small magnification of the CD163 (Figure S4E) and IL-6 (Figure S4F) mRNA. The CD163 expression was detectable only with higher magnification in the periosteum (Figure S4G). IL-6 mRNA was present in the bone marrow (Figure S4H).

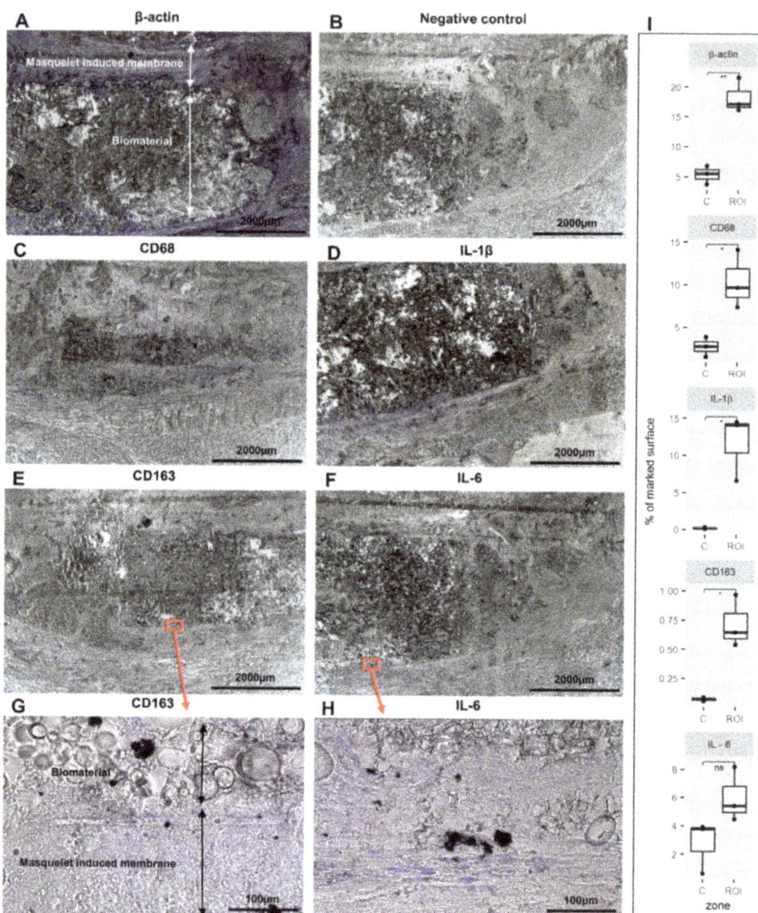

Figure 9. In situ hybridization in the operated rat femur: (**A**) expression of β-actin mRNA (positive control); (**B**) negative control; (**C**) CD68 mRNA; (**D**) IL-1β mRNA; (**E**) CD163 mRNA; (**F**) IL-6 mRNA; (**G**) expanded view: high magnification image of the area within the red rectangle in image (**E**); (**H**) expanded view: high magnification image of the area within the red rectangle in image (**F**); (**I**) the stained area of the picture was quantitatively analyzed using ImageJ. The Masquelet induced membrane region was compared with the control region (**C**) (n = 3). ** $p < 0.01$, * $p < 0.5$, ns = not significant, compared to control region.

Because co-labeling was not possible with the in situ DIG technique, in situ hybridization coupled with hybridization-chain-reaction detection (in situ HCR) was performed on cryosections of the bone. Indeed, three different mRNA expressions were co-detected, taking into account the rapid loss of signal of fluorophores within two weeks after labeling. Using the in situ DIG method, a high expression of IL-1β and IL-6 was detected in the regenerating bone. To examine whether M2-like macrophages were expressing these two cytokines, in situ HCR was performed with CD163, IL-1β, and IL-6 probes altogether (Figure 10A). Macrophages co-expressing the three signals were identified in the regenerated region of the operated femur, while no expression was detected in the negative control (Figure 10B). By random examination of the fluorescence of 100 cells labeled with a CD163 probe, the percentage of the M2b-like macrophages was approximately 68% of the M2-like macrophages.

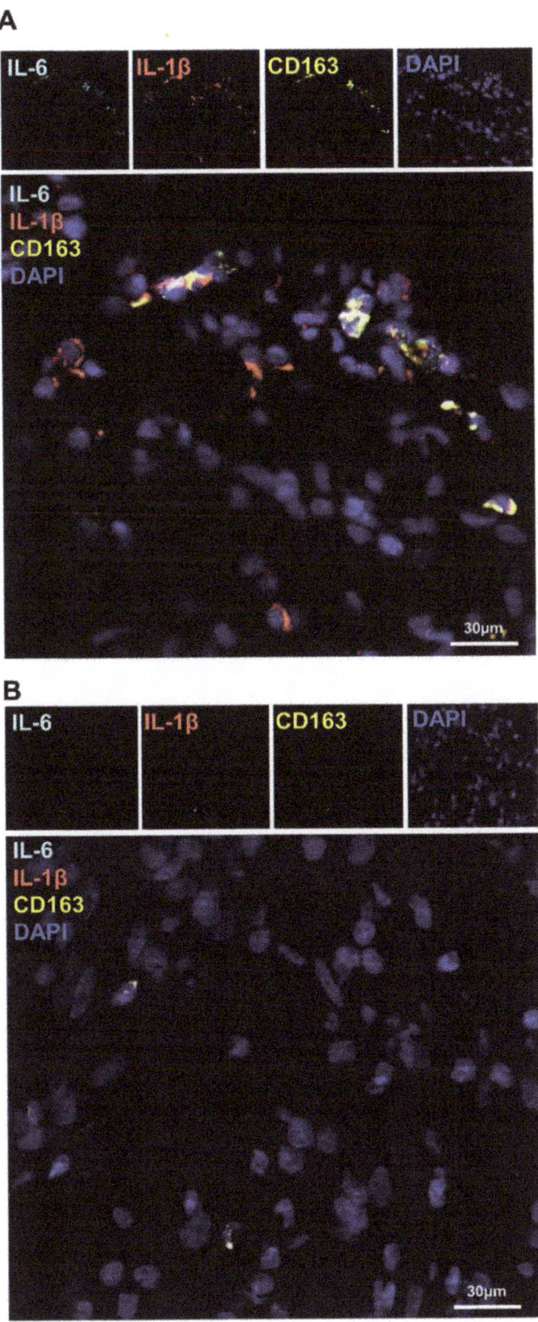

Figure 10. Identification of the M2b-like macrophages in the operated femur by in situ hybridization combined with hybridization-chain-reaction detection (in situ HCR): (**A**) in situ hybridization combined with hybridization-chain-reaction detection (in situ HCR); (**B**) negative control: Probe-IL-6 (Alexa488, turquoise fluorescence), Probe-IL-1β (Alexa546, red fluorescence), Probe-CD163 (Alexa647, yellow fluorescence), nuclear staining with DAPI (blue fluorescence).

4. Discussion

Bone and muscle tissues constantly interact with each other by secreting cytokines and other soluble factors [69]. Analysis of cytokine expressions during bone regeneration is very important for developing new therapeutic approaches. We aimed to analyze the different macrophage-specific cytokine expressions during bone regeneration in the sections of the entire femur surrounded by muscle. Before sectioning, the tissues must be fixed and embedded in paraffin. In our case, paraffin embedding was not possible because the muscle significantly slowed down (around 5 weeks) the decalcifying solution (ethylene diamine tetra acetic acid (EDTA)) from penetration and did not allow the use of ISH techniques in these sections (data not shown) due to the mRNA degradation. To avoid this problem, an improved version of the CryoJane tape transfer system was developed in our laboratory to obtain the cryosections of the entire femur surrounded by muscle, which was suitable for ISH techniques [49].

The role of macrophages in tissue development, homeostasis, and wound healing is essential [17,19,21,22,37–39]. Bone regeneration is a very complex process. The continuous interaction between bone-forming cells and macrophages is essential for successful bone healing [1,70,71]. However, their role in bone regeneration remains poorly understood. Subsequently, a deeper understanding of their involvement after an injury is essential to developing new therapeutic strategies for bone repair. A broad spectrum of biomaterials, differing in their structure, porosity, composition, and chemistry, is used as therapeutic strategies for tissue repair. It is known from in vitro studies that macrophage polarization remains biomaterial dependent. Data in the literature suggest that the host response in the in vivo condition is more complex which limits the interpretation of in vivo results compared to in vitro observations [72–74]. Previously, it was speculated that M1 and M2 macrophages subsequently participate in a different stage of bone regeneration [1,75–77]. Three weeks after lesion/surgery, both M1-like and M2-like types of macrophages were detected in the injured bone tissue. Mostly M2-like macrophages were identified, while few but detectable M1-like macrophages were observed. Our results suggest that bone regeneration is a complex process, and in contrast to our expectation, it requires the continuous presence of two types of macrophages. Several publications suggested that macrophages exert their effects on bone regeneration through secreted cytokines [1,8,36,38]. Vi and collaborators (2015) demonstrated that even the conditioned medium of the macrophages was able to activate the "mineral deposition and bone formations" [8].

Macrophage subtypes produce diverse cytokines at different stages of regeneration. Identification of cytokine-expressing cells by immunolabeling is quite complicated because these proteins and peptides are secreted, making it very difficult to determine which cells have produced them. In this work, we demonstrated the possibility to identify macrophages in bone after the detection of cytokines expressed by macrophages with in situ hybridization coupled with in situ HCR detection methods. This tool allows in situ characterization of macrophages during bone regeneration, which was not possible earlier. Fluorescence in situ hybridization (FISH) appeared in the 1980s [78–80]. The advantage of this method compared with in situ DIG was that the expression of several mRNAs could be analyzed in the same section [47,81]. However, it was only suitable for detecting highly expressing mRNA. Unfortunately, the expression of cytokines is usually low. The in situ HCR [49–53] method is a promising new method to study gene expression in bone tissues. This technique allows the detection of multiple mRNAs also in the same section and is suitable for the visualization of mRNA with low expression.

Other groups exhibited that at the bone damage location, M1 macrophages secreted pro-inflammatory cytokines (TNF, IL-1β, and IL6) and induced the recruitment of the mesenchymal stem cells (MSC) and other progenitor cells at the local site. Cytokine expression was investigated in the operated and non-operated tissues by qPCR and in situ DIG methods. Interestingly, very high expression of IL-6 and IL-1β was detected by qPCR, and in situ DIG showed that these two cytokines were localized around the biomaterial. M1-like and M2-like macrophages were present in regenerating bone although the M2-like macrophages were more abundant. The M2b macrophages are the only M2-like macrophages that co-secrete these two cytokines. We found that the in situ HCR method is a good compromise in macrophage identification and allows in situ determination of the presence of the M2b-like macrophages.

5. Conclusions

In summary, at least five observations were highlighted: (1) There was a significant difference between non-operated and operated bone tissues. With immunostaining, in the operated bone tissue we observed mostly M2-like macrophages with smaller quantities of M1-like macrophages. No macrophages were found in the non-operated bone tissue. (2) The localization of the macrophages was limited to the wounded area. (3) The operated tissue showed a strong IL-1β and IL-6 cytokine expression as measured by RT-qPCR detection. (4) In addition, the in situ HCR method has been proved useful as it allows selectively exploring the RNA expressions of the cytokines and macrophage markers in the cells of the wounded bone tissue. With this technique, it was possible to identify the in vivo M2b-like macrophages in the wounded bone tissue. (5) It was pointed out that microspectroscopical analysis is very important in macrophage characterization because it can differentiate between fluorescence and autofluorescence.

Finally, this work highlighted the limits of immunostaining and the interest in in situ HCR method providing valuable information for in vivo macrophage characterization. This technique is an important approach for analyzing the biological mechanism of bone regeneration and offers a new perspective in the field of regenerative medicine.

Supplementary Materials: The following supporting information can be downloaded at: https://www.mdpi.com/article/10.3390/biomedicines10061385/s1. Figure S1: Identification of M1-like and M2-like macrophages in the rat femurs; (A–D) immunolabeling with anti-Iba1 and CD206 antibodies of the operated femurs; (B–D) expanded view: high magnification image of the area within the red rectangle in image A; anti-Iba1 (Alexa488, turquoise fluorescence), labeling the M1 and M2 macrophages; anti-CD206 (Alexa568, red fluorescence) labeling the M2 macrophages and satellite cells; nuclear staining with DAPI (blue fluorescence); thin arrow: M1-like macrophages; thick arrow: M2-like macrophages. Figure S2: Identification of M2-like macrophages and satellite cells in the rat femurs: (A) immunolabeling with anti-CD163 and CD206 antibodies of the operated femurs; (B) negative control; anti-CD163 (Alexa488, turquoise fluorescence), labeling the M2-like macrophages; anti-CD206 (Alexa568, red fluorescence) labeling the M2-like macrophages and satellite cells; nuclear staining with DAPI (blue fluorescence); thin arrow: M2-like macrophages; thick arrow: satellite cells. Figure S3: Identification of macrophages M1-like and M2-like in the femurs of the rat; (A) negative control of Figure 4B; (B) negative control of Figure 4C; (C) negative control of Figure 4D; (D) negative control of Figure 4E; (E) negative control of Figure 4F; (F) negative control of Figure 4G; anti-CD68 (Alexa488, turquoise fluorescence) labeling the M1-like and M2-like macrophages.; anti-CD206 (Alexa568, red fluorescence) labeling the M2-like macrophages and satellite cells; nuclear staining with DAPI (blue fluorescence). Figure S4: In situ hybridization in the non-operated femur of the rat; (A) expression of β-actin mRNA (positive control); (B) negative control; (C) CD68 mRNA; (D) IL-1β mRNA; (E) CD163 mRNA; (F) IL-6 mRNA; (G) expanded view: high magnification image of the area within the red rectangle in image E; (H) expanded view: high magnification image of the area within the red rectangle in image F. Table S1: Oligos used for RT-qPCR: FW (forward primer); RV (reverse primer). Table S2. Oligos are used for in situ hybridization experiments: T3 (T3 promoter); T7 (T7 promoter); FW (forward primer); and RV (reverse primer).

Author Contributions: K.N. and A.-L.F. designed the experiments; C.C. performed the cryo-coupe; K.N. and E.S. performed the immunofluorescence and in situ hybridization; J.B. performed the spectrum analysis; M.O. helped with image preparation; K.N., X.H. and A.-L.F. designed the analysis approaches; X.H. and A.-L.F. devised and conducted the analyses; M.D. and J.-M.C. provided femurs of the rat samples; K.N., X.H. and A.-L.F. wrote the manuscript with input from all the authors. All authors discussed the results. All authors have read and agreed to the published version of the manuscript.

Funding: Work was supported by the Délégation Générale de l'Armement (DGA, Ministry of French Army) (PDH2-NRBC-4-NR-4306 and PDH-SAN-1-217/206).

Institutional Review Board Statement: This study was approved by the French Army Animal Ethics Committee (N°2011/22.1). All rats were treated in compliance with the European legislation (dir 2010/63/EU) implemented into French law (decree 2013-118) regulating animal experimentation.

Informed Consent Statement: Not applicable.

Data Availability Statement: Not applicable.

Acknowledgments: We are very grateful to Zsolt Kelemen for his helpful advice and critical reviewing our manuscript. We are also grateful to Xavier Butigieg for helping with image preparations.

Conflicts of Interest: The authors declare no conflict of interest.

References

1. Pajarinen, J.; Lin, T.; Gibon, E.; Kohno, Y.; Maruyama, M.; Nathan, K.; Lu, L.; Yao, Z.; Goodman, S.B. Mesenchymal Stem Cell-Macrophage Crosstalk and Bone Healing. *Biomaterials* **2019**, *196*, 80–89. [CrossRef] [PubMed]
2. Sun, Y.; Li, J.; Xie, X.; Gu, F.; Sui, Z.; Zhang, K.; Yu, T. Macrophage-Osteoclast Associations: Origin, Polarization, and Subgroups. *Front. Immunol.* **2021**, *12*, 778078. [CrossRef] [PubMed]
3. Smith, J.K. Osteoclasts and Microgravity. *Life* **2020**, *10*, 207. [CrossRef] [PubMed]
4. Gibon, E.; Lu, L.; Goodman, S.B. Aging, Inflammation, Stem Cells, and Bone Healing. *Stem. Cell Res. Ther.* **2016**, *7*, 44. [CrossRef]
5. Weitzmann, M.N. Bone and the Immune System. *Toxicol. Pathol.* **2017**, *45*, 911–924. [CrossRef]
6. Horwood, N.J. Macrophage Polarization and Bone Formation: A Review. *Clin. Rev. Allergy Immunol.* **2016**, *51*, 79–86. [CrossRef]
7. Kaur, S.; Raggatt, L.J.; Batoon, L.; Hume, D.A.; Levesque, J.-P.; Pettit, A.R. Role of Bone Marrow Macrophages in Controlling Homeostasis and Repair in Bone and Bone Marrow Niches. *Semin. Cell Dev. Biol.* **2017**, *61*, 12–21. [CrossRef]
8. Vi, L.; Baht, G.S.; Whetstone, H.; Ng, A.; Wei, Q.; Poon, R.; Mylvaganam, S.; Grynpas, M.; Alman, B.A. Macrophages Promote Osteoblastic Differentiation In-Vivo: Implications in Fracture Repair and Bone Homeostasis. *J. Bone Miner. Res.* **2015**, *30*, 1090–1102. [CrossRef]
9. Heideveld, E.; Horcas-Lopez, M.; Lopez-Yrigoyen, M.; Forrester, L.M.; Cassetta, L.; Pollard, J.W. Methods for Macrophage Differentiation and in Vitro Generation of Human Tumor Associated-like Macrophages. *Methods Enzym.* **2020**, *632*, 113–131. [CrossRef]
10. Orecchioni, M.; Ghosheh, Y.; Pramod, A.B.; Ley, K. Macrophage Polarization: Different Gene Signatures in M1(LPS+) vs. Classically and M2(LPS-) vs. Alternatively Activated Macrophages. *Front. Immunol.* **2019**, *10*, 1084. [CrossRef]
11. Chanput, W.; Mes, J.J.; Wichers, H.J. THP-1 Cell Line: An in Vitro Cell Model for Immune Modulation Approach. *Int. Immunopharmacol.* **2014**, *23*, 37–45. [CrossRef]
12. Martinez, F.O.; Sica, A.; Mantovani, A.; Locati, M. Macrophage Activation and Polarization. *Front. Biosci.* **2008**, *13*, 453–461. [CrossRef]
13. Mosser, D.M.; Edwards, J.P. Exploring the Full Spectrum of Macrophage Activation. *Nat. Rev. Immunol.* **2008**, *8*, 958–969. [CrossRef]
14. Mantovani, A.; Sica, A.; Sozzani, S.; Allavena, P.; Vecchi, A.; Locati, M. The Chemokine System in Diverse Forms of Macrophage Activation and Polarization. *Trends Immunol.* **2004**, *25*, 677–686. [CrossRef]
15. Graff, J.W.; Dickson, A.M.; Clay, G.; McCaffrey, A.P.; Wilson, M.E. Identifying Functional MicroRNAs in Macrophages with Polarized Phenotypes. *J. Biol. Chem.* **2012**, *287*, 21816–21825. [CrossRef]
16. Rőszer, T. Understanding the Mysterious M2 Macrophage through Activation Markers and Effector Mechanisms. *Mediat. Inflamm.* **2015**, *2015*, 816460. [CrossRef]
17. Abdelaziz, M.H.; Abdelwahab, S.F.; Wan, J.; Cai, W.; Huixuan, W.; Jianjun, C.; Kumar, K.D.; Vasudevan, A.; Sadek, A.; Su, Z.; et al. Alternatively Activated Macrophages; a Double-Edged Sword in Allergic Asthma. *J. Transl. Med.* **2020**, *18*, 58. [CrossRef]
18. Locati, M.; Curtale, G.; Mantovani, A. Diversity, Mechanisms, and Significance of Macrophage Plasticity. *Annu. Rev. Pathol.* **2020**, *15*, 123–147. [CrossRef]
19. Arora, S.; Dev, K.; Agarwal, B.; Das, P.; Syed, M.A. Macrophages: Their Role, Activation and Polarization in Pulmonary Diseases. *Immunobiology* **2018**, *223*, 383–396. [CrossRef]
20. Huang, X.; Xiu, H.; Zhang, S.; Zhang, G. The Role of Macrophages in the Pathogenesis of ALI/ARDS. *Mediat. Inflamm.* **2018**, *2018*, 1264913. [CrossRef]
21. Murray, P.J. Macrophage Polarization. *Annu. Rev. Physiol.* **2017**, *79*, 541–566. [CrossRef]
22. Van den Bossche, J.; O'Neill, L.A.; Menon, D. Macrophage Immunometabolism: Where Are We (Going)? *Trends Immunol.* **2017**, *38*, 395–406. [CrossRef]
23. Wynn, T.A.; Chawla, A.; Pollard, J.W. Macrophage Biology in Development, Homeostasis and Disease. *Nature* **2013**, *496*, 445–455. [CrossRef]

24. Wang, L.-X.; Zhang, S.-X.; Wu, H.-J.; Rong, X.-L.; Guo, J. M2b Macrophage Polarization and Its Roles in Diseases. *J. Leukoc. Biol.* **2019**, *106*, 345–358. [CrossRef]
25. Minutti, C.M.; Knipper, J.A.; Allen, J.E.; Zaiss, D.M.W. Tissue-Specific Contribution of Macrophages to Wound Healing. *Semin. Cell Dev. Biol.* **2017**, *61*, 3–11. [CrossRef]
26. Viniegra, A.; Goldberg, H.; Çil, Ç.; Fine, N.; Sheikh, Z.; Galli, M.; Freire, M.; Wang, Y.; Van Dyke, T.E.; Glogauer, M.; et al. Resolving Macrophages Counter Osteolysis by Anabolic Actions on Bone Cells. *J. Dent. Res.* **2018**, *97*, 1160–1169. [CrossRef]
27. Shapouri-Moghaddam, A.; Mohammadian, S.; Vazini, H.; Taghadosi, M.; Esmaeili, S.-A.; Mardani, F.; Seifi, B.; Mohammadi, A.; Afshari, J.T.; Sahebkar, A. Macrophage Plasticity, Polarization, and Function in Health and Disease. *J. Cell Physiol.* **2018**, *233*, 6425–6440. [CrossRef]
28. Selders, G.S.; Fetz, A.E.; Radic, M.Z.; Bowlin, G.L. An Overview of the Role of Neutrophils in Innate Immunity, Inflammation and Host-Biomaterial Integration. *Regen. Biomater.* **2017**, *4*, 55–68. [CrossRef]
29. Mantovani, A. Wandering Pathways in the Regulation of Innate Immunity and Inflammation. *J. Autoimmun.* **2017**, *85*, 1–5. [CrossRef]
30. Snyder, R.J.; Lantis, J.; Kirsner, R.S.; Shah, V.; Molyneaux, M.; Carter, M.J. Macrophages: A Review of Their Role in Wound Healing and Their Therapeutic Use. *Wound Repair Regen.* **2016**, *24*, 613–629. [CrossRef]
31. Batoon, L.; Millard, S.M.; Wullschleger, M.E.; Preda, C.; Wu, A.C.-K.; Kaur, S.; Tseng, H.-W.; Hume, D.A.; Levesque, J.-P.; Raggatt, L.J.; et al. CD169[+] Macrophages Are Critical for Osteoblast Maintenance and Promote Intramembranous and Endochondral Ossification during Bone Repair. *Biomaterials* **2019**, *196*, 51–66. [CrossRef] [PubMed]
32. Miron, R.J.; Bosshardt, D.D. OsteoMacs: Key Players around Bone Biomaterials. *Biomaterials* **2016**, *82*, 1–19. [CrossRef] [PubMed]
33. Tan, B.; Tang, Q.; Zhong, Y.; Wei, Y.; He, L.; Wu, Y.; Wu, J.; Liao, J. Biomaterial-Based Strategies for Maxillofacial Tumour Therapy and Bone Defect Regeneration. *Int. J. Oral. Sci.* **2021**, *13*, 9. [CrossRef] [PubMed]
34. Lebaudy, E.; Fournel, S.; Lavalle, P.; Vrana, N.E.; Gribova, V. Recent Advances in Antiinflammatory Material Design. *Adv. Healthc. Mater.* **2021**, *10*, e2001373. [CrossRef]
35. Witherel, C.E.; Abebayehu, D.; Barker, T.H.; Spiller, K.L. Macrophage and Fibroblast Interactions in Biomaterial-Mediated Fibrosis. *Adv. Healthc. Mater.* **2019**, *8*, e1801451. [CrossRef]
36. Sadtler, K.; Sommerfeld, S.D.; Wolf, M.T.; Wang, X.; Majumdar, S.; Chung, L.; Kelkar, D.S.; Pandey, A.; Elisseeff, J.H. Proteomic Composition and Immunomodulatory Properties of Urinary Bladder Matrix Scaffolds in Homeostasis and Injury. *Semin. Immunol.* **2017**, *29*, 14–23. [CrossRef]
37. Klopfleisch, R. Macrophage Reaction against Biomaterials in the Mouse Model-Phenotypes, Functions and Markers. *Acta Biomater.* **2016**, *43*, 3–13. [CrossRef]
38. Boersema, G.S.A.; Grotenhuis, N.; Bayon, Y.; Lange, J.F.; Bastiaansen-Jenniskens, Y.M. The Effect of Biomaterials Used for Tissue Regeneration Purposes on Polarization of Macrophages. *Biores. Open Access* **2016**, *5*, 6–14. [CrossRef]
39. Ogle, M.E.; Segar, C.E.; Sridhar, S.; Botchwey, E.A. Monocytes and Macrophages in Tissue Repair: Implications for Immunoregenerative Biomaterial Design. *Exp. Biol. Med.* **2016**, *241*, 1084–1097. [CrossRef]
40. Huffman, L.K.; Harris, J.G.; Suk, M. Using the Bi-Masquelet Technique and Reamer-Irrigator-Aspirator for Post-Traumatic Foot Reconstruction. *Foot Ankle Int.* **2009**, *30*, 895–899. [CrossRef]
41. Masquelet, A.C. Free Vascularized Corticoperiosteal Grafts. *Plast. Reconstr. Surg.* **1991**, *88*, 1106. [CrossRef]
42. Masquelet, A.C. Induced Membrane Technique: Pearls and Pitfalls. *J. Orthop. Trauma* **2017**, *31* (Suppl. 5), S36–S38. [CrossRef]
43. Masquelet, A.C. The Induced Membrane Technique. *Orthop. Traumatol. Surg. Res.* **2020**, *106*, 785–787. [CrossRef]
44. Muñoz, J.; Akhavan, N.S.; Mullins, A.P.; Arjmandi, B.H. Macrophage Polarization and Osteoporosis: A Review. *Nutrients* **2020**, *12*, 2999. [CrossRef]
45. Eggold, J.T.; Rankin, E.B. Erythropoiesis, EPO, Macrophages, and Bone. *Bone* **2019**, *119*, 36–41. [CrossRef]
46. Michalski, M.N.; McCauley, L.K. Macrophages and Skeletal Health. *Pharmacol. Ther.* **2017**, *174*, 43–54. [CrossRef]
47. Nikovics, K.; Favier, A.-L. Macrophage Identification in Situ. *Biomedicines* **2021**, *9*, 1393. [CrossRef]
48. Nikovics, K.; Morin, H.; Riccobono, D.; Bendahmane, A.; Favier, A. Hybridization-chain-reaction Is a Relevant Method for in Situ Detection of M2d-like Macrophages in a Mini-pig Model. *FASEB J.* **2020**, *34*, 15675–15686. [CrossRef]
49. Nikovics, K.; Castellarin, C.; Holy, X.; Durand, M.; Morin, H.; Bendahmane, A.; Favier, A. In Situ Gene Expression in Native Cryofixed Bone Tissue. *Biomedicines* **2022**, *10*, 484–498. [CrossRef]
50. Choi, H.M.T.; Beck, V.A.; Pierce, N.A. Next-Generation in Situ Hybridization Chain Reaction: Higher Gain, Lower Cost, Greater Durability. *ACS Nano* **2014**, *8*, 4284–4294. [CrossRef]
51. Choi, H.M.T.; Calvert, C.R.; Husain, N.; Huss, D.; Barsi, J.C.; Deverman, B.E.; Hunter, R.C.; Kato, M.; Lee, S.M.; Abelin, A.C.T.; et al. Mapping a Multiplexed Zoo of MRNA Expression. *Development* **2016**, *143*, 3632–3637. [CrossRef]
52. Choi, H.M.T.; Schwarzkopf, M.; Fornace, M.E.; Acharya, A.; Artavanis, G.; Stegmaier, J.; Cunha, A.; Pierce, N.A. Third-Generation in Situ Hybridization Chain Reaction: Multiplexed, Quantitative, Sensitive, Versatile, Robust. *Development* **2018**, *145*, dev165753. [CrossRef]
53. Choi, H.M.T.; Schwarzkopf, M.; Pierce, N.A. Multiplexed Quantitative in Situ Hybridization with Subcellular or Single-Molecule Resolution Within Whole-Mount Vertebrate Embryos: QHCR and DHCR Imaging (v3.0). *Methods Mol. Biol.* **2020**, *2148*, 159–178. [CrossRef]

54. Song, S.H.; Kim, J.E.; Lee, Y.J.; Kwak, M.H.; Sung, G.Y.; Kwon, S.H.; Son, H.J.; Lee, H.S.; Jung, Y.J.; Hwang, D.Y. Cellulose Film Regenerated from Styela Clava Tunics Have Biodegradability, Toxicity and Biocompatibility in the Skin of SD Rats. *J. Mater. Sci. Mater. Med.* **2014**, *25*, 1519–1530. [CrossRef]

55. Betjes, M.G.; Haks, M.C.; Tuk, C.W.; Beelen, R.H. Monoclonal Antibody EBM11 (Anti-CD68) Discriminates between Dendritic Cells and Macrophages after Short-Term Culture. *Immunobiology* **1991**, *183*, 79–87. [CrossRef]

56. Chistiakov, D.A.; Killingsworth, M.C.; Myasoedova, V.A.; Orekhov, A.N.; Bobryshev, Y.V. CD68/Macrosialin: Not Just a Histochemical Marker. *Lab. Investig.* **2017**, *97*, 4–13. [CrossRef]

57. Kosmac, K.; Peck, B.D.; Walton, R.G.; Mula, J.; Kern, P.A.; Bamman, M.M.; Dennis, R.A.; Jacobs, C.A.; Lattermann, C.; Johnson, D.L.; et al. Immunohistochemical Identification of Human Skeletal Muscle Macrophages. *Bio-Protoc.* **2018**, *8*, e2883. [CrossRef]

58. Tidball, J.G.; Villalta, S.A. Regulatory Interactions between Muscle and the Immune System during Muscle Regeneration. *Am. J. Physiol. Regul. Integr. Comp. Physiol.* **2010**, *298*, R1173–R1187. [CrossRef]

59. Fujimori, M.; Kimura, Y.; Ueshima, E.; Dupuy, D.E.; Adusumilli, P.S.; Solomon, S.B.; Srimathveeravalli, G. Lung Ablation with Irreversible Electroporation Promotes Immune Cell Infiltration by Sparing Extracellular Matrix Proteins and Vasculature: Implications for Immunotherapy. *Bioelectricity* **2021**, *3*, 204–214. [CrossRef]

60. Zhang, X.; Thompkins-Johns, A.; Ziober, A.; Zhang, P.J.; Furth, E.E. Hepatic Macrophage Types Cluster with Disease Etiology in Chronic Liver Disease and Differ Compared to Normal Liver: Implications for Their Biologic and Diagnostic Role. *Int. J. Surg. Pathol.* **2022**. [CrossRef]

61. Oh, T.; Do, D.T.; Vo, H.V.; Kwon, H.-I.; Lee, S.-C.; Kim, M.H.; Nguyen, D.T.T.; Le, Q.T.V.; Tran, T.M.; Nguyen, T.T.; et al. The Isolation and Replication of African Swine Fever Virus in Primary Renal-Derived Swine Macrophages. *Front. Vet. Sci.* **2021**, *8*, 645456. [CrossRef] [PubMed]

62. Fauch, L.; Palander, A.; Dekker, H.; Schulten, E.A.; Koistinen, A.; Kullaa, A.; Keinänen, M. Narrowband-Autofluorescence Imaging for Bone Analysis. *Biomed. Opt. Express* **2019**, *10*, 2367–2382. [CrossRef] [PubMed]

63. Capasso, L.; D'Anastasio, R.; Guarnieri, S.; Viciano, J.; Mariggiò, M. Bone Natural Autofluorescence and Confocal Laser Scanning Microscopy: Preliminary Results of a Novel Useful Tool to Distinguish between Forensic and Ancient Human Skeletal Remains. *Forensic. Sci. Int.* **2017**, *272*, 87–96. [CrossRef] [PubMed]

64. Monici, M. Cell and Tissue Autofluorescence Research and Diagnostic Applications. *Biotechnol. Annu. Rev.* **2005**, *11*, 227–256. [CrossRef] [PubMed]

65. Vedeswari In Vivo Autofluorescence Characteristics of Pre- and Post-Treated Oral Submucous Fibrosis: A Pilot Study-PubMed. Available online: https://pubmed.ncbi.nlm.nih.gov/19884705/ (accessed on 17 February 2021).

66. Weiss, G.; Schaible, U.E. Macrophage Defense Mechanisms against Intracellular Bacteria. *Immunol. Rev.* **2015**, *264*, 182–203. [CrossRef]

67. Altan-Bonnet, G.; Mukherjee, R. Cytokine-Mediated Communication: A Quantitative Appraisal of Immune Complexity. *Nat. Rev. Immunol.* **2019**, *19*, 205–217. [CrossRef]

68. Chu, Y.-H.; Hardin, H.; Zhang, R.; Guo, Z.; Lloyd, R.V. In Situ Hybridization: Introduction to Techniques, Applications and Pitfalls in the Performance and Interpretation of Assays. *Semin. Diagn. Pathol.* **2019**, *36*, 336–341. [CrossRef]

69. Herrmann, M.; Engelke, K.; Ebert, R.; Müller-Deubert, S.; Rudert, M.; Ziouti, F.; Jundt, F.; Felsenberg, D.; Jakob, F. Interactions between Muscle and Bone-Where Physics Meets Biology. *Biomolecules* **2020**, *10*, 432. [CrossRef]

70. Sandberg, O.H.; Tätting, L.; Bernhardsson, M.E.; Aspenberg, P. Temporal Role of Macrophages in Cancellous Bone Healing. *Bone* **2017**, *101*, 129–133. [CrossRef]

71. Jablonská, E.; Horkavcová, D.; Rohanová, D.; Brauer, D.S. A Review of in Vitro Cell Culture Testing Methods for Bioactive Glasses and Other Biomaterials for Hard Tissue Regeneration. *J. Mater. Chem. B* **2020**, *8*, 10941–10953. [CrossRef]

72. Bonnardel, J.; Guilliams, M. Developmental Control of Macrophage Function. *Curr. Opin. Immunol.* **2018**, *50*, 64–74. [CrossRef]

73. Eme-Scolan, E.; Dando, S.J. Tools and Approaches for Studying Microglia In Vivo. *Front. Immunol.* **2020**, *11*, 583647. [CrossRef]

74. Sica, A.; Mantovani, A. Macrophage Plasticity and Polarization: In Vivo Veritas. *J. Clin. Investig.* **2012**, *122*, 787–795. [CrossRef]

75. Loi, F.; Córdova, L.A.; Pajarinen, J.; Lin, T.; Yao, Z.; Goodman, S.B. Inflammation, Fracture and Bone Repair. *Bone* **2016**, *86*, 119–130. [CrossRef]

76. Nich, C.; Takakubo, Y.; Pajarinen, J.; Ainola, M.; Salem, A.; Sillat, T.; Rao, A.J.; Raska, M.; Tamaki, Y.; Takagi, M.; et al. Macrophages-Key Cells in the Response to Wear Debris from Joint Replacements. *J. Biomed. Mater. Res. A* **2013**, *101*, 3033–3045. [CrossRef]

77. Claes, L.; Recknagel, S.; Ignatius, A. Fracture Healing under Healthy and Inflammatory Conditions. *Nat. Rev. Rheumatol.* **2012**, *8*, 133–143. [CrossRef]

78. Bauman, J.G.; Wiegant, J.; Borst, P.; van Duijn, P. A New Method for Fluorescence Microscopical Localization of Specific DNA Sequences by in Situ Hybridization of Fluorochromelabelled RNA. *Exp. Cell Res.* **1980**, *128*, 485–490. [CrossRef]

79. Bayani, J.; Squire, J.A. Fluorescence in Situ Hybridization (FISH). *Curr. Protoc. Cell Biol.* **2004**, *23*, 22–24. [CrossRef]

80. Bayani, J.; Squire, J.A. Comparative Genomic Hybridization. *Curr. Protoc. Cell Biol.* **2005**, *25*, 22–26. [CrossRef]

81. Pothos, A.; Plastira, K.; Plastiras, A.; Vlachodimitropoulos, D.; Goutas, N.; Angelopoulou, R. Comparison of Chromogenic in Situ Hybridisation with Fluorescence in Situ Hybridisation and Immunohistochemistry for the Assessment of Her-2/Neu Oncogene in Archival Material of Breast Carcinoma. *Acta Histochem. Cytochem.* **2008**, *41*, 59–64. [CrossRef]

Review

Critical Role of Inflammation and Specialized Pro-Resolving Mediators in the Pathogenesis of Atherosclerosis

Subhapradha Rangarajan [1], Davit Orujyan [1], Patrida Rangchaikul [1] and Mohamed M. Radwan [2,*]

1   College of Osteopathic Medicine of the Pacific, Western University of Health Sciences, Pomona, CA 91766, USA
2   Department of Translational Research, College of Osteopathic Medicine of the Pacific, Western University of Health Sciences, Pomona, CA 91766, USA
*   Correspondence: mradwanahmed@westernu.edu; Tel.: +1-909-469-7041

Abstract: Recent research on how the body resolves this inflammation is gaining traction and has shed light on new avenues for future management of cardiovascular diseases. In this narrative review, we discuss the pathophysiological mechanisms of atherosclerosis, the recent development in the understanding of a new class of molecules called Specialized Pro-resolving Mediators (SPMs), and the impact of such findings in the realm of cardiovascular treatment options. We searched the MEDLINE database restricting ourselves to original research articles as much as possible on the complex pathophysiology of atherosclerosis and the role of SPMs. We expect to see further research in translating these findings to bedside clinical trials in treating conditions with a pathophysiological basis of inflammation, such as coronary artery disease, asthma, and periodontal disease.

Keywords: cardiovascular disease; atherosclerosis; efferocytosis; inflammation; resolution; immunotherapy

Citation: Rangarajan, S.; Orujyan, D.; Rangchaikul, P.; Radwan, M.M. Critical Role of Inflammation and Specialized Pro-Resolving Mediators in the Pathogenesis of Atherosclerosis. *Biomedicines* **2022**, *10*, 2829. https://doi.org/10.3390/biomedicines10112829

Academic Editor: Krisztina Nikovics

Received: 25 September 2022
Accepted: 3 November 2022
Published: 6 November 2022

Publisher's Note: MDPI stays neutral with regard to jurisdictional claims in published maps and institutional affiliations.

1. Introduction

Cardiovascular diseases have been ranked as one of the top causes of death in the United States since 1980, with over 600,000 deaths recorded in 2018, irrespective of ethnicity [1].

Atherosclerosis, a disease of the arteries set off initially by fat deposition, is a major cause of life-threatening cardiovascular events. It was long thought to be a passive process caused by the accumulation of cholesterol within the lumen of arteries resulting in ischemia and an eventual complete blockage. However, in recent decades studies proved that plaque inflation and rupture are the events that lead to the life-threatening consequences of atherosclerosis [2,3].

Arteries are composed of endothelial cells (EC), elastin, collagen, and smooth muscle cells [4]. ECs line the lumen of vessels and are subject to physical demands, such as shear stress, imposed by the flow of blood. Such stressors fluctuate and vary through the length of the artery, owing to the rheological properties of the blood and vulnerable areas of the arteries, such as branching points. These factors contributed to the initial focus of the pathogenesis of atherosclerosis and amplified through environmental factors such as age-related arterial degeneration, lifestyle choices of diet and exercise, and other risk factors such as obesity, hypertension, hyperlipidemia, diabetes mellitus, and smoking [2,5,6].

Recently the concept of inflammation resolution has garnered attention, as studies showed that the end of acute inflammation is an active concerted effort by a class of molecules termed SPMs and not a passive process that fizzles away in time. Thus, allowing us to look at treating inflammation by enhancing its resolution. Since atherosclerosis is a chronic inflammation that takes years, a temporal dependency dictates the efficacy and efficiency of such processes [7,8]. This temporal dependency poses a challenge in identifying risk groups with current diagnostic tools since not all individuals with the same cholesterol level end up with the same stage of atherosclerosis, and the lumens of such vulnerable arteries are hard to access at times [2].

Current treatments for atherosclerosis include reducing blood cholesterol levels and surgical upkeep of the arterial lumen—based on the knowledge that cholesterol accumulation is the precursor of its pathogenesis; some developments have been made in treating atherosclerosis with anti-inflammatory medications [9].

2. Pathogenesis of Atherosclerosis

The stressors discussed above subject the EC to injury. Intact ECs cannot regenerate the nearby wounded site or migrate distally to an injured site to repair, causing the incessant injury to exhaust EC's turnover capacity. Lack of a repair mechanism results in perpetually dysfunctional ECs. These cells lose their tight junctions and become more permeable to molecules that otherwise would not be able to enter the intima. The ECs undergo alterations in their adhesive characteristics, becoming 'sticky,' leading to more monocyte and T cell attachment in early plaque formation and during plaque growth, respectively. Impaired ECs also exhibit growth-stimulatory characteristics, enabling the entry of LDL molecules and monocytes into the intimal layer, triggering the formation of a fatty streak, the first step in the long process of plaque formation [6,10].

Apo B-100 receptors of the LDL particles bind to the proteoglycan molecules of the extracellular matrix, enter the subintimal layer, and get oxidized. While their oxidation process is multifactorial and not fully understood, it is postulated that Nitric Oxide Synthase generated by the activated macrophages might have a significant role in it. Myeloperoxidase, 15-lipooxygenase, hypochlorous acid, and phenoxy radical intermediates also bring about such oxidation [2]. Oxidized molecular species modify the Lysine residues of Apo B 100 [11].

In atherosclerosis, the oxidized LDL (ox-LDL) induces the release of chemical mediators by the cells in its vicinity and promotes the accumulation of macrophages. This process initiates plaque formation and recruits inflammatory cells, beginning the process of chronic inflammation in the arterial wall [11,12].

Within the intima layer, the damaged ECs release macrophage colony-stimulating factor (M-CSF), which converts the initial set of monocytes into macrophages. These, in turn, generate monocyte chemoattractant protein (MCP-1), which increases the number of immunocompetent cells in the region. Studies have shown that MCP$-/-$ and LDLR$-/-$ mice do not have a risk of atherosclerosis [13]. MCP-1 and the ox-LDL particle are the essential chemokines involved in atherosclerosis [3]. In addition, LDL oxidation generates reactive aldehydes and truncated lipids that trigger a pro-inflammatory cascade in ECs and the expression of adhesion molecules such as VCAM-1, E-Selectin, and P-Selectin. Receptors for MCP-1 on monocytes are heavily upregulated during early plaque formation and are expressed by endothelial cells, smooth muscle cells, and macrophages [2,3,5,6]. Transient blockage of P-selectin, one of the receptors on EC or its ligand, in an apoe$-/-$, cholesterol-fed mice before the incident of vascular injury resulted in a substantial reduction of neointima formation [14].

Macrophages phagocytose ox-LDL molecules through the scavenger receptors SR-A and CD36. An increase in ox-LDL intake does not downregulate these receptors, so macrophages can potentially keep intaking these particles until they undergo apoptosis [2]. In healthier conditions or with high HDL presence, such macrophages can transfer the LDL species to the HDL molecules to be circulated back to the liver. Apoptosis of macrophages spills out the ingested ox-LDL giving it the color and the name fatty streak [2,3,5,6].

Vascular Smooth Muscle Cells (VSMC) are called into the growing plaque within the intima and generate collagen, elastin, and other extracellular proteoglycans that give the plaque its fibrous cap [5,6]. Even as the fibrous cap provides uniformity and stability to the plaque, at its core is a soft lipid and necrotic material that, with enough growth, occludes the lumen. On the other hand, if this lesion is not uniform, its stability is compromised, allowing the possibility of rupture from shear stress [6].

3. Plaque Stability

It has been established that it is the "stability" of the plaque rather than its formation, the culprit behind life-threatening cardiovascular events (Figure 1). The process in which a "fatty streak" morphs into an unstable plaque that ruptures, involves a complex interplay of biochemicals between subsets of different types of immune-modulating cells and the local environment. The most detrimental effect of a stable plaque could be ischemia of a distal organ due to lumen occlusion. In contrast, an unstable plaque with its risk of rupture and resulting thrombo-embolism will most definitely result in life-threatening cardiovascular events such as infarction or stroke.

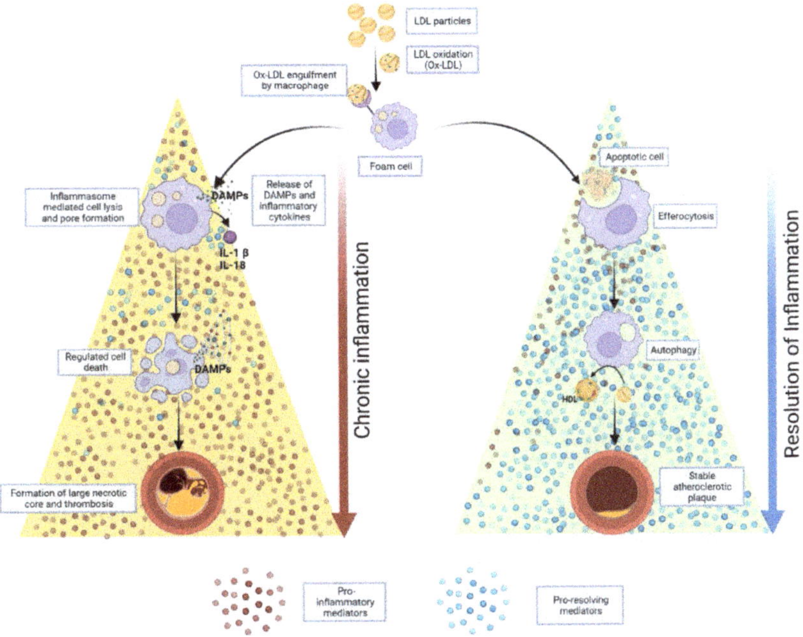

Figure 1. The proportion of pro-inflammatory and pro-resolving mediators in its microenvironment decides the stability and fate of a growing plaque. An abundance of pro-inflammatory mediators, results in inflammasome mediated pore formation leading to inflamed cell death and further release of proinflammatory mediators such as DAMPs, prolonging the inflammation cycle and rendering the plaque unstable and prone to thrombus formation. On the contrary, with an abundance of SPMs in the milieu leads to efferocytosis, a non-phlogistic clearance of any cellular debris, and regulated autophagy leading to a stable plaque.

As discussed above, fibrous cap comprises primarily of connective tissue and VSMCs, that prevents macrophage-derived tissue factor from encountering various coagulation factors in blood. Plaque instability rises from thinning of the fibrous cap due to the various cytokines, apoptosis of VSMC, and the reduction of collagen production [6,15]. Reduction of collagen production seems to be one of the major milestones in the cascade of events that leads to the rupture of the fibrous cap. Studies in mice with various impaired {collagen pathways, such as, enhanced INF-γ signaling, or genetically induced scurvy, have shown to develop plaques susceptible to rupture due to a weak fibrous cap [9]. Molecular processes such as cell adhesion, cytoskeletal restructuring, and migration play important role behind the scenes in enabling various cellular players that define the characteristics of a stable plaque. A ubiquitously expressed molecule Talin-1 aids in intercellular communications through integrin activation, and crosstalk. While the exact role of it, if any, in plaque stability needs to be determined, its expression has been shown to be downregulated in plaques vs. control arteries and has been found to exhibit a positive correlation with a stable vs. an unstable plaque. Further, miRNA-330-5p has been identified as a potential positive regulator of Talin-1 [16]. This is an example that the establishment of a plaque's stability is multifactorial involving molecular basis as well as genetic expressions. Some studies have shown a positive correlation and a predominance of M1 subtype of macrophages in ruptured plaques; however, any causal association is under debate, and our understanding is still evolving [17].

4. VSMCs

Within the growing stable plaque, an interaction between the T cells and macrophages secrete a wide array of chemokines and growth factors that target circulating monocytes, local endothelial cells, and smooth muscle cells. This proliferation can lead to lesion expansion. As discussed above, the stability of the plaque is derived from the presence of a thick fibrous cap produced in part by the VSMCs. Activated VSMCs switch their phenotype from contractile to synthetic and synthesize fibrotic proteins such as collagen [18]. Typically, TGF-β is a potent activator of collagen synthesis by VSMCs; however, T-cells secrete INF-γ, which inhibits collagen synthesis, especially types I and III, that are majorly found in extracellular matrices of arteries [9,19]. It has been shown that INF-γ inhibit collagen production by VSMCs even in the presence of TGF-β [19]. The source of INF-γ is T-cells, the presence of which in atherosclerotic lesions helps to tie the link between adaptive immune response to the stability of a plaque.

VSMCs also express scavenger receptors such as LOX-1 and CD-36 that internalize ox-LDL, generating foam cells out of VSMCs, as they would with macrophages. This generation of foam cells from VSMCs is most likely due to VSMCs' phenotypic change to resemble monocytes and mesenchymal stem cells. Endothelial and VSMCs, under the conditions of acute inflammation, start to produce pro-inflammatory molecules such as TNF-α, IL-1β, and MCP-1, among many others, which attract neutrophils to the inflammatory site. Neutrophils, in turn, act on these cells to increase their pro-inflammatory effects, thus producing a snowball effect that prolongs the inflammatory response and intimal hyperplasia [20]. Studies have shown that a knockout of the transcription factor, Klf-4, which possibly mediates the phenotypic change in VSMCs, results in a reduction of VSMC-derived macrophage-like cells, a smaller lesion size, and increased fibrous cap thickness. These changes consequently increase plaque stability [21].

5. Macrophages, T Cells, Platelets

Macrophages, classically have been categorized into M1, involved in pro-inflammatory pathways, and M2, participating in anti-inflammatory events. As noted above, monocytes recruited to the site of plaque formation are activated by toll-like receptors (TLRs), and INF-γ present in the micro-environment of a growing plaque to M1 subtype [17,22,23]. These M1 type macrophages go on to secrete other pro-inflammatory cytokines such as IL-1b, IL-6, TNF, IL-12, and IL-23 along with molecules such as reactive oxygen and nitrogen species that sustain the ongoing inflammation [17]. M1 type macrophage also recruit Th1 and natural killer cells to the growing plaque further exacerbating the injury due to uncontrolled inflammation [24].

While historically, atherosclerotic plaques were characterized mainly by M1 type macrophages, recent studies have brought to fore other subtypes of macrophages, such as M4, Mox, M(Hb), Mhem. These subtypes vary in their local milieu and therefore their activation, their phenotypic markers, and in their participation or role in the pathogenesis as well as the characteristics of an atherosclerotic plaque. Recent studies with mouse models and carotid plaques have shown the presence of M2 subtype in advanced lesions [25]. However, since the cause of their presence could not be narrowed down, and since macrophage phenotypic switching based on their micro-environment is well-known, such representation is challenging and studies are still ongoing to demonstrate the presence of distinct types of macrophages and their role in plaque stability and severity [17].

The reduction in collagen levels in a plaque is partly due to decreased production mediated by T cells and increased levels of breakdown by macrophages, neutrophils, endothelial cells, and SMCs. Studies have shown that macrophages secrete several matrix metalloproteinases (MMPs), such as MMP-1, MMP-2, MMP-8, MMP-9, and MMP-13, which are structurally identical to enzymes that degrade fibrillar collagen types I and III, proteoglycans, and elastin. Such MMPs also activate platelet aggregation and adhesion. Ox-LDL increases MMP-14 expression in endothelial cells, which in turn activates MMP-2, a potent gelatinase that acts on collagen IV, a constituent of the basement membrane. T-cells can bring such secretions by macrophages through the CD40 ligand (CD40L). The finding that the lack of MMPs resulted in increased and better-organized collagen in such plaques proved the role of MMPs in plaque stability. Elastases such as cathepsins S and K, and neutrophil elastase, have been found in plaques, which lower plaque stability possibly by degrading the extra-cellular matrix [19]. CD40L by T cells not only increases the secretion of MMPs but has also been attributed to the expression of tissue factor (TF) expression by macrophages. TF is the major activator of thrombosis when the cells encounter coagulation factors in blood. CD40L are also derived from platelets, which can cause its aggregation when the local environment is conducive. This can lead to local small arteries that feed the plaque to rupture leading to an intraplaque hemorrhage. This loop can continue whereby the exposed CD40L can activate TF in the local environment, leading to the growth of the thrombus, which in turn can cause further inflammation. While rupture of a plaque is the climax of the negative cardiac events, superficial erosion of the endothelial cells lining the plaque contributes to the weakening of its cap [19].

Not all small-scale changes that occur regularly within a plaque that result in mural thrombus formation lead to negative cardiac events, as seen in the thrombus formed within the vasculature of patients that did not die of heart conditions. These mural thrombi have platelets as its participants and platelet derived growth factor (PDGF) thus secreted might have led to the fibrosis of the plaque conferring it with stability. This 'healing' of the minor plaque ruptures lead to the 'expansive remodeling' of the artery rather than the 'constrictive remodeling' observed in an immature growing inflamed plaque [19].

6. Apoptosis

TNF family of cytokines from platelets has a pronounced effect on the apoptosis of VSMCs. In addition, chemokines such as IL-1beta, TNF-alpha, and IFN-gamma by activated macrophages and T cells in the immediate environment, induce apoptosis in VSMC and block collagen production [26]. Myeloperoxidase produces hypochlorous acid in the event of increased oxidative stress, which induces apoptosis in EC. Increased caspase-3 and DNA ladders in ECs support the theory that oxidative stress inherent in inflammation-induced plaque formation leads to a cyclic deterioration of plaque stability [19].

Loss of collagen through apoptosis of VSMCs, increased MMPs, and the accumulation of necrotic debris by apoptotic macrophages, all cumulatively result in an unstable plaque, increasing the chance for rupture. Risk of plaque rupture also increases with superficial erosion of endothelial monolayer, the risk of which is increased with apoptosis of endothelial cells due to inflammation [19]. Apoptosis is beneficial in clearing the cells that help eliminate oxidized elements in the initial lesion, but the same processes, if left unchecked, are detrimental during the later stages. Removing the injurious agent or resolving the inflammation could reverse the progression of the lesion from a fatty streak to an unstable plaque.

7. Resolution of Inflammation

Resolution of inflammation has been established as an active process that begins with and is characterized by a reduction in neutrophil recruitment and an increase in efferocytosis, a non-phlogistic clearance of cellular debris by macrophages [8].

As with the active inflammation process, its resolution involves a myriad of chemical modulators from a variety of cell populations through complex chemical pathways that are interconnected. Resolution of an ongoing inflammation is kicked off by the "class switching" of prostaglandins and leukotrienes to lipoxins, which are also derived from arachidonic acid [27].

The resolution of acute inflammation was studied by analyzing the exudate from inflamed tissues and was found to be mediated by molecules derived from essential fatty acids, Eicosapentaenoic acid, and Docosahexaenoic acid (DHA). These molecules are termed SPMs and are classified into further subdivisions—resolvins (E and D series), protectins, lipoxins, and maresins. E-series resolvins are generated from EPA, lipoxins from arachidonic acid, while the rest, D-series resolvins, protectins, and Maresins, are derived from DHA [21,28]. These SPMs assist in reducing inflammation via several mechanisms, including increasing efferocytosis, as shown in (Figure 2) [29].

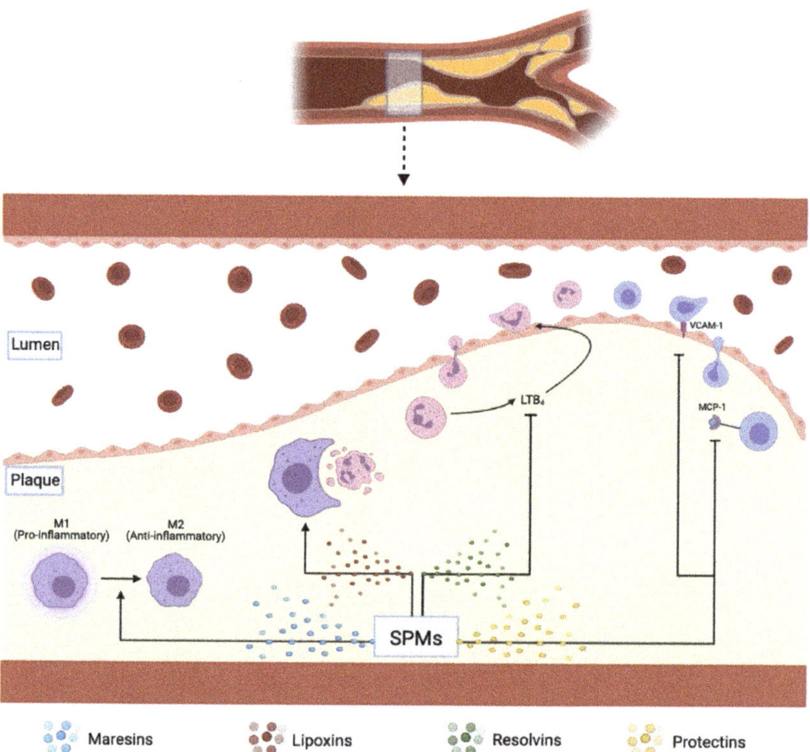

Figure 2. SPMs including maresins, lipoxins, resolvins, and protectins, assist in plaque resolution through various pathways: converting pro-inflammatory M1 to anti-inflammatory M2 macrophages, increasing effective efferocytosis, downregulating pro-inflammatory LTB4, VCAM-1, and MCP-1.

Resolution of an ongoing inflammation stimulates the increased formation of lipoxins, which facilitate resolution by stopping further recruitment of neutrophils, inducing non-phlogistic migration, and induction of macrophages to clear apoptotic neutrophils [27].

8. Efferocytosis

The process in which apoptotic cells are cleared is called efferocytosis. While mainly managed by macrophages, vascular smooth muscle cells and neighboring cells may also have efferocytotic roles. As discussed above, apoptosis is a significant phenomenon that defines the progression of a plaque: EC, VSMC, and foamy macrophages all undergo apoptosis fueled by the growing plaque's chemokine environment. Tabas et al. have concluded that clearance of such apoptotic cells is the real issue within an atherosclerotic plaque than apoptosis itself [10]. Martinet et al. have shown that efferocytosis is reduced by approximately 20-fold in a plaque relative to normal [30,31]. Interestingly, macrophages that become foam cells do not unload the engulfed ox-LDL to HDLs due to their defective efferocytosis, suppressing reverse cholesterol pathways' normal functioning. Defective efferocytosis also triggers macrophages to secrete pro-inflammatory signals such as TGF-β or IL-10. When efferocytosis is not complete, macrophages undergo cell membrane lysis spilling out necrotic chemicals such as proteases, thrombogenic tissue factors, and angiogenesis-promoting cytokines, creating more pro-inflammatory pathways (Figure 2). It is also thought that phenotype switching of macrophages from anti-inflammatory M2 to pro-inflammatory M1 has a role in diminishing the efficiency of efferocytosis [32].

Several classes of cellular molecules highly regulate efferocytosis: "find me" ligands that recruit phagocytes to the site of apoptosis, bridging molecules that link phagocytes to their targets, and "eat me" ligands on the apoptotic cell surface. These "eat me" ligands on cell surfaces bind and activate engulfment receptors on phagocytes. A counter molecule class called "don't eat me" ligand is present on viable cells but is downregulated in apoptotic cells. Studies have shown that ox-LDL in plaques induces auto-antibody generation within macrophages and other phagocytes. These autoantibodies mask the cell-surface "eat me" ligands on the dying cell. Ox-LDL also seems to act as a competitive inhibitor of scavenger receptors, making them less efficient in clearing apoptotic cells, as shown in (Figure 3). Calreticulin (Calr) is one of the key "eat me" ligands that binds to LDL Receptor-Related Protein (LRP1) on phagocytic cells and induces engulfment. Carriers of a risk allele at chromosome 9p21 are shown to express less Calr due to an inherited defect in TGF-β signaling, resulting in a more extensive lesion formation. Plaques of such mice models (that lack one of the 9p21 candidate genes) have been shown to exhibit plaque destabilizing features. VSMCs deficient in Calr have been shown to resist phagocytosis in vitro, induce pro-inflammatory foam-cell phenotype on cocultured macrophages, and suppress reverse cholesterol transport. It was shown that an exogenous introduction of Calr reversed these effects in-vitro [32].

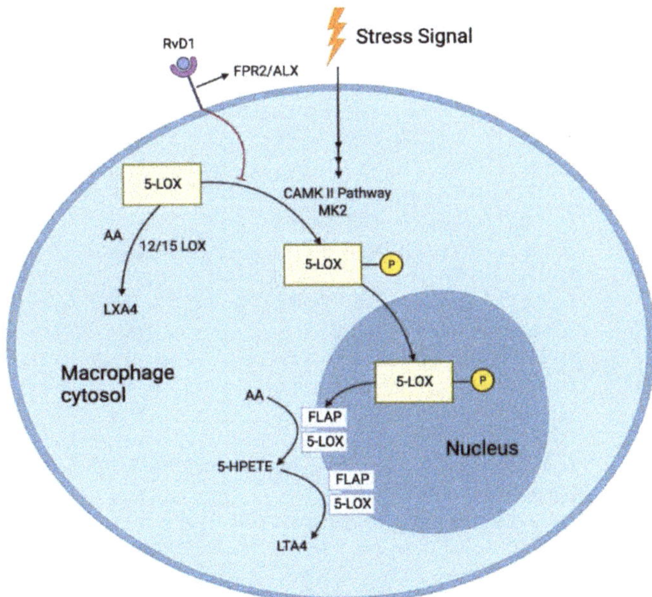

Figure 3. RvD1 prevents the nuclear location of 5-LOX, increasing the production of pro-resolving LXA4 and reducing the production of pro-inflammatory LTA4 in response to stress signals.

As mentioned above, the "don't eat me" ligands maintain the balance of efferocytosis and protect healthy cells from being phagocytosed. CD47 is one such ligand on healthy cells that interacts with the alpha receptor on phagocytes, shutting off the efferocytotic pathways within the phagocytes. TNF-α weakens the downregulation of CD47 in atherosclerotic plaque cells and renders them resistant to efferocytosis. CD47 blocking antibodies have shown to have beneficial effects in mouse models by preventing atherosclerotic progression, regressing the necrotic core, and preventing the plaque from rupturing [32].

A few other molecules have been implicated in the failure of efferocytosis in atherosclerosis, including Milk fat globule epidermal growth factor 8 (Mfge8) and Mer receptor tyrosine kinase (Mertk). Mfge8 is a bridging molecule between $\alpha v \beta 3$ integrin on the macrophages and externalized phosphatidyl serine on the apoptotic body [14]. This molecule seems to be expressed less in atherosclerotic plaque. In mouse models that were created with both LDLR$-/-$, and transplanted Mfge8$-/-$ bone marrow had advanced atherosclerosis with larger necrotic core and systemic inflammation. It is thought that Mfge8 might also have a role in reverse cholesterol transport by binding transglutaminase 2 [32].

Mouse models with both LDLR$-/-$, and transplanted Mertk$-/-$ bone marrow showed similar plaque properties to those with absent Mfge8 and LDLR. Furthermore, a mouse with a defective kinase form of Mertk resulted in more plaque necrosis than those found in ApoE$-/-$. Metalloproteinases, generally found in abundance in pro-inflammatory settings, cleave Mertk into a soluble inactive form. This inactive molecule provides a decoy receptor to Growth Arrest Specific 6 and leads to competitive inhibition of efferocytosis [32].

Weissman and colleagues have found that cancer cells upregulate "don't eat me" ligands to evade phagocytosis. Antibodies and decoy molecules that inhibit such processes and restore normal phagocytosis have been developed and are under study. If such treatments prove effective, they will have immense potential in treating atherosclerosis. Similarly, antibodies to TNF-α have been shown to reduce the expression of CD47. Anti-TNF-α antibodies are used in patients with rheumatological conditions. Such patients appear to be protected from myocardial infarction or adverse cardiovascular effects. Mouse treated with a combination of anti-CD47 and anti-TNF-α antibodies showed a better reduction in atherosclerosis than anti-CD47 alone. Given that there is a genetic susceptibility in reducing the efficiency of efferocytosis through the example of Calr expression, a genotype-driven therapy will most benefit such individuals [32].

9. Resolvins

9.1. Resolvin D Series (RvD)

RvD1 binds to two receptors, ChemR23 and BLT1, and increases macrophage phagocytosis and PMN apoptosis, respectively. RvD1 also upregulates anti-inflammatory IL-10 and downregulates pro-inflammatory LTB4 [20]. Overexpression of the enzyme 15-Lipoxygenase (15-LOX) reduced atherosclerotic plaques in rabbits [8]. Lack of resolution of inflammation and the increased ratio of LTB4 to RvD1 have been implicated as the real culprit behind what starts as a host-beneficial process to life-threatening events [33,34]. Since SPMs act in their local environments, the microclimate of the area of tissue necrosis is vital in determining their viability and action; thus, a 5-LOX closer to the cell periphery can interact with 12/15 LOX to produce RvD1 from DHA. Furthermore, RvD1 prevents the nuclear location of 5-LOX, increasing the production of pro-resolving LXA4 and reducing the production of LTA4 (Figure 3) [35]. Liquid Chromatography tandem Mass Spectrometry of human carotid plaques revealed LXA4 and RvD1 as the major SPMs, requiring 5-LOX. Furthermore, these pro-resolving molecules were much less in vulnerable plaques than the stable ones. However, the intermediates such as 5-HEPE, 15-HEPE, 17, and 14-HDHA through the actions of 5/15/12-LOX were high in vulnerable plaques, indicating that the enzymes (5-LOX and 15-LOX) themselves were bioactive even in vulnerable plaques [33]. One mechanism by which there could be a reduction in RvD1 was proposed to be the relocation of 5-LOX to the nucleus via the persistent activation of Ca^{2+}/Calmodulin Dependent Protein Kinase II (CAMKII) caused by the oxidative stress from substances such as 7-ketocholesterol (7-KC) found in the atherosclerotic plaque. This was proven by the reduction in the nuclear localization of 5-LOX once CAMKII expression was suppressed in human macrophages [35]. Inhibition of CAMKII also prevented the reduction of RvD1 production by 7-KC [33]. RvD1 also blocked the synthesis of LTB4 from AA. Such pathways were mediated by the receptors formyl peptide receptor 2/ lipoxin A4 receptor (FPR2/ALX), as both a receptor blocking antibody, as well as an antagonist, blocked this reduction of LTB4 by RvD1 [35]. Nuclear localization of 5-LOX produces LTA4, which is then transformed to LTB4 through the action of LTA4 hydrolase. This nuclear localization is brought about by the phosphorylation of LOX-5 at Ser271 by p38MAPK-activated protein kinase 2 (MK2). RvD1 blocks this phosphorylation through the receptors FPR2/ALX and G-protein-coupled receptor (GPCR) Gi (and GPR32 in humans). LXA4, which shares the same receptors, also has been found to reduce the phosphorylation. RvD1 could not block the synthesis of LTB4 when incubated with large quantities of LTA4, demonstrating that RvD1 cannot alter the pathway once LTA4 is formed [35]. Fredman et al. showed that the decrease in RvD1 was associated with the progression of the atherosclerotic plaque in the aortic arch of Ldlr−/− mice fed with a western diet for 8 or 17 weeks. Analysis of the lipid mediators in early vs. late plaques showed an approximately 87-fold decrease in RvD1 in advanced plaques but no significant change in LTB4 [33]. An increase in the lesional RvD1 levels was noticed when RvD1 was administered within the physiologic range. This increase resulted in the reduction of LTB4 levels, showing that RvD1 could have facilitated this reduction by preventing the nuclear localization of 5-LOX. This simultaneous increase in RvD1 (through external administration) and reduction in LTB4 (through RvD1 mediated reduction in the nuclear localization of 5-LOX) resulted in the ratio of RvD1:LTB4 being reverted to its early plaque levels. In addition, RvD1 increased other p SPMs, reduced oxidized CAMKII, reduced oxidative stress in the plaques, enhanced efferocytosis of macrophages, reduced the size of necrotic cores, and reduced the levels of collagenase and MMP9 without a concomitant reduction in the number of macrophages or VSMCs, resulting in a thickened fibrous cap. All these effects contribute to the stability of the plaque and in slowing its progression to an advanced type [33]. RvD1 was shown to act on human PMN, in-vitro, through a GPCR receptor that was inhibited by Pertussis toxin (PTX) and reduced their actin polymerization. They also blocked β2 integrin molecules on human PMN that were regulated by LTB4. Enhancement of phagocytosis by macrophages was also observed through the interaction of the receptors ALX, GPR32, and RvD1 [36]. RvD1 was also found to limit monocyte adhesion, reactive oxygen species (ROS), and pro-inflammatory cytokine production in VSMCs derived from the saphenous vein in vitro and in rabbit arteries that underwent balloon angioplasty [20]. It was found to alter the cytoskeletal properties of arterial smooth muscle cells (ASMC) in rats, thereby inhibiting their migration. Furthermore, it reduces their proliferation, oxidative stress, and translocation of p65, a molecule vital in NF-κB stimulation, which is implicated widely in the inflammation processes. All these beneficial effects were observed without damaging the viability of ASMCs [18]. Interestingly, a positive effect of RvD1 on reducing neutrophil infiltration comes from the analysis done by Recchiuti et al. They found that RvD1, possibly through its upregulation of certain micro RNAs (miRNAs) in humans, brought about a reduction in the resolution interval by ~4 h. These miRNAs were found to target immune-competent proteins such as the NF-κB pathway and 5-LOX (in the leukotriene pathway). By blocking these pathways, the concentration of pro-inflammatory mediators is reduced [37].

Akagi et al. demonstrated that pretreatment of ASMCsin vitro with DHA-derived SPMs, RvD2, and maresin-1(MaR1), impaired their migration towards PDGF in a dose-dependent manner by 74% and 80%, respectively at a 100 nM concentration [38]. They also showed that GPCR could have mediated this response since the reduction in migration was attenuated in the presence of PTX, which inhibits GPCR proteins [20,36,38]. Of the many immunologically active cytokines, TNFα induces the NFκB pathway by nuclear translocation of p65 (Figure 4). This pathway results in the transcription of many pro-inflammatory cytokines such as TNFα, Interleukin (IL)-1, IL-6, and IL-8. An in vitro study showed that RvD2 or MaR1 treatment of mouse ASMCs reduced p65 nuclear translocation by 24% and 28%, respectively, at a concentration of 500 nM. At the same concentration, such treatment also has reduced TNF-α induced superoxide production by 46% and 53%, respectively [38]. Furthermore, RvD2 reduced the cultured VSMCs production of VCAM-1 and ICAM-1 induced by TNF-α [20]. Such effects of these two SPMs demonstrate their anti-inflammatory and pro-resolving characteristics in their local environment. This pro-resolving action reduced the neo-intimal hyperplasia (neointima: media area ratio), which has been proven to be a result of chronic inflammation, by 67% and 71% by RvD2 and MaR1, respectively. The same study also showed a decrease in migration of neutrophils and monocytes to the area of injury, achieved by the suppression of MCP-1 expression by activated VSMCs and an increase in the M2 phenotype of macrophages which promoted resolution of the ongoing inflammation [38].

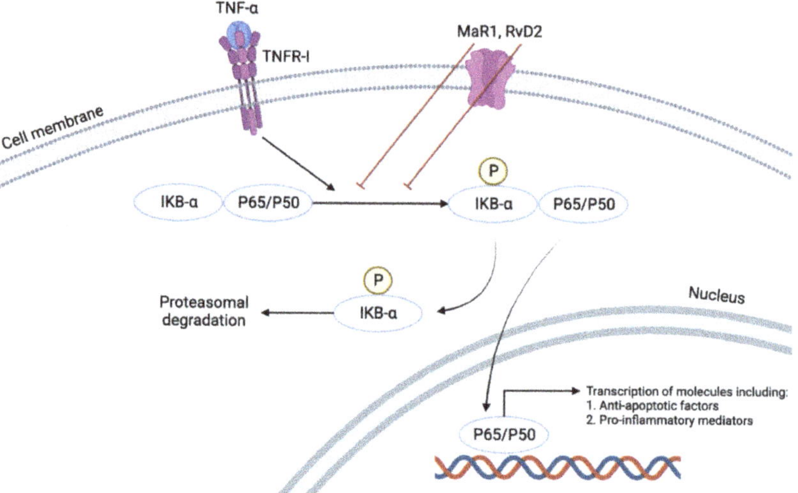

Figure 4. TNFα induces the NFκB pathway by nuclear translocation of p65. This pathway results in the transcription of many pro-inflammatory cytokines such as TNFα, Interleukin (IL)-1, IL-6, and IL-8. RvD2 or MaR1 treatment reduces p65 nuclear translocation, demonstrating their anti-inflammatory and pro-resolving characteristics in their local environment.

Aspirin-triggered RvD1 (AT-RvD1) has been shown to activate the nuclear factor erythropoietin 2 related factor 2 (Nrf2), increasing the expression of genes such as heme oxygenase-1 (HMOX1), and NAD(P)H quinone oxidoreductase 1 (NQO-1), aiding in combating oxidative stress in mice lung injuries [39].

9.2. Resovlin E Series (RvE)

Resolvin E (RvE) series (E1 and E2) molecules are produced from EPA through the action of 5-lipoxygenase (5-LOX) [40]. In addition, RvE1 receptors have been found to be GPCRs [41].

RvE series mediate the resolution of inflammation through the following means:

1. Reduction in chemotaxis of PMNs by affecting changes in their actin polymerization [42].
2. Increase in non-phlogistic phagocytic activity of macrophages [42].
3. Dose-dependent increase in IL-10 (an anti-inflammatory cytokine) by macrophages [42].
4. Downregulation of leukocyte integrin activation, reducing their response to platelet activation factor (PAF), a potent pro-inflammatory cytokine. This downregulation was postulated to be mediated through the interaction of RvE2 and leukotriene B4

receptor, BLT1. RvE1 binds BLT1 and ChemR23 equally, but RvE2 is a weak agonist of the ChemR23 receptor [42]. Platelet aggregation is one of the hallmarks of acute inflammation, brought about in part by ADP, which activates other platelets and leukocytes through intracellular signaling pathways. These pathways ultimately end with activation of platelet receptor GP IIb/IIIa, and granule secretion. RvE1 has been shown to regulate an ADP-mediated pathway that results in P-selectin surface mobilization through the ChemR23 receptor [43].

These SPMs have proven to be strong local modulators of acute inflammation through these actions. One of the major risks in atherosclerosis—thrombus formation- is avoided by preventing platelet aggregation. GP IIb/IIIa is a receptor for fibrinogen that is activated by ADP within platelets. RvE1, at concentrations of ~100 nM, has reduced fibrinogen binding to platelets. The lack of complete blockage of ADP stimulation of platelets by RvE1 is beneficiary as platelet aggregation, and thrombus formation are required for hemostasis [43].

10. Lipoxins

Lipoxins (LXs), as mentioned above, are endogenously made eicosanoids with anti-inflammatory properties [44]. They are produced via two reactions mediated by a set of lipoxygenases (LOX) using arachidonic acid (AA) as a precursor. AA is initially acted on by LOX-12/15 to produce an intermediate that proceeds to generate LXs via LOX-5. In humans, two types of LXs are made through this pathway, LX A4 and B4 [45]. LX A4 acts as an endogenous anti-inflammatory mediator by interacting with different players in the inflammatory immune process. One of the receptors it interacts with is a G-protein coupled ALX receptor found in many tissues and cells in the body, including neutrophils, macrophages, and endothelial cells [46]. When LX A4 interacts with the ALX receptor on neutrophils, it causes a reduction in the concentration of free oxygen radicals and production of pro-inflammatory cytokines and chemokines. It also inhibits the transmigration of neutrophils through the endothelium and induces apoptosis [46]. These changes on neutrophils is beneficial as the increase in ROS by neutrophils has been shown to play a part in plaque formation and rupture [47]. In addition, LX A4 when interacting with ALX receptor on macrophages induces phagocytosis of apoptotic leukocytes, hence ameliorating inflammation and inducing resolution [46]. Interestingly, an experiment performed in animal models with LX A4 injections demonstrated inhibition of the production of pro-inflammatory cytokines such as IL-1b, IL-6, and IL-8 and reduced infiltration of neutrophils and levels of TNF-α [46].

Similarly, LX B4 as shown to be a strong mediator in resolving inflammation during atherosclerosis [47]. Kraft et al. has recently performed an experiment depicting the effect of lipoxins in healthy individuals and those with atherosclerotic disease. The results were interesting, as they depicted lipoxins having opposing effects on neutrophils between healthy individuals and those with atherosclerosis. In patient with atherosclerosis, LX B4 seemed to inhibit the oxidative burst in neutrophils; and they describe the process of neutrophil oxidative burst as being a key player in the atherosclerotic process. On the other hand, in healthy individual LX B4 was shown to increase oxidative burst in neutrophils [47].

Furthermore, another player in clot formation and atherosclerosis are CD-11b integrins, that have been shown to regulate chemotaxis of neutrophils and neutrophil-platelet aggregation [47,48]. However, in the experiment by Kraft et al., they also illustrated that LX B4 in atherosclerotic patients caused a decrease in the synthesis of CD-11b integrin; thus, further decreasing neutrophil chemotaxis and the harmful effects of neutrophils in the atherosclerotic process [47]. Although both LX A4 and B4 have some similar effects in decreasing the detrimental effects of neutrophils, it was shown that LX B4 seemed to have a more potent effect that LX A4 [47].

In a study of the temporal relationship between peptide-derived and lipid-derived resolution compounds on patients that underwent abdominal aortic aneurysm surgery by Pillai et al., pro-inflammatory and pro-resolving mediators were closely assessed both before and after surgery. Two distinct groups of profiles emerged from this study that displayed either pro-inflammatory or pro-resolving milieu after surgery. The early resolving group (Group B) was named such, as their average LXA4 levels rose steadily from just after 5 min to 72 h post-surgery, while in the late resolving group (Group A), LXA4 peaked at 5 min post-surgery and declined significantly thereafter. Group A also had overall lower levels of ATL, higher levels of TXB2, and significantly high levels of LTB4 immediately pre- and post-unclamping of the aorta. At the same time, group B showed high levels of ATL at 5 min and 6 h post-surgery, overall low levels of TXB2 with a slight increase at 72 h post-surgery, and an overall low level of LTB4 with a mild increase at 24 h timeline. While group A patients exhibited what might look like a pro-inflammatory group, all the patients survived. Although the resolution mediating molecules showed up late, they helped recover from inflammation induced by surgeries [41]. Lastly, a standard OTC medicine, aspirin (and others of its class), which performs its action by inhibiting enzymes cyclooxygenase 1/2, allows for the action of LOX enzymes to increase as both their enzyme activities act on AA, resulting in increased LXs [45].

Our body's innate mechanisms to deal with and curtail inflammation beyond its stipulated duration, are numerous with complex interplays within the participating agents. One such mechanism involves the nuclear factor Nrf2, a basic leucine zipper transcription factor that increases the transcription of genes that code for antioxidant proteins such as HMOX1, NQO-1, superoxide dismutase (SOD), and thioredoxin (TXN) involved in the reduction of ROS. While ROS are essential for homeostasis, elevated ROS associated with atherosclerosis induce and exacerbate endothelial dysfunction. HMOX1 has been shown to reduce atherosclerosis in mouse models [39], and upregulation of Nrf2/HMOX1 protected the human endothelial cells against TNF-α activation [49].

LXA4 prevented vascular endothelial cell (EC) damage due to oxidative stress through Nrf2 and increased the production of HMOX1. The LXA4/FPR2 receptor agonist BML-111 has shown to increase Nrf2 signaling and prevent oxidative stress in autoimmune myocarditis mouse model [39].

11. Maresins

Maresins, macrophage mediators in resolving inflammation are made from the ω-3 fatty acid DHA. The key enzyme in the synthesis of maresins is 12-LOX and is synthesized mainly by M2 macrophages [50,51]. Maresins appear to be tightly linked with macrophage dependent cardiac tissue regeneration and act as pro-resolving mediators by augmenting the secretion of TGF-β and decreasing concentrations of IL-6 and TNF-α [52]. MaR1 synthesized by macrophages act on BLT1 and Leucine-rich repeat -containing G-protein-coupled receptor 6 (LGR6) receptors to stimulate the phenotypic transition of macrophages from pro-inflammatory M1 to pro-resolving M2 [52,53]. In addition, MaR1 acts on retinoic acid-related orphan receptor-α (RORα) and LGR6 to enhance efferocytosis and phagocytosis by stimulating phosphorylation of several proteins such as extracellular signal-regulated kinase (ERK) and cAMP responsive element-binding protein (CREB1) [50,54,55]. In a study conducted on human saphenous vein EC and VSMC in vitro, Chatterjee and colleagues found that MaR1 weakened TNF-α induced monocyte adhesion by downregulating the cell surface adhesion molecule E-selectin. However, VCAM-1 and ICAM-1 expressions remained unchanged. MaR1 also reduced ROS generation in both EC and VSMC by downregulating NADPH oxidases (NOX4, NOX1, NOX2). Because cell adhesion, and the creation of ROS, are a couple of the hallmark events in inflammation, by minimizing them, MaR1 is a SPM that stop polymorphonuclear infiltration and inhibit ROS product. Blockage of TNF-α induction was discovered to be through inhibition of I-κ Kinase (IKK) phosphorylation and, eventually, the reduction in the nuclear translocation of the p65 subunit of NF-κB as mentioned above. Phosphorylation of IKK, in turn, phosphorylates and subjects I-kappa α to proteasomal degradation, resulting in the release of p65 from its I-κ complex, which then migrates to the nucleus to act as a transcription factor. NF-κB has been well established as a key transcription factor in synthesizing many pro-inflammatory molecules that act in a paracrine way to stimulate local inflammation [56]. It has been shown that Aspirin enables the production of Aspirin-Triggered Lipoxin (ATL), a Lipoxin A4 epimer, by interacting with the receptors FPR2/ALX found on VSMCs and macrophages in atherosclerotic plaques. The presence of these receptors correlated negatively with the clinical manifestation of the disease, implying a more stable plaque, possibly through increased collagen and decreased collagenases [57]. However, MaR1 has been shown to increase collagen synthesis leading to plaque stabilization by reducing the expression of Arginase-2 (ARG2) in endothelial cells and nitric oxide synthase 2 (NOS2) in macrophages, while increasing the expression of TGF-β1 and ARG1 [52].

Similar to AT-RvD1, MaR1 has also been found to induce Nrf2, increasing cytoplasmic HMOX1, thus reducing the levels of ROS and improved pulmonary ischemia/reperfusion injury [39].

While these studies shed light on SPM (LXA4, AT-RvD1, and MaR1)-Nrf2 relationship in increasing antioxidative proteins within pulmonary physiology, there is still debate on the overall cardioprotective effects of Nrf2, with studies showing that mice with Nrf2−/−developed less atherosclerosis [58] and HMOX1 was seen highest within human plaques with characteristics of high instability [59]. Roles of Nrf2 in lipid metabolism [58], in reduction of scavenger receptor CD36 resulting in reduced foam cell formation [60], and in NOD-, LRR- and pyrin domain-containing protein 3 (NLRP3) inflammasome induction by cholesterol crystals within the atherosclerotic plaque have been attributed to these counterintuitive results of increased atherosclerosis with complete absence of Nrf2 expression [61].

12. Protectins

Protectins, of which PD1 is one of the most studied, are another class of SPMs, generated from DHA, through enzymatic action on an epoxide intermediate [50,62]. PD1 issynthesized by polymorphonuclear cells, macrophages, and eosinophils [63–65]. The effect of PD1 is also correlated with the specific stereochemistry of the molecule; it has been shown that the R-epimer of PD1 is much more effective in its anti-inflammatory properties compared to the S-epimer [62,66]. Protectins' anti-inflammatory activities include inhibiting neutrophil migration, as well as reducing the concentrations of TNF-α and IFN-γ by acting on GPCR37 or parkin-associated endothelin receptor-like receptor (PAELR) [67,68]. Among many other cell-protective and immunoregulatory actions, it has been shown that PD1 reduces the production of pro-inflammatory cytokines, leukocyte accumulation, and T-cell migration following an ischemic injury [67,69]. It has also been shown to downregulate the expression of VCAM-1 and MCP-1 in human aortic endothelial cells [20]. PD1 acts as an anti-inflammatory agent by regulating C-C chemokine receptor type 5 (CCR5) expression on neutrophils and decreases neutrophil infiltration of tissues [70,71]. In addition, it augments phagocytosis and efferocytosis of macrophages, which in turn clear apoptotic neutrophils [71]. Interestingly, PD1 expression has been shown to increase in the first few hours of a myocardial infarct, a correlation that further suggests a possible pro-resolving role of PD1 during early stages of inflammation [50].

13. Therapeutic Perspective of SPMs in Atherosclerosis

A continuous evolution of our understanding of the complex pathophysiology of atherosclerosis and the emergence of the existence of SPMs pave way for individualized and targeted pharmacotherapy in the treatment of atherosclerosis. As detailed above, this involves a complete understanding of all the pathways each of these molecules participate in, the consequences of altering such events, spatially and temporally, on both near and remotely associated structures. It would also demand from the scientific community, a feasible way to replicate in-vivo microcosms, so the bioavailability, pharmokinetics, and pharmacodynamics behave as observed and anticipated in the numerous studies leading up to it [50]. One challenge for example is the rapid metabolic inactivation of in vivo LXA4 and LXB4 by prostaglandin dehydrogenase [72].

For instance, with the knowledge of FPR2 receptor as a master switch in promoting resolution leading the cascade of events that blocks the phosphorylation and nuclear colocalization of 5-LOX, resulting in attenuation of pro-inflammatory cytokines, has led to the development of the molecule BMS986235, a FPR2 agonist by Bristol-Meyers Squibb. They had recently concluded a Phase1 trial. FPR2 has been shown to be a receptor to both LXA4, and another anti-inflammatory protein, annexin A1 (AnxA1), the interaction that leads to recruitment and polarization of macrophages to M2 phenotype. It would be interesting to watch whether the agonists of FPR2 deliver the same results as reducing pro-inflammatory molecules, and in stimulating macrophages to switch to their anti-inflammatory, M2 phenotype [73]. Another potent FPR1 and FPR2 agonist, called Compound 43 was developed by Bristol Myers Squibb. Compound 43 has been shown to induce phagocytic and chemotactic activities in mouse models, and later was patented to treat myocardial infarction [74]. Compound 17b, another agonist of FRP1/FPR2 had a similar effect on myocardial injury in mice models [75].

Another receptor involved in resolvin pathways, ChemR23, could be exploited to induce cascades leading to pro-resolution by utilizing its agonists. chemerin-9 is one such agonist, an adipokine highly expressed in white adipose tissue. Infusion of chemerin-9 resulted in decreased concentrations of TNF-α, and size of atherosclerotic lesions, and improved vascular functions [76].

G. Bannenberg et al., have shown that stable 3-oxa-ATL analogs that were resistant to β-oxidation (ZK-142/ZK-996) or the corresponding trienyne analog (ZK-990/ZK-994) exhibited anti-inflammatory effects in terms of inhibiting leukocytes and myeloperoxidase activity following oral, intravenous, or topical administration [77].

In a study by Tang et al., involving mouse models, a metabolically stable analog of aspirin-triggered resolvin D1, termed p-RvD1 (17R-hydroxy-19-para-fluorophenoxy-resolvin D1 methyl ester) has been shown to reduce damage to vascular endothelial cells resulting in markedly reduced vascular permeability in lung injury. Benzo-RvD1 (BRvD), a synthetic analog of resolvin, 17R-RvD1, reduced the migration of VSMC and inhibited NF-κB translocation in cytokine stimulated endothelial cells by 12% to 21% in a model of rat carotid angioplasty [78].

Construction of novel nano particles derived from neutrophil-derived endogenous microparticles, opens a promising door into the stable delivery of any SPM analogues. Study using these nanoparticles enriched with aspirin-triggered resolvin D1 or a LXA4 analog reduced neutrophil influx, shortened resolution timelines, and demonstrated pro-resolving actions in murine peritonitis [79].

The enzyme 5-LOX converts AA to LTA4 through the intermediates 5-hydroperoxyeicosatetraenoic acid (5-HPETE), and 5-hydroxyeicosatetraenoic acid (5-HETE) with the help of the protein 5-Lipoxygenase Activating Protein (FLAP) [80]. If the protein FLAP is inhibited, then the "class switch" from prostaglandins and leukotrienes to lipoxins could be achieved, leading to the cascade of resolution. As such, a few FLAP inhibitors/antagonists, or molecules that interfere in the FLAP mechanisms, AZD5718, BRP-201, BRP-187 have been developed that are currently being studied to understand the complexity of their consequences in inflammation and resolution.

14. Conclusions

As we gain more insights into the molecular mechanisms of atherosclerosis, our understanding of its pathophysiology leading to cardiovascular disease has begun to include hitherto incompletely characterized immunoactive SPMs (Figure 1). Their complex temporal and functional interdependences seem to forge a path of resolution that begins with the start of inflammation, which opens up an entirely novel way of treating the world's number one cause of death. While the current treatment modalities are mainly damage-control with some passive preventive measures, further exploration of this new understanding will undoubtedly lead to targeted, individualized medicine that has the potential to be both preventative and curative.

15. Limitations of the Study

As depicted in this review, SPMs each play their role in an intertwined manner to resolve inflammation and thus curtail the pathogenesis of atherosclerosis in its early stages. Through this article we offer a simple introduction and a bird's eye view of the role of SPMs in atherosclerosis and overview of their cellular and molecular mechanisms in resolution of inflammation, that has been proven to be implicated in the pathogenesis of the atherosclerotic cardiovascular disease. The exact molecular structure of each of the SPMs, detailed review of the studies that led to their discovery, an in-depth analysis of the interplay between various signal molecules, their receptors, and cell types in the pathogenesis of atherosclerosis are beyond the scope of this article. Also limited is the availability of a comprehensive list and analysis of any ongoing clinical trials with SPMs.

Author Contributions: Conceptualization, M.M.R. and S.R.; figures and formatting, S.R. and P.R.; writing, S.R., D.O. and M.M.R.; review and editing, S.R., M.M.R. and D.O. All authors have read and agreed to the published version of the manuscript.

Funding: This research received no external funding.

Institutional Review Board Statement: Not applicable.

Informed Consent Statement: Not applicable.

Data Availability Statement: Not applicable.

Conflicts of Interest: The authors declare no conflict of interest.

References

1. CDC. Available online: https://www.cdc.gov/nchs/fastats/leading-causes-of-death.htm (accessed on 31 December 2021).
2. Falk, E. Pathogenesis of atherosclerosis. *J. Am. Coll. Cardiol.* **2006**, *47* (Suppl. 8), C7–C12. [CrossRef] [PubMed]
3. Hansson, G.K.; Hermansson, A. The immune system in atherosclerosis. *Nat. Immunol.* **2011**, *12*, 204–212. [CrossRef] [PubMed]
4. Alberts, B.; Jonson, A.; Lewis, J.; Morgan, D.; Raff, M.; Roberts, K.; Walter, P. *Molecular Biology of the Cell*, 6th ed.; Garland Science, Taylor & Francis Group: New York, NY, USA, 2014; pp. 1235–1238.
5. Halvorsen, B.; Otterdal, K.; Dahl, T.B.; Skjelland, M.; Gullestad, L.; Øie, E.; Aukrust, P. Atherosclerotic Plaque Stability—What Determines the Fate of a Plaque? *Prog. Cardiovasc. Dis.* **2008**, *51*, 183–194. [CrossRef] [PubMed]
6. Ross, R. Cell Biology of Atherosclerosis. *Annu. Rev. Physiol.* **1995**, *57*, 791–804. [CrossRef]
7. Bäck, M.; Yurdagul, A., Jr.; Tabas, I.; Öörni, K.; Kovanen, P.T. Inflammation and its resolution in atherosclerosis: Mediators and therapeutic opportunities. *Nat. Rev. Cardiol.* **2019**, *16*, 389–406. [CrossRef] [PubMed]
8. Serhan, C.N. Pro-resolving lipid mediators are leads for resolution physiology. *Nature* **2014**, *510*, 92–101. [CrossRef]
9. Libby, P.; Aikawa, M. Stabilization of atherosclerotic plaques: New mechanisms and clinical targets. *Nat. Med.* **2002**, *8*, 1257–1262. [CrossRef]
10. Hansson, G.K.; Libby, P.; Tabas, I. Inflammation and plaque vulnerability. *J. Intern Med.* **2015**, *278*, 483–493. [CrossRef]
11. Glass, C.K.; Witztum, J.L. Atherosclerosis: The Road Ahead. *Cell* **2001**, *104*, 503–516. [CrossRef]
12. Fu, P.; Birukov, K.G. Oxidized phospholipids in control of inflammation and endothelial barrier. *Transl. Res.* **2009**, *153*, 166–176. [CrossRef]
13. Linton, M.F.; Fazio, S. Macrophages, inflammation, and atherosclerosis. *Int. J. Obes. Relat. Metab. Disord.* **2003**, *27* (Suppl. S3), S35–S40. [CrossRef] [PubMed]
14. Phillips, J.W.; Barringhaus, K.G.; Sanders, J.M.; Hesselbacher, S.E.; Czarnik, A.C.; Manka, D.; Vestweber, D.; Ley, K.; Sarembock, I.J. Single Injection of P-Selectin or P-Selectin Glycoprotein Ligand-1 Monoclonal Antibody Blocks Neointima Formation After Arterial Injury in Apolipoprotein E-Deficient Mice. *Circulation* **2003**, *107*, 2244–2249. [CrossRef] [PubMed]
15. Libby, P.; Geng, Y.J.; Aikawa, M.; Schoenbeck, U.; Mach, F.; Clinton, S.K.; Sukhova, G.K.; Lee, R.T. Macrophages and atherosclerotic plaque stability. *Curr. Opin. Lipidol.* **1996**, *7*, 330–335. [CrossRef] [PubMed]
16. Wei, X.; Sun, Y.; Han, T.; Zhu, J.; Xie, Y.; Wang, S.; Wu, Y.; Fan, Y.; Sun, X.; Zhou, J.; et al. Upregulation of miR-330-5p is associated with carotid plaque's stability by targeting Talin-1 in symptomatic carotid stenosis patients. *BMC Cardiovasc. Disord.* **2019**, *19*, 149. [CrossRef] [PubMed]
17. Jinnouchi, H.; Guo, L.; Sakamoto, A.; Torii, S.; Sato, Y.; Cornelissen, A.; Kuntz, S.; Paek, K.H.; Fernandez, R.; Fuller, D.; et al. Diversity of macrophage phenotypes and responses in atherosclerosis. *Cell. Mol. Life Sci.* **2020**, *77*, 1919–1932. [CrossRef]
18. Wu, B.; Mottola, G.; Schaller, M.; Upchurch, G.R.; Conte, M.S. Resolution of vascular injury: Specialized lipid mediators and their evolving therapeutic implications. *Mol. Asp. Med.* **2017**, *58*, 72–82. [CrossRef]
19. Libby, P. The molecular mechanisms of the thrombotic complications of atherosclerosis. *J. Intern. Med.* **2008**, *263*, 517–527. [CrossRef]

20. Miyahara, T.; Runge, S.; Chatterjee, A.; Chen, M.; Mottola, G.; Fitzgerald, J.M.; Serhan, C.N.; Conte, M.S. D-series resolvin attenuates vascular smooth muscle cell activation and neointimal hyperplasia following vascular injury. *FASEB J.* **2013**, *27*, 2220–2232. [CrossRef]

21. Shankman, L.S.; Gomez, D.; Cherepanova, O.A.; Salmon, M.; Alencar, G.F.; Haskins, R.M.; Swiatlowska, P.; Newman, A.A.C.; Greene, E.S.; Straub, A.C.; et al. KLF4-dependent phenotypic modulation of smooth muscle cells has a key role in atherosclerotic plaque pathogenesis. *Nat. Med.* **2015**, *21*, 628–637. [CrossRef]

22. Verreck, F.A.; de Boer, T.; Langenberg, D.M.; Hoeve, M.A.; Kramer, M.; Vaisberg, E.; Kastelein, R.; Kolk, A.; de Waal-Malefyt, R.; Ottenhoff, T.H. Human IL-23-producing type 1 macrophages promote but IL-10-producing type 2 macrophages subvert immunity to (myco)bacteria. *Proc. Natl. Acad. Sci. USA* **2004**, *101*, 4560–4565. [CrossRef]

23. Wang, N.; Liang, H.; Zen, K. Molecular mechanisms that influence the macrophage m1-m2 polarization balance. *Front. Immunol.* **2014**, *5*, 614. [CrossRef] [PubMed]

24. Mantovani, A.; Sica, A.; Sozzani, S.; Allavena, P.; Vecchi, A.; Locati, M. The chemokine system in diverse forms of macrophage activation and polarization. *Trends Immunol.* **2004**, *25*, 677–686. [CrossRef] [PubMed]

25. Bouhlel, M.A.; Derudas, B.; Rigamonti, E.; Dièvart, R.; Brozek, J.; Haulon, S.; Zawadzki, C.; Jude, B.; Torpier, G.; Marx, N.; et al. PPARgamma activation primes human monocytes into alternative M2 macrophages with anti-inflammatory properties. *Cell Metab.* **2007**, *6*, 137–143. [CrossRef]

26. Geng, Y.J.; Libby, P. Evidence for apoptosis in advanced human atheroma. Colocalization with interleukin-1 beta-converting enzyme. *Am. J. Pathol.* **1995**, *147*, 251–266.

27. Serhan, C.N.; Savill, J. Resolution of inflammation: The beginning programs the end. *Nat. Immunol.* **2005**, *6*, 1191–1197. [CrossRef] [PubMed]

28. Heinz, J.; Marinello, M.; Fredman, G. Pro-resolution therapeutics for cardiovascular diseases. *Prostaglandins Other Lipid Mediat.* **2017**, *132*, 12–16. [CrossRef]

29. Fredman, G.; Spite, M. Specialized pro-resolving mediators in cardiovascular diseases. *Mol. Asp. Med.* **2017**, *58*, 65–71. [CrossRef]

30. Schrijvers, D.M.; De Meyer, G.R.; Herman, A.G.; Martinet, W. Phagocytosis in atherosclerosis: Molecular mechanisms and implications for plaque progression and stability. *Cardiovasc. Res.* **2007**, *73*, 470–480. [CrossRef]

31. Schrijvers, D.M.; De Meyer, G.R.; Kockx, M.M.; Herman, A.G.; Martinet, W. Phagocytosis of apoptotic cells by macrophages is impaired in atherosclerosis. *Arterioscler. Thromb. Vasc. Biol.* **2005**, *25*, 1256–1261. [CrossRef]

32. Kojima, Y.; Weissman, I.L.; Leeper, N.J. The Role of Efferocytosis in Atherosclerosis. *Circulation* **2017**, *135*, 476–489. [CrossRef]

33. Fredman, G.; Hellmann, J.; Proto, J.D.; Kuriakose, G.; Colas, R.A.; Dorweiler, B.; Connolly, E.S.; Solomon, R.; Jones, D.M.; Heyer, E.J.; et al. An imbalance between specialized pro-resolving lipid mediators and pro-inflammatory leukotrienes promotes instability of atherosclerotic plaques. *Nat. Commun.* **2016**, *7*, 12859. [CrossRef] [PubMed]

34. Thul, S.; Labat, C.; Temmar, M.; Benetos, A.; Bäck, M. Low salivary resolvin D1 to leukotriene B(4) ratio predicts carotid intima media thickness: A novel biomarker of non-resolving vascular inflammation. *Eur. J. Prev. Cardiol.* **2017**, *24*, 903–906. [CrossRef] [PubMed]

35. Fredman, G.; Ozcan, L.; Spolitu, S.; Hellmann, J.; Spite, M.; Backs, J.; Tabas, I. Resolvin D1 limits 5-lipoxygenase nuclear localization and leukotriene B4 synthesis by inhibiting a calcium-activated kinase pathway. *Proc. Natl. Acad. Sci. USA* **2014**, *111*, 14530–14535. [CrossRef]

36. Krishnamoorthy, S.; Recchiuti, A.; Chiang, N.; Yacoubian, S.; Lee, C.H.; Yang, R.; Petasis, N.A.; Serhan, C.N. Resolvin D1 binds human phagocytes with evidence for proresolving receptors. *Proc. Natl. Acad. Sci. USA* **2010**, *107*, 1660–1665. [CrossRef] [PubMed]

37. Recchiuti, A.; Krishnamoorthy, S.; Fredman, G.; Chiang, N.; Serhan, C.N. MicroRNAs in resolution of acute inflammation: Identification of novel resolvin Dl-miRNA circuits. *FASEB J.* **2011**, *25*, 544–560. [CrossRef] [PubMed]

38. Akagi, D.; Chen, M.; Toy, R.; Chatterjee, A.; Conte, M.S. Systemic delivery of proresolving lipid mediators resolvin D2 and maresin 1 attenuates intimal hyperplasia in mice. *FASEB J.* **2015**, *29*, 2504–2513. [CrossRef]

39. Kang, G.J.; Kim, E.J.; Lee, C.H. Therapeutic Effects of Specialized Pro-Resolving Lipids Mediators on Cardiac Fibrosis via NRF2 Activation. *Antioxidants* **2020**, *9*, 1259. [CrossRef] [PubMed]

40. Freire, M.O.; Van Dyke, T.E. Natural resolution of inflammation. *Periodontology 2000* **2013**, *63*, 149–164. [CrossRef] [PubMed]

41. Pillai, P.S.; Leeson, S.; Porter, T.F.; Owens, C.D.; Kim, J.M.; Conte, M.S.; Serhan, C.N.; Gelman, S. Chemical mediators of inflammation and resolution in post-operative abdominal aortic aneurysm patients. *Inflammation* **2012**, *35*, 98–113. [CrossRef]

42. Oh, S.F.; Dona, M.; Fredman, G.; Krishnamoorthy, S.; Irimia, D.; Serhan, C.N. Resolvin E2 Formation and Impact in Inflammation Resolution. *J. Immunol.* **2012**, *188*, 4527–4534. [CrossRef]

43. Fredman, G.; Van Dyke Thomas, E.; Serhan Charles, N. Resolvin E1 Regulates Adenosine Diphosphate Activation of Human Platelets. *Arterioscler. Thromb. Vasc. Biol.* **2010**, *30*, 2005–2013. [CrossRef] [PubMed]

44. Maderna, P.; Godson, C. Lipoxins: Resolutionary road. *Br. J. Pharmacol.* **2009**, *158*, 947–959. [CrossRef] [PubMed]

45. Ryan, A.; Godson, C. Lipoxins: Regulators of resolution. *Curr. Opin. Pharmacol.* **2010**, *10*, 166–172. [CrossRef] [PubMed]

46. Tułowiecka, N.; Kotlęga, D.; Bohatyrewicz, A.; Szczuko, M. Could Lipoxins Represent a New Standard in Ischemic Stroke Treatment? *Int. J. Mol. Sci.* **2021**, *22*, 4207. [CrossRef] [PubMed]

47. Kraft, J.D.; Blomgran, R.; Bergström, I.; Soták, M.; Clark, M.; Rani, A.; Rajan, M.R.; Dalli, J.; Nyström, S.; Quiding-Järbrink, M.; et al. Lipoxins modulate neutrophil oxidative burst, integrin expression and lymphatic transmigration differentially in human health and atherosclerosis. *FASEB J.* **2022**, *36*, e22173. [CrossRef]

48. Sheikh, S.; Nash, G.B. Continuous activation and deactivation of integrin CD11b/CD18 during de novo expression enables rolling neutrophils to immobilize on platelets. *Blood* **1996**, *87*, 5040–5050. [CrossRef]

49. Satta, S.; Mahmoud, A.M.; Wilkinson, F.L.; Yvonne Alexander, M.; White, S.J. The Role of Nrf2 in Cardiovascular Function and Disease. *Oxidative Med. Cell. Longev.* **2017**, *2017*, 9237263. [CrossRef]

50. Kotlyarov, S.; Kotlyarova, A. Molecular Pharmacology of Inflammation Resolution in Atherosclerosis. *Int. J. Mol. Sci.* **2022**, *23*, 4808. [CrossRef]

51. Tang, S.; Wan, M.; Huang, W.; Stanton, R.C.; Xu, Y. Maresins: Specialized Proresolving Lipid Mediators and Their Potential Role in Inflammatory-Related Diseases. *Mediat. Inflamm* **2018**, *2018*, 2380319. [CrossRef]

52. Salazar, J.; Pirela, D.; Nava, M.; Castro, A.; Angarita, L.; Parra, H.; Durán-Agüero, S.; Rojas-Gómez, D.M.; Galbán, N.; Añez, R.; et al. Specialized Proresolving Lipid Mediators: A Potential Therapeutic Target for Atherosclerosis. *Int. J. Mol. Sci.* **2022**, *23*, 3133. [CrossRef]

53. Viola, J.R.; Lemnitzer, P.; Jansen, Y.; Csaba, G.; Winter, C.; Neideck, C.; Silvestre-Roig, C.; Dittmar, G.; Döring, Y.; Drechsler, M.; et al. Resolving Lipid Mediators Maresin 1 and Resolvin D2 Prevent Atheroprogression in Mice. *Circ. Res.* **2016**, *119*, 1030–1038. [CrossRef] [PubMed]

54. Chiang, N.; Libreros, S.; Norris, P.C.; de la Rosa, X.; Serhan, C.N. Maresin 1 activates LGR6 receptor promoting phagocyte immunoresolvent functions. *J. Clin. Investig.* **2019**, *129*, 5294–5311. [CrossRef] [PubMed]

55. Im, D.S. Maresin-1 resolution with RORα and LGR6. *Prog. Lipid Res.* **2020**, *78*, 101034. [CrossRef]

56. Chatterjee, A.; Sharma, A.; Chen, M.; Toy, R.; Mottola, G.; Conte, M.S. The pro-resolving lipid mediator maresin 1 (MaR1) attenuates inflammatory signaling pathways in vascular smooth muscle and endothelial cells. *PLoS ONE* **2014**, *9*, e113480. [CrossRef] [PubMed]

57. Petri, M.H.; Laguna-Fernandez, A.; Tseng, C.-N.; Hedin, U.; Perretti, M.; Bäck, M. Aspirin-triggered 15-epi-lipoxin A4 signals through FPR2/ALX in vascular smooth muscle cells and protects against intimal hyperplasia after carotid ligation. *Int. J. Cardiol.* **2015**, *179*, 370–372. [CrossRef] [PubMed]

58. Barajas, B.; Che, N.; Yin, F.; Rowshanrad, A.; Orozco, L.D.; Gong, K.W.; Wang, X.; Castellani, L.W.; Reue, K.; Lusis, A.J.; et al. NF-E2–Related Factor 2 Promotes Atherosclerosis by Effects on Plasma Lipoproteins and Cholesterol Transport That Overshadow Antioxidant Protection. *Arterioscler. Thromb. Vasc. Biol.* **2011**, *31*, 58–66. [CrossRef] [PubMed]

59. Cheng, C.; Noordeloos, A.M.; Jeney, V.; Soares, M.P.; Moll, F.; Pasterkamp, G.; Serruys, P.W.; Duckers, H.J. Heme Oxygenase 1 Determines Atherosclerotic Lesion Progression Into a Vulnerable Plaque. *Circulation* **2009**, *119*, 3017–3027. [CrossRef] [PubMed]

60. Sussan, T.E.; Jun, J.; Thimmulappa, R.; Bedja, D.; Antero, M.; Gabrielson, K.L.; Polotsky, V.Y.; Biswal, S. Disruption of Nrf2, a key inducer of antioxidant defenses, attenuates ApoE-mediated atherosclerosis in mice. *PLoS ONE* **2008**, *3*, e3791. [CrossRef]

61. Freigang, S.; Ampenberger, F.; Spohn, G.; Heer, S.; Shamshiev, A.T.; Kisielow, J.; Hersberger, M.; Yamamoto, M.; Bachmann, M.F.; Kopf, M. Nrf2 is essential for cholesterol crystal-induced inflammasome activation and exacerbation of atherosclerosis. *Eur. J. Immunol.* **2011**, *41*, 2040–2051. [CrossRef]

62. Hansen, T.V.; Vik, A.; Serhan, C.N. The Protectin Family of Specialized Pro-resolving Mediators: Potent Immunoresolvents Enabling Innovative Approaches to Target Obesity and Diabetes. *Front. Pharmacol.* **2018**, *9*, 1582. [CrossRef]

63. Katakura, M.; Hashimoto, M.; Inoue, T.; Mamun, A.A.; Tanabe, Y.; Arita, M.; Shido, O. Chronic Arachidonic Acid Administration Decreases Docosahexaenoic Acid- and Eicosapentaenoic Acid-Derived Metabolites in Kidneys of Aged Rats. *PLoS ONE* **2015**, *10*, e0140884. [CrossRef] [PubMed]

64. Perretti, M. The resolution of inflammation: New mechanisms in patho-physiology open opportunities for pharmacology. *Semin. Immunol.* **2015**, *27*, 145–148. [CrossRef] [PubMed]

65. Serhan, C.N.; Gotlinger, K.; Hong, S.; Lu, Y.; Siegelman, J.; Baer, T.; Yang, R.; Colgan, S.P.; Petasis, N.A. Anti-inflammatory actions of neuroprotectin D1/protectin D1 and its natural stereoisomers: Assignments of dihydroxy-containing docosatrienes. *J. Immunol.* **2006**, *176*, 1848–1859. [CrossRef] [PubMed]

66. Serhan, C.N.; Fredman, G.; Yang, R.; Karamnov, S.; Belayev, L.S.; Bazan, N.G.; Zhu, M.; Winkler, J.W.; Petasis, N.A. Novel proresolving aspirin-triggered DHA pathway. *Chem. Biol.* **2011**, *18*, 976–987. [CrossRef]

67. Ariel, A.; Li, P.L.; Wang, W.; Tang, W.X.; Fredman, G.; Hong, S.; Gotlinger, K.H.; Serhan, C.N. The docosatriene protectin D1 is produced by TH2 skewing and promotes human T cell apoptosis via lipid raft clustering. *J. Biol. Chem.* **2005**, *280*, 43079–43086. [CrossRef]

68. Bannenberg, G.L.; Chiang, N.; Ariel, A.; Arita, M.; Tjonahen, E.; Gotlinger, K.H.; Hong, S.; Serhan, C.N. Molecular Circuits of Resolution: Formation and Actions of Resolvins and Protectins. *J. Immunol.* **2005**, *174*, 4345–4355. [CrossRef]

69. Marcheselli, V.L.; Hong, S.; Lukiw, W.J.; Tian, X.H.; Gronert, K.; Musto, A.; Hardy, M.; Gimenez, J.M.; Chiang, N.; Serhan, C.N.; et al. Novel docosanoids inhibit brain ischemia-reperfusion-mediated leukocyte infiltration and pro-inflammatory gene expression. *J. Biol. Chem.* **2003**, *278*, 43807–43817. [CrossRef]

70. Ariel, A.; Fredman, G.; Sun, Y.P.; Kantarci, A.; Van Dyke, T.E.; Luster, A.D.; Serhan, C.N. Apoptotic neutrophils and T cells sequester chemokines during immune response resolution through modulation of CCR5 expression. *Nat. Immunol.* **2006**, *7*, 1209–1216. [CrossRef]

71. Schwab, J.M.; Chiang, N.; Arita, M.; Serhan, C.N. Resolvin E1 and protectin D1 activate inflammation-resolution programmes. *Nature* **2007**, *447*, 869–874. [CrossRef]
72. Andrews, D.; Godson, C. Lipoxins and synthetic lipoxin mimetics: Therapeutic potential in renal diseases. *Biochim. Biophys. Acta Mol. Cell Biol. Lipids* **2021**, *1866*, 158940. [CrossRef]
73. Perretti, M.; Godson, C. Formyl peptide receptor type 2 agonists to kick-start resolution pharmacology. *Br. J. Pharmacol.* **2020**, *177*, 4595–4600. [CrossRef] [PubMed]
74. Maciuszek, M.; Cacace, A.; Brennan, E.; Godson, C.; Chapman, T.M. Recent advances in the design and development of formyl peptide receptor 2 (FPR2/ALX) agonists as pro-resolving agents with diverse therapeutic potential. *Eur. J. Med. Chem.* **2021**, *213*, 113167. [CrossRef] [PubMed]
75. Qin, C.X.; May, L.T.; Li, R.; Cao, N.; Rosli, S.; Deo, M.; Alexander, A.E.; Horlock, D.; Bourke, J.E.; Yang, Y.H.; et al. Small-molecule-biased formyl peptide receptor agonist compound 17b protects against myocardial ischaemia-reperfusion injury in mice. *Nat. Commun.* **2017**, *8*, 14232. [CrossRef]
76. Xie, Y.; Liu, L. Role of Chemerin/ChemR23 axis as an emerging therapeutic perspective on obesity-related vascular dysfunction. *J. Transl. Med.* **2022**, *20*, 141. [CrossRef] [PubMed]
77. Bannenberg, G.; Moussignac, R.-L.; Gronert, K.; Devchand, P.R.; Schmidt, B.A.; Guilford, W.J.; Bauman, J.G.; Subramanyam, B.; Daniel Perez, H.; Parkinson, J.F.; et al. Lipoxins and novel 15-epi-lipoxin analogs display potent anti-inflammatory actions after oral administration. *Br. J. Pharmacol.* **2004**, *143*, 43–52. [CrossRef]
78. Werlin, E.C.; Kim, A.; Kagaya, H.; Chen, M.; Wu, B.; Mottola, G.; Spite, M.R.; Sansbury, B.; Conte, M.S. A Synthetic Resolvin Analogue (Benzo-Rvd1) Attenuates Vascular Smooth Muscle Cell Migration and Neointimal Hyperplasia. *JVS-Vasc. Sci.* **2020**, *1*, 247–248. [CrossRef]
79. Norling, L.V.; Spite, M.; Yang, R.; Flower, R.J.; Perretti, M.; Serhan, C.N. Cutting Edge: Humanized Nano-Proresolving Medicines Mimic Inflammation-Resolution and Enhance Wound Healing. *J. Immunol.* **2011**, *186*, 5543–5547. [CrossRef] [PubMed]
80. Bäck, M.; Weber, C.; Lutgens, E. Regulation of atherosclerotic plaque inflammation. *J. Intern. Med.* **2015**, *278*, 462–482. [CrossRef]

Review

Wound Healing versus Metastasis: Role of Oxidative Stress

Tatiana Lopez [1,2], Maeva Wendremaire [1,2], Jimmy Lagarde [1,2], Oriane Duquet [2,3], Line Alibert [4], Brice Paquette [4], Carmen Garrido [1,2,5] and Frédéric Lirussi [1,2,3,*]

[1] UMR 1231, Lipides Nutrition Cancer, INSERM, 21000 Dijon, France
[2] UFR des Sciences de Santé, Université Bourgogne Franche-Comté, 25000 Besançon, France
[3] Plateforme PACE, Laboratoire de Pharmacologie-Toxicologie, Centre Hospitalo-Universitaire Besançon, 25000 Besançon, France
[4] Service de Chirurgie, Centre Hospitalo-Universitaire Besançon, 25000 Besançon, France
[5] Centre Georges François Leclerc, 21000 Dijon, France
* Correspondence: frederic.lirussi@univ-fcomte.fr

Abstract: Many signaling pathways, molecular and cellular actors which are critical for wound healing have been implicated in cancer metastasis. These two conditions are a complex succession of cellular biological events and accurate regulation of these events is essential. Apart from inflammation, macrophages-released ROS arise as major regulators of these processes. But, whatever the pathology concerned, oxidative stress is a complicated phenomenon to control and requires a finely tuned balance over the different stages and responding cells. This review provides an overview of the pivotal role of oxidative stress in both wound healing and metastasis, encompassing the contribution of macrophages. Indeed, macrophages are major ROS producers but also appear as their targets since ROS interfere with their differentiation and function. Elucidating ROS functions in wound healing and metastatic spread may allow the development of innovative therapeutic strategies involving redox modulators.

Keywords: wound healing; metastasis; oxidative stress; macrophage; Hypoxia Induced Factor; Nuclear Factor Kappa B; nuclear factor erythroid-2-related factor 2

Citation: Lopez, T.; Wendremaire, M.; Lagarde, J.; Duquet, O.; Alibert, L.; Paquette, B.; Garrido, C.; Lirussi, F. Wound Healing versus Metastasis: Role of Oxidative Stress. *Biomedicines* **2022**, *10*, 2784. https://doi.org/10.3390/biomedicines10112784

Academic Editor: Krisztina Nikovics

Received: 20 September 2022
Accepted: 29 October 2022
Published: 2 November 2022

Publisher's Note: MDPI stays neutral with regard to jurisdictional claims in published maps and institutional affiliations.

1. Introduction

The process of wound healing is a successive well-organized cascade of events involving specific cellular and molecular actors with the intent to restore tissue homeostasis and protect it from infection. On the contrary, unsuccessful healing is associated with severe clinical outcomes such as tumor development. It is also now well documented that some wounds like diabetic wounds or septic injury are associated with tumor progression and/or with an increased risk of cancer relapse.

Excessive and prolonged inflammation during inadequate wound healing process creates a microenvironment that shares strong similarities with tumor stroma. Notably, these microenvironments are characterized by hypoxia that generates a neovascularization for nourishment, influx of leukocytes sustaining the inflammatory response and breakdown/remodeling of the extracellular matrix. The tight similarity between wounds and tumor stroma generation have been first proposed by Rudolph Virchow in 1858 with his 'irritation theory', in which he concluded that irritation and its subsequent inflammation were the essential factors that led to the formation of neoplastic tissues [1]. Over a century later, these same similarities have led Harold Dvorak to state that tumors are 'wounds that do not heal' [2]. The strong similarities between the wound healing process and metastasis dissemination have been extensively and recently reviewed elsewhere in particular concerning the pivotal role of inflammation. Nevertheless, inflammation is also associated with oxidative stress either via Reactive Oxygen Species (ROS) and/or via Reactive Nitrogen Species (RNS) production that may play a key role in the clearing repairing process. Surprisingly, while these mechanisms are well known for wound healing, it remains poorly studied concerning metastasis. In this mini-review, we will focus on the specific contribution of oxidative stress in these two (physio)-pathological processes, in particular by describing the molecular and cellular actors involved. Studying pathophysiological mechanism of wound healing may help to better understand the metastasis process and lead to new therapies and vice versa.

2. Good and Bad Wound Healing: Acute versus Chronic and Chronology of the Cellular Actors

Four distinct and overlapping steps are needed in the physiological wound healing process: (1) hemostasis, (2) inflammation, (3) new tissue formation and (4) tissue remodeling [3–5].

Hemostasis consists of the formation of platelet plug, blood clot and consequent local hypoxia. Then, during the inflammation phase, neutrophils and tissue resident macrophages are the first immune responding cells to the wound [6–9]. This stage also induces immune cells invasion particularly monocytes recruited from the bone marrow and differentiated into mature inflammatory macrophages (named M1) [10]. Proteolytic enzymes, pro-inflammatory cytokines, growth factors and ROS are secreted [11] to protect organism against bacterial or other micro-organisms invasion. After this step, the levels of pro-inflammatory cytokines and oxidative stress decrease to return to a basal state [11]. Resolving anti-inflammatory macrophages (named M2) contribute to remove cells and bacteria debris by efferocytosis or phagocytosis [12]. Keratinocytes, fibroblasts and endothelial cells migrate to the wound and proliferate to initiate new tissue formation stage. Finally, tissue remodeling macrophages promote matrix metalloproteinase (MMP) expression in order to restore functional and anatomical integrity of tissue (Figure 1) [13–16].

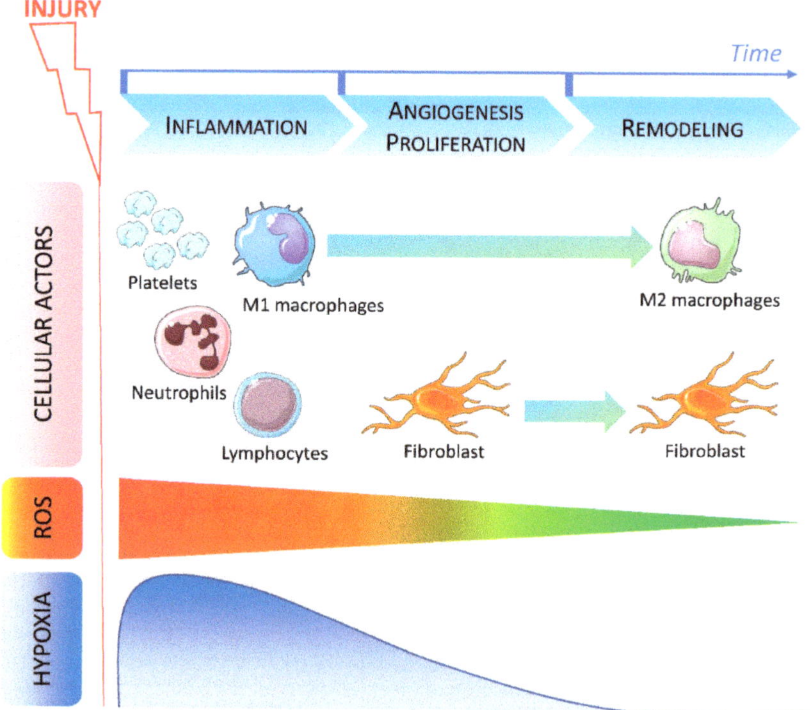

Figure 1. Timeline of cellular actors, ROS and hypoxia involved in wound healing.

After injury, various cells are recruited during the early phase of wound healing. During the inflammation stage, platelets first migrate to the site of the injury to induce coagulation followed by neutrophils. At the same time, ROS level increases while local oxygen concentration decreases leading to hypoxia. Lymphocytes and M1 macrophages are then recruited and promote inflammation. During the angiogenesis and proliferation stage, fibroblasts migrate to the wound and macrophages polarization is modified. Hypoxia is reduced and ROS level decreases indicating the beginning of the late phase. During the remodeling phase, macrophages are polarized in an M2 resolving phenotype and fibroblasts are still present. ROS return to a physiological low level and hypoxia is abolished.

Disturbance at any point in the wound healing process can contribute to pathological wound like fibrosis or non-healing wound.

2.1. Macrophages in Wound Healing Process

Macrophages are major contributing cells in the wound healing process following organ damage either induced by infection, autoimmune disorders, mechanical or toxic injuries. Evidence demonstrates that macrophages depletion reduces inflammatory responses whereas macrophages activation reduces recovery responses [13,17]. Beside tissue resident macrophages, bone marrow-derived macrophages, along with neutrophils, are among the first cells recruited to the site of injury. Their role, widely reviewed within the past years, is described at each step of tissue repair allowing them to be grouped into three types of activation. Firstly, early research highlighted their pro-inflammatory and scavenging contribution to the inflammatory stage [18,19]. Cellular response is then initiated by secreted inflammatory mediators (chemokines, ROS, matrix metalloproteases) leading to pathogens killing and phagocytosis [20,21]. At this stage, macrophages are mainly described with a pro-inflammatory 'classical' M1 phenotype. Secondly, in response to microenvironment stimuli, the predominant macrophage population can mature to an anti-inflammatory healing phenotype depicted to remove dead cells and dampen inflammation [22,23]. These M2 resolving macrophages promote cellular proliferation and blood vessel development through growth factors (Platelet-Derived Growth Factor [PDGF], insulin-like growth factor-1, Vascular Endothelial Growth Factor [VEGF]) and reduce local hypoxia following injury [24,25]. They secrete Transforming Growth Factor-β1 (TGF-β1), which will allow fibroblasts differentiation, stromal cells migration and expansion, wound contraction and closure. In the final stage, a specific subtype of macrophages, called tissue-remodeling macrophages, instruct tissue repair suppressing immune response and subsequently resolving inflammation.

These three functional phenotypes involve an activation continuum that evolves, according to cellular ontogenesis and environmental stimuli, from a pro-inflammatory to a remodeling phenotype [26,27]. Each stage of wound healing must be carefully regulated, especially by different macrophage phenotypes whose roles are unique and critical [28].

2.2. ROS in Wound Healing

ROS (superoxide anion [$O_2^{\bullet-}$] and hydrogen peroxide [H_2O_2]) act in the early phase of wound healing to induce vasoconstriction, platelet activation and defend host from bacterial invasion [11,29,30]. They play a pivotal role in orchestrating wound healing owing to the function of their signaling mediators in immune and stromal cells. ROS allow the recruitment of neutrophils, macrophages and/or lymphocytes to the site of injury [31] and promotes endothelial migration and division. Oxidative stress indicators include glutathione oxidation, modulation of redox-sensitive kinases, or transcription factors such as Nuclear Factor-Kappa B (NF-κB) [32].

ROS level is finely controlled by small anti-oxidant molecules (vitamin C, vitamin E, α-tocopherol, Nicotinamide adenine dinucleotide phosphate [NADPH]) or by an endogenous anti-oxidant and pro-oxidant specialized group of enzymes [33]. Anti-oxidant enzymes (catalase [CAT], glutathione peroxidase [GPx], superoxide dismutase [SOD], NADPH quinone oxidoreductase-1 [NQO-1], Heme-oxygenase-1 [HO-1]) are designed to detoxify

ROS and thereby eliminate their deleterious effects. Contrariwise, NADPH oxidases (NOXs) are a family of major ROS-producing enzymes. The seven transmembrane isoforms (NOX1, NOX2, NOX3, NOX4, NOX5, Duox1, and Duox2) have tissue- and cell type-specific expression profiles and are involved in ROS production as NOX2 and NOX4 mRNA are overexpressed in injury [34]. In addition to the control of the redox state, these enzymes are implicated in a wide range of cellular processes, which includes apoptosis, cellular signal transduction, host defense, angiogenesis and oxygen sensing [35].

A precise homeostatic control of oxidative state is essential for normal tissue repair while extreme (low or high) levels of ROS can impair wound healing [36–39]. Indeed, several studies indicated that reduced ROS level, by magnetic field or pro-oxidant enzyme deficiency, improved wound healing in a model of diabetic mice [34,40]. Furthermore, it has been well documented that non-healing wounds, due to diabetes, or chronic wounds characteristic of pathologies such as inflammatory bowel diseases, are associated with a higher ROS level [32,41–43]. Elevated and sustained ROS are, in these cases, due to excessive or uncontrolled oxidant production or decreased anti-oxidants level (Vitamin E, glutathione) or enzymes activity (CAT, GPx or SOD) [44,45]. This results in a prolonged inflammation process. It therefore appears important to be able to modulate ROS production in healing and to redirect the therapeutic strategy towards their control.

3. Macrophages Polarization: Role of ROS and NOXs

Because macrophages and ROS play major roles in the process of tissue repair, we will focus this review on the mechanisms underlying ROS production during macrophages differentiation and polarization in an oxidative microenvironment.

Based on hydroxyl radical (HO$^\bullet$) imaging, macrophages differentiation stage and HO$^\bullet$ formation are closely interlinked and involve NADPH and consequently NOXs [46]. Furthermore, macrophages polarization towards the pro-inflammatory M1 phenotype resulted in an increased $O_2^{\bullet-}$ and H_2O_2 production compared to M2-polarized macrophages [47]. These results suggest the implication of NOX enzymes in this process.

Further evidence identified the pro-oxidants enzymes NOX1, NOX2 and NOX4 in phagocytes [47,48]. NOX1 and NOX2 are the main isotypes expressed in both bone marrow monocytes and bone marrow-derived macrophages [49]. NOX2 is the most well-characterized enzyme for its role in phagocytic function and is the highest expressed in both human and murine immature macrophages, followed by NOX4 and NOX1 [47,49].

3.1. ROS in Macrophages Differentiation/Polarization and Function

Macrophages produce ROS, which can modulate macrophages function at various stages. Firstly, ROS are essential for the monocytes to macrophages differentiation. Indeed, previous studies indicated that chemically inhibition of ROS generation may affect the monocyte-macrophage differentiation process. Treatment with butylated hydroxyanisole (BHA), a ROS inhibitor, during differentiation blocked the increase in the expression of the macrophage marker CD11b, the induction of $O_2^{\bullet-}$ production and the specific macrophage morphology features [50,51]. This loss of morphology was partially recovered by low concentrations of H_2O_2 [50]. In the context of healing, ROS produced by neutrophils allow bone marrow monocytes to differentiate into macrophages [52].

Secondly, ROS are required for M2 differentiation. ROS inhibitors have been reported to block the overexpression of the M2 marker CD163, the M2 cytokine interleukin-10 (IL-10) and the chemokines CCL17, CCL18 and CCL24 [50,51]. ROS inhibition only acts during the polarization stage and has no effect on the phenotype and function once the macrophage is mature. Indeed, decreased M2 ROS production, after lipopolysaccharide (LPS) treatment, do not affect the expression of M2 markers such as CD163 or CD200R [53]. With regard to the pro-inflammatory macrophages, this treatment had no effect on the CD86 marker and little effect on the secretion of M1 cytokines, Tumor Necrosis Factor-α (TNF-α) and IL-6 [50]. Other studies concluded that depletion of H_2O_2 by catalase or inhibition of ROS favors the expression of M1 markers on bone marrow-derived macrophages [47] and a function on T cell proliferation comparable to M1 macrophages [51]. It therefore seems that ROS are required for macrophages differentiation and polarization.

3.2. NOXs in Macrophages Polarization

At the molecular level, NOX1 and NOX2 are implicated in this process. Indeed, in monocytes from NOX1/2 double knockout mice, ROS generation was largely blocked and affected macrophages differentiation resulting in more rounded and less differentiated cells [49]. NOX2 and its product $O_2^{\bullet-}$ specifically promote an M1 phenotype with phagocytic activity and pro-inflammatory properties [54,55]. Accordingly, NOX2 deficiency reduced pro-inflammatory M1 macrophages and promoted M2 macrophages polarization in a mouse model of brain injury [56]. In contrast, M2 polarization of macrophages is characterized by both reduced NOX2 activity and reduced $O_2^{\bullet-}$ production. Loss of NOX1 and NOX2 affects the differentiation of monocytes to macrophages and the polarization of M2 macrophages. The M2 populations from NOX1/2 double knockout mice were substantially reduced compared with the wild-type mice [49]. In a wound healing model, NOX1/2 double knockout mice had less infiltration of M2-type macrophages in the wound edge and a delayed wound healing compared with wild-type mice [49,57]. These results may indicate a defect in macrophages polarization or recruitment to the site of injury. Another study, on in vitro murine macrophages, indicated that loss of NOX2 induced a small but significant reduction in M1 polarization with no effect on M2 polarization [47,49].

Because NOX4 expression is increased during phorbol myristate acetate (PMA)-induced monocytes to macrophages differentiation, several studies analyzed the contribution of this enzyme on this process. Data showed that NOX4 expression remained upregulated in the PMA-induced differentiating macrophages, while treatment with apocynin downregulated NOX4 in an in vitro system [46]. When NOX4 was chemically inhibited, TNF-α and IL-1β expression was increased in human macrophages, derived from peripheral blood monocytes, indicating M1 polarization. This was accompanied by a significant downregulation in M2 markers [47]. On the contrary, other studies focused on murine intestinal macrophages abundantly found in inflammatory bowel diseases and expressing various phenotypes. They revealed that NOX4 inhibitor suppressed the M1 polarization of intestinal macrophages, reducing the proportion of F4/80$^+$ CD11c$^+$ macrophages and inflammatory cytokines levels [58]. We can assume that these divergent results of NOX4 inhibition relate with the macrophages lineage and that NOX4 may act on distinct differentiation and polarization stages. Furthermore, the absence of NOX4 increased ROS formation in M1-polarized macrophages. Because the major source of ROS in M1 macrophages is NOX2, studies revealed that its expression was elevated in NOX4-deficient M1 polarized macrophages [47].

3.3. Molecular Events and Signaling Pathways Involved in Wound Healing

Wound healing stages involve specific molecular hallmarks such as hypoxia, inflammation and oxidative stress. These markers are regulated, among other things, by numerous transcription factors. Activation of these transcription factors is a key event for the hypoxic or inflammatory signaling cascades and the oxidative stress response. We will describe here the main signaling targets identified in macrophages (i.e., Hypoxia Induced Factor [HIF], NF-κB and nuclear factor erythroid-2-related factor 2 [Nrf2]) and their functional interrelation (Figure 2).

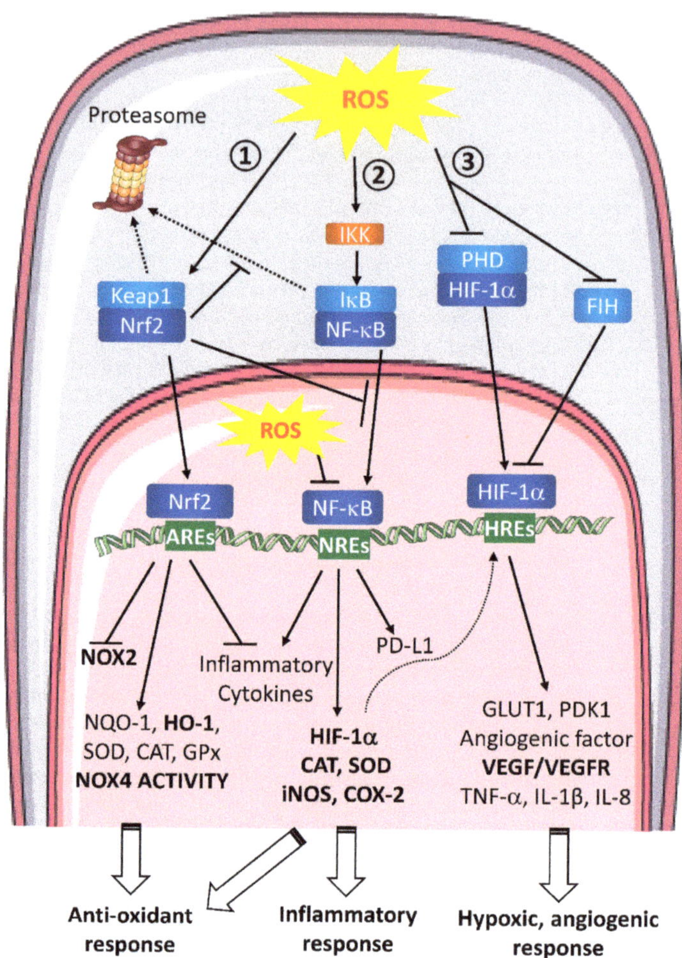

Figure 2. ROS signaling pathways involved in wound healing and metastasis. Extracellular ROS activate intracellular signaling pathways. ① Nrf2 pathway. In unstressed conditions, Keap1 retains Nrf2 in the cytoplasm. When ROS are produced, Keap1 is oxidized and ubiquitinated thereby leading to its proteasomal degradation. Consequently, Nrf2 is free to translocate to the nucleus and binds to the anti-oxidant response elements (AREs). This binding inhibits NOX2 and pro-inflammatory cytokines transcription and enhances the anti-oxidant defense response expression. ② NF-κB pathway. In normal conditions, NF-κB is associated with IκB and retained in the cytoplasm. In the presence of ROS, IKK is activated and can phosphorylate IκB to induce its dissociation with NF-κB and its proteasomal degradation. Then, free NF-κB translocates to the nucleus, binds to NF-κB Response Elements (NREs) and induces target genes transcription leading to a global inflammatory response. ROS are able to directly act in the nucleus inhibiting NF-κB binding to the NREs. ③ HIF pathway. In homeostatic conditions, HIF-1α is hydroxylated by PHDs and targeted for proteasomal degradation. HIF-1α is also regulated by FIH, which blocks the interaction between HIF-1α transactivation domain and coactivators on HREs. During hypoxia or when ROS are produced, PHDs are inactivated which stabilizes HIF-1α and FIH is inhibited. HIF-1α then translocates into the nucleus where it binds to HIF Response Elements (HREs). Transcription of target genes is induced leading to hypoxic and angiogenic response. Nrf2, NF-κB and HIF pathways are closely interlinked. Nrf2 can inhibit IκB proteasomal degradation and NF-κB nuclear translocation, NF-κB pathway induces anti-oxidant response regulating iNOS and COX-2 transcription and HIF-1α expression is regulated by NF-κB.

- HIF (Figure 2 ①)

In a wound, local oxygen level is reduced due to blood vessel destruction [59]. A change in oxygen concentration regulates transcription factors, the main being HIF. This local hypoxia implicates macrophages and induces ROS production, among others, as signaling molecules to restore normoxia [42,60]. As oxidative stress and macrophages are closely related during wound healing, the role of ROS on HIF activation, in macrophages, has been investigated.

HIF are a family of 3 transcription factors (HIF-1, HIF-2 and HIF-3). These heterodimers of β-subunits and hypoxia-induced α-subunits (HIF-1α, HIF-2α HIF-3α bind to hypoxia-responsive elements and activate target genes transcription. HIF-1α induces the expression of glucose transporter 1 (GLUT1), and pyruvate dehydrogenase kinase isoform 1 (PDK1) in macrophages [61,62]. In cancer cells, HIF-1α increases Programmed death-ligand 1 (PD-L1) expression and cytokines secretion (i.e., VEGF) thereby promoting tumor associated macrophages (TAM) accumulation and immune escape.

In homeostatic conditions, HIF activation is regulated by proteasomal degradation. HIF is hydroxylated by prolyl hydroxylases (PHDs) and subsequently ubiquitinated by the E3 ubiquitin ligase von Hippel-Lindau. These modifications direct HIF to the ubiquitin-proteasome system for degradation. Another layer of regulation involves the interaction between proteins from the signaling pathway. This level involves Factor inhibiting HIF (FIH), which blocks interactions between the HIF-α transactivation domain and coactivators. When oxygen concentration decreases, PHDs are inactive and HIF is stabilized in the cytoplasm. This accumulation allows the transcription factor to translocate in the nucleus and to regulate target genes expression [63].

Several studies have focused on oxidative stress and HIF during hypoxia or normoxia [64,65]. They revealed that ROS contribute to HIF transcriptional activity by stabilizing HIF-1α and inhibiting FIH. Indeed, Chandel et al. demonstrate that catalase abolishes HIF-1α stabilization under hypoxic conditions [64]. Conversely, high concentration of H_2O_2 can induce HIF-1α stabilization in normoxia [65].

In macrophages, stimuli like LPS or pathogenic microorganisms' infection, can upregulate HIF-1α expression and activity through NF-κB signaling [65–67]. Indeed, Li et al. demonstrate the critical role of HIF-1α during macrophages polarization towards pro-inflammatory phenotype. They also found that HIF-1α is necessary for macrophages responses when these cells are challenged with pathogens. Furthermore, in HIF-1α deficient macrophages, mRNA expression, production and secretion of several pro-inflammatory cytokines (TNF-α and IL-6) or VEGF are inhibited independently of oxygen level [67,68]. In inflammatory bowel diseases, on the contrary, effects of HIF knockout in myeloid cells depend on the type of transcription factor studied. Finally, in an intestinal context, HIF-1 has been reported to promote inflammation while HIF-2 protects against chemically induced inflammation [69].

- NF-κB (Figure 2 ②)

NF-κB plays a crucial role in inflammatory and immune responses and is subject to complex regulation. It participates in a plethora of macrophages regulatory mechanisms and is associated with extensive ROS production. Its role in healing is therefore important at all stages of the process, whether it is at the early inflammatory phase or at the later phase of tissue formation and remodeling [70,71].

NF-κB is a homo- and hetero-dimeric complex resulting from the five monomers in mammals (RelA/p65, RelB, cRel, NF-κB1 p50, and NF-κB2 p52) [72]. The heterogeneity of NF-κB targets is further increased by interactions of NF-κB dimers with other transcription factors. The most well characterized heterodimer during inflammatory response is the p50/p65 complex. NF-κB is kept inactive in the cytosol by binding to the inhibitory protein IκBα (nuclear factor of kappa light polypeptide gene enhancer in B-cells inhibitor, alpha). Under various stimuli (inflammation, cytosolic ROS), the IκB kinase (IKK) complex, which is constituted of two catalytic subunits IKKα and IKKβ and a regulatory subunit IKKγ

(or NEMO), is phosphorylated. This complex, thus activated, phosphorylates IκBα thereby targeting the protein for proteasomal degradation. NF-κB is then free to translocate to the nucleus and initiate the transcription of several genes [70,71,73]. ROS can also oxidize NF-κB cysteines and inhibit its DNA binding, reducing its activity. In addition to its major role in inflammation, an immunosuppressive one has been described in the context of tumor microenvironment where ROS induce PD-L1 expression through NF-κB binding to its promoter [74]. In the same way, in an inflammatory bowel disease model, ROS activate NF-κB signaling leading to the recruitment and the polarization of intestinal macrophages to an M2 phenotype [75]. In metabolic disorders such as obesity and type 2 diabetes, Luo et al. demonstrate that celastrol, a natural anti-oxidant, is able to suppress M1 macrophage polarization and enhance M2 polarization through inhibition of NF-κB nuclear translocation. This M1 polarization is mediated by the Nrf2 activation pathway [76].

- Nrf2 (Figure 2 ③)

Transcription factor Nrf2 is a basic leucine zipper (bZIP), which is a major sensor for oxidative stress [77]. It has been described to maintain redox homeostasis and to attenuate inflammation and thereby to be involved in wound healing [78].

Under unstressed conditions, Nrf2 is retained in the cytoplasm by Kelch-like ECH-associated protein 1 (Keap1) that functions as an Nrf2 Inhibitor [79]. Keap1 is an adapter protein of the E3 ubiquitin ligase Cul3-Ring-box 1, which is responsible for the ubiquitination and proteasomal degradation of Nrf2.

Upon oxidative stress, several cysteine residues on Keap1 are subjected to oxidation which induced a conformational change in the protein and prevents Nrf2 ubiquitination and subsequent degradation. As a consequence, Nrf2 is released from Keap1 and accumulates in the cytoplasm. Nrf2 then translocates into the nucleus and forms a heterodimer with bZIP proteins. On one hand, the heterodimer Nrf2 binds to anti-oxidant response elements of target genes and regulates the expression of cytoprotective anti-oxidant genes and detoxifying enzymes implicated in NADPH, glutathione and thioredoxin systems (HO-1 and NQO-1) [80]. Nrf2 is also implicated in NOX expression as its deletion in fibroblast induces an upregulation of NOX4 [81]. As Nrf2 is essential to maintain redox homeostasis, its inhibition in fibroblasts reduces specific NADPH ROS production during treatment with ionomycin (a calcium ionophore agent) while it does not interfere with ROS levels in basal conditions. In Nrf2 knockout mice, ROS level is increased compared to wild-type mice [81].

On the other hand, Nrf2 is described to regulate gene expression of pro-inflammatory cytokines independently of ROS level [82]. In this case, evidence suggested that Nrf2 can bind to the proximity of the pro-inflammatory gene (not only on anti-oxidant response elements) and interferes with the polymerase II thereby inhibiting the transcription initiation step [82].

Furthermore, in macrophages, a high level of Nrf2 decreases LPS-induced cytokines while, in its absence, pro-inflammatory cytokines are upregulated [83,84]. Microarrays analyses, on bone marrow derived macrophages from Nrf2 knockout mice, indicated that genes induced during M1 polarization are downregulated [82,83]. Other indirect evidence suggests that Nrf2 induces M2 macrophages polarization. Overexpression of HO-1, a Nrf2 target gene, induces an anti-inflammatory response in cultured macrophages [85]. In a model of delayed diabetic wound healing, Nrf2 activation accelerates the wound process while Nrf2 inhibition mimics the effects of diabetes and the delayed process [84]. In inflammatory bowel diseases, Nrf2 has been reported to protect against colitis. The first study describing this role, performed by Khor et al., reveals that Nrf2 knockout mice are more sensitive to chemically induced colitis [86]. Further studies indicate that Nrf2 prevents the early stages of carcinogenesis associated with colitis [87].

Nevertheless, Nrf2 has been also described to favor the progression of cancer cells. In TAM, nuclear translocation of Nrf2 is increased and its targeted anti-oxidant genes are overexpressed. In Nrf2 knockdown macrophages, treatment with cancer cell medium blocked the induced over-expression of M2 markers and down-regulation of M1 markers [88]. Controversially, in macrophages exposed to the tumor fluid, data indicate that Nrf2 nuclear localization is reduced, indicating an alteration in the oxidative status [89].

In wound healing, signaling pathways are closely linked and are activated at different stages of the process. Their activation is not stage-specific but presents a continuum. Their roles are in some cases redundant and allow the activation of the same target genes, therefore having an identical overall effect. In some cases, transcription of target genes from one signaling pathway will activate another pathway. It is therefore difficult to know exactly the role of each signaling pathway in wound healing.

4. Metastasis

Many wound healing cellular actors, molecular mechanisms and signaling pathways are also implicated in metastasis [2]. Therefore, elucidating the link between wound healing and metastatic cancer progression may allow the development of better therapeutic strategies against these two pathologies.

4.1. Metastasis Hallmarks

Metastatic spread comprises a complex succession of cellular biological events leading to the dissemination of cancer cells from the tumor to the surrounding tissues and to distant organs, through blood and lymphatic vessels [90]. Furthermore, it also involves crosstalk between cancer cells and components of the tumor microenvironment [91].

The metastatic process begins with the hypoxia at the primary tumor site due to excessive cell proliferation [92]. Reduced oxygen level induces HIF-1α stabilization and its nuclear translocation, which promotes the expression of various genes involved among others in angiogenesis, glucose metabolism, extracellular matrix remodeling, epithelial-mesenchymal transition, metastasis, cancer stem cell maintenance and immune invasion [93]. In parallel, hypoxia-induced necrosis results in a continuous release of cellular debris, notably High Mobility Group Box protein-1 (HMGB1) by dying tumor cells [94]. HMGB1 has been shown to be up-regulated in tissue biopsies from cancer patients [95]. Interestingly, HMGB1 plays opposite roles depending on its redox state. Oxidized HMGB1 induces the production of pro-inflammatory cytokines whereas the reduced form interacts with TAM therefore regulating monocyte recruitment, angiogenesis and immune suppression [96]. In fine, altering the redox status of HMGB1 may be considered as a therapeutic approach to combat metastasis and favor wound healing.

Angiogenesis provides oxygen and nutrients supply essential for cancer cells to dissociate from the basal membrane delineating the epithelial compartment from the stroma. This requires the degradation of the extracellular matrix (ECM), through the activation of matrix metalloproteinases [97]. Under normal circumstances, cells detachment from the ECM leads to the induction of an apoptosis called anoikis, a form of programmed cell death that occurs in anchorage-dependent cells [98]. However, cancer cells develop a trans-differentiation program known as epithelial–mesenchymal transition (EMT), which render the cells resistant to anoikis [99]. Anoikis plays an important role in the prevention of metastasis and promoting its induction might be an interesting therapeutic strategy. Finally, cells acquire stemness properties. Stemness is the ability of a cell to perform self-renewal and is capable of pluripotency. This is an important feature for supplying material for wound closure and for the establishment of cancer cells at the metastatic sites [100].

4.2. ROS and Metastasis

One of the principal mechanisms underlying metastasis in human cells is the disruption of the redox balance. This imbalance in redox homeostasis is induced by an increase in free radicals, mainly ROS [101]. Cancer cells have elevated expression levels of NOXs (NOX1, NOX2, NOX4, NOX5), leading to high levels of ROS [101,102]. Consequently, cancer cells have been shown to be more tolerant to oxidative stress via increased expression of catalase and superoxide dismutase. However, the lack of robust anti-oxidant defenses may have detrimental consequences in the tumor microenvironment and in the adjacent normal cells [103].

- Dual effect of ROS

Although several processes of metastasis are redox-sensitive, it is still controversial whether ROS have oncogenic/metastatic or tumor suppressive functions. The answer appears to depend on ROS levels and the cancer stage, leading many authors to consider ROS as a "double-edged sword" [101]. Low to moderate ROS levels can promote survival of cancer cells by inducing EMT and stem cell differentiation, enhancing angiogenesis and switching to glycolytic metabolism. Conversely, excessive production of ROS induced by chemotherapy and radiotherapy is detrimental to the survival of cancer cells and causes cellular damage [104,105]. Concerning the stage of the disease, it has been reported that in the early stages of cancer, ROS promote cancer initiation by inducing base pair substitution mutations in pro-oncogenes such as Ras and tumor suppressor genes such as p53 [106]. As cancer progresses, an intracellular excess of ROS triggers apoptosis of tumor cells. To escape this ROS-induced apoptosis, tumor cells produce high levels of anti-oxidants [106]. In the last stages of tumor development, ROS have a pro-metastatic role promoting the spread of cancer cells.

- ROS and angiogenesis

Additionally, ROS are involved in angiogenesis. Angiogenesis is mainly mediated by VEGF whose expression can be regulated by nutrient deprivation and hypoxia, both of which increase levels of ROS [107,108]. Activation of angiogenesis by ROS can involve different signaling pathways. Firstly, ROS have been shown to activate PI3K/Akt/mTOR signaling cascade in different cancer cell lines (MCF-7, HepG2, H-1299, PC-3), enhancing HIF-1α and VEGF expression and ultimately angiogenesis [109,110]. The role of ROS has been confirmed by several studies showing that catalase and glutathione peroxidase overexpression or NOX4 knockdown lead to a decrease in VEGF and HIF-1α levels and inhibit angiogenesis in human ovarian cancer cells [111,112]. Further, oxidative stress can induce angiogenesis in a VEGF-independent manner through the activation of the TLR/NF-κB pathway. West et al. demonstrated the proangiogenic effects of TLR1/2 stimulation by oxidative stress, represented by lipid oxidation products, in murine and human melanoma [113]. In addition, angiogenesis is also mediated by matrix metalloproteinases and upregulated by ROS [114].

- ROS, EMT and anoikis resistance

Several studies have proven that ROS are a major cause of EMT. ROS-induced EMT has been reported to be NOX4-dependent in human metastatic breast epithelial cells [115] and in lung cancer cells [116]. NOX4 is an important source of ROS induced by TGF-β and under hypoxia, two important mediators in cancer metastasis [117,118]. Furthermore, NOX4 inhibition significantly attenuated the distant metastasis of breast cancer cells to lung and bone [119].

Resistance to anoikis seems to concern not only the field of cancer but also this phenomenon may be interesting in wound healing. Indeed, ROS are considered as one of the key players in anoikis sensitivity. In recent studies, ROS generation induced by NOX4 has been involved in anoikis resistance of gastric [120] and lung cancer cells [121]. ROS promote EMT by inducing the expression and activity of MMPs that mediate proteolytic degradation of ECM components [122,123]. TGF-β1, a well-established player of EMT induction, regulates MMP-9 to facilitate cell migration and invasion via the activation of NF-κB through a ROS-dependent mechanism [123]. Similarly, ROS production induced MMP-2 secretion and activation results in pancreatic cells invasion [122]. In colorectal cancer, the EMT process is highly regulated through some of the classic tumorigenic signaling pathways, such as the NF-κB, HIF-1, and TGF-β1 pathways [124]. Intriguingly, TGF-β1 induces EMT through Nrf2 activation as well as ROS production in lung adenocarcinoma cells [116]. Indeed, Nrf2 is a key transcriptional regulator that drives anti-oxidant gene expression and protection from oxidative damages. Oxidative stress plays a critical regulatory role in these pathways by degrading inhibitors or inducing nuclear translocation and consequent transcription [124].

- ROS and stemness

Cancer stem cells possess a particular redox status, since they have lower ROS levels and increased anti-oxidant capacity than differentiated cancer cells [125,126]. Increasing evidence shows that these low amounts of ROS are actually needed to maintain the quiescence and self-renewal potential of cancer stem cells (CSC). Previous studies have demonstrated that ROS contribute to reduce stemness and to enhance differentiation of CSC. For example, glioblastoma stem cells have potent anti-oxidant defense mechanisms and H_2O_2 has been shown to inhibit their self-renewal and induce their differentiation [127]. ROS have been reported to promote hematopoietic stem cell differentiation with a progressive increase in ROS levels with the advancing differentiation stages. Moreover, inhibition of ROS production has been found to attenuate the differentiation of hematopoietic stem cells [128]. In summary, hypoxia-associated increase in ROS in tumor cells promotes stemness. Although oxidative stress promotes the development of CSC, ROS level declines after this acquisition of stemness, allowing the maintenance of the sub-population.

4.3. Oxidative Stress and Metastasis: Cellular Actors Involved

Macrophages, neutrophils and fibroblasts are major ROS producers in the tumor microenvironment [92]. Here, we will focus on macrophages and fibroblasts since neutrophils activation in wound healing and metastasis has been already extensively reviewed [129].

- Macrophages

In cancer, macrophages present in the tumor are known as TAM and can represent up to 50% of the tumor mass [130]. ROS can be both beneficial and detrimental for the anti-cancer immune function. Therefore, they may indirectly impact cancer progression by altering cancer immune surveillance [131]. Although macrophages have anti-tumor effects as immune cells, experimental and clinical evidence have revealed that TAM contribute to tumor progression and metastasis. High levels of TAM are associated with weak prognosis and decreased overall survival in various cancers [132–135]. The effect of ROS in TAM polarization toward a M1 or M2 phenotype has been discussed, as several studies showed that ROS can stimulate both activation statuses in TAM [49,50,136,137]. M1 and M2 macrophages are two extremes in a continuum of macrophage functional states, which reflect the different effects that can be observed on tumor cells [138].

$O_2^{\bullet-}$ production promotes M2 polarization through activation of ERK and JNK signaling pathways [49,50]. Moreover, administration of the anti-oxidant BHA blocked TAM infiltration and tumor progression, which suggests a beneficial effect of ROS inhibition in tumor therapy [50]. Indeed, another ROS scavenger, oligo-fucoidan, has been reported to inhibit M2 polarization and TAM infiltration in subcutaneous colorectal tumors [139]. Conversely, Wu et al. demonstrated that increased NOX-dependent ROS production by irradiation of macrophages promotes a pro-inflammatory M1 phenotype that is associated with improved response to radiotherapy in rectal cancer [137]. Similarly, iron overload has been reported to polarize macrophages towards an M1 phenotype by increasing ROS production and reduction in ROS levels by N-Acetyl-Cysteine repressed M1 polarization [136]. These results confirm a link between ROS generation and M1 polarization of macrophages. Apart from polarization, ROS also govern TAM apoptosis. For example, inhibition of autophagy in macrophages increases ROS levels, provokes TAM apoptosis and leads to regression of the primary tumor [140]. TAM are also major players in the regulation of tumor angiogenesis in colorectal cancer [141]. They have been demonstrated to enhance the expression of angiogenic proteins in the tumor microenvironment in an oxidative stress-dependent manner by regulating the activity of NOXs [142].

- Fibroblasts

In wound healing, fibroblast's function includes renewal of ECM, the regulation of epithelial differentiation and the regulation of inflammation. Cancer-Associated Fibroblasts (CAFs) are the most predominant stromal cell type in the tumor microenvironment [143]. They are major producers of ROS [144], which facilitates metastasis through the activation of angiogenesis [145]. Moreover, cancer cells induce ROS overproduction in CAFs contributing to a pro-oxidative tumor microenvironment [146]. Conversely, ROS produced by CAFs enhance ROS generation in cancer cells, increasing tumor aggressiveness [147]. CAF-mediated ROS production are involved in the increased metastasis potential of prostate carcinoma. CAF drive cancer cells to secrete cyclooxygenase-2 (COX-2)-mediated ROS, which is mandatory for EMT, stemness and dissemination of metastatic cells [148]. Finally, CAFs, in a mouse model of squamous skin carcinogenesis, promote macrophage recruitment and neovascularization in close association with NF-κB [149].

5. Conclusion and Future Perspectives

Although the intertwining of wound healing and metastasis have already been well described in the literature, this review highlights the molecular and cellular similarities between these two processes. Notably, accumulating evidence designates ROS and macrophages as major regulators of these pathologies, in which disturbance can lead to either pathological wounds or cancer cells spread. These two actors are intrinsically linked since macrophages are the main source of oxidative stress and, at the same time, their differentiation and polarization require ROS. In this context, both appear as potential therapeutic targets.

As recapitulated in Figure 3, a high level of ROS is a common feature in the development of non-healing wound and metastasis. Controlling oxidative stress level in wound and tumor cells environment can be an interesting strategy both to promote wound healing and to prevent metastatic spread. The excessive ROS accumulation could be managed by (1) scavenging agents, (2) limiting its production and/or (3) increasing anti-oxidant defenses. ROS-scavenging hydrogel showed enhanced wound healing abilities by down-regulating pro-inflammatory cytokines, up-regulating the M2 phenotype of macrophages and promoting angiogenesis and the production of collagen [150]. Secondly, the production of ROS can be limited through NOXs inhibition. To date, few studies have focused on this area due to the lack of specificity and pharmacological knowledge on NOXs inhibitors [151]. Nevertheless, a dual protective effect against oxidative stress has been demonstrated by beta3-adrenergic receptor stimulation on macrophages. Indeed, it results in the inhibition of NOXs activity, a decreased NOX2 level and an increased catalase expression [152]. Although this study was conducted for preterm birth management, the use of beta3-adrenergic receptor agonists can be applied to other pathologies associated with excessive oxidative stress production. Finally, the use of anti-oxidants such as vitamins, polyphenols and flavonoids has been widely studied [102,153]. Unfortunately, when used as monotherapy, clinical studies did not provide any therapeutic benefit. Along with the tremendous rise of the immune-checkpoint modulators as anti-cancer drugs, this led researchers to investigate the potential synergistic effects of ROS blockade and immunotherapy. For example, recent studies reported that vitamin C supplementation improved anti-cancer immunotherapies efficiency in various murine tumor models [154,155].

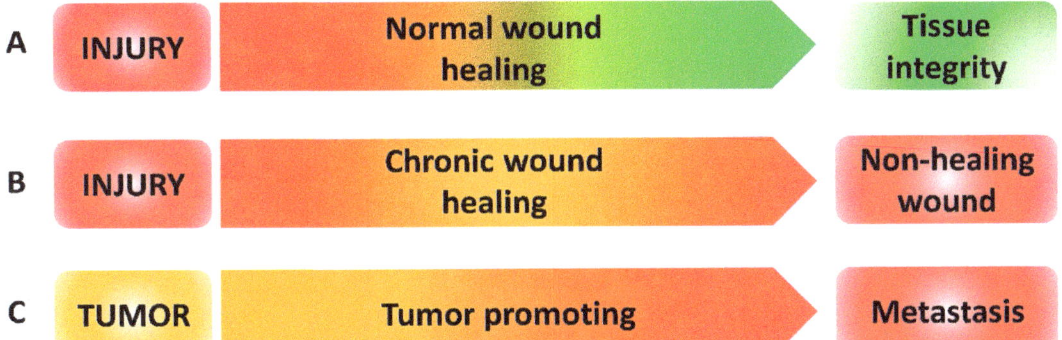

Figure 3. ROS levels during wound healing and metastasis. (**A**) ROS level during normal wound healing. After injury, high levels of ROS (red) are produced and then decreased to low level (green) over time to restore tissue integrity. (**B**) ROS level during chronic wound healing. After injury, high levels of ROS (red) are produced and failed to be reduced inducing non-healing wound. (**C**) ROS level during tumor progression. In tumor, ROS are produced in an intermediate level (orange). When ROS level increased, tumor progression is promoted leading to metastasis.

Reprogramming of macrophages appears as the second target for the management of cancer metastasis and, by extension, of wound healing. Indeed, since macrophages are also involved in wound pathophysiology, this therapeutic approach can also be interesting in wound healing. Administration of the anti-oxidant BHA blocked M2 macrophage differentiation resulting in suppression of tumorigenesis in three different mouse cancer models [50]. Similarly, another ROS scavenger, oligo-fucoidan, induced monocyte polarization toward M1-like macrophages and repolarized M2 macrophages into M1 phenotypes; therefore, inhibiting colorectal tumor progression [139].

It is worth mentioning that some limitations of targeting oxidative stress as a promising treatment in wound healing and metastasis relies on the balance needed between beneficial and harmful effects of ROS. As a double-faceted agent, ROS also play a pivotal role in orchestrating wound healing mechanisms [156] and as potent genotoxic agents causing DNA damage in cancer cells [102]. As proof, radiotherapy and chemotherapy induce oxidative stress necessary for their anti-tumoral activity [104,105]. Furthermore, due to some disparities in the mechanisms of these two diseases, questions arise as to the modalities and timing of administration of therapies. Defective wound healing would require local treatment while systemic treatment seems more suitable to prevent and treat metastases.

In summary, this review offers a compilation that may provide a better understanding of the pivotal role of oxidative stress in both wound healing and metastasis, encompassing the contribution of macrophages. Although the treatment of metastases or chronic wounds is a real challenge, new therapeutic approaches involving administration of redox modulators need to be considered.

6. Methods

This literature review was based on searches on PubMed, Web of science, Springer and Wiley databases, with no time limit but giving preference to recent articles.

Author Contributions: All of the authors contributed equally to all aspects of the article. All authors have read and agreed to the published version of the manuscript.

Funding: This work was supported by a French Government grant managed by the French National Research Agency under the program "Investissements d'Avenir" with reference ANR-11-LABX-0021

(LabEX LipSTIC). This work was also supported by the Agence Nationale de la Recherche (grant ANR-17-CE17-0023 to FL).

Institutional Review Board Statement: Not applicable.

Informed Consent Statement: Not applicable.

Data Availability Statement: The data that support the findings of this study are available from the corresponding author upon reasonable request.

Conflicts of Interest: The authors declare no conflict of interest.

References

1. Virchow, R. *Die Cellularpathologie in Ihrer Begründung Auf Physiologische Und Pathologische Gewebelehre*; Hirschwald: Berlin, Germany, 1858; Volume 16.
2. Dvorak, H.F. Tumors: Wounds That Do Not Heal. Similarities between Tumor Stroma Generation and Wound Healing. *N. Engl. J. Med.* **1986**, *315*, 1650–1659. [CrossRef]
3. Singer, A.J.; Clark, R.A. Cutaneous Wound Healing. *N. Engl. J. Med.* **1999**, *341*, 738–746. [CrossRef] [PubMed]
4. Martin, P. Wound Healing–Aiming for Perfect Skin Regeneration. *Science* **1997**, *276*, 75–81. [CrossRef] [PubMed]
5. Sorg, H.; Tilkorn, D.J.; Hager, S.; Hauser, J.; Mirastschijski, U. Skin Wound Healing: An Update on the Current Knowledge and Concepts. *Eur. Surg. Res.* **2017**, *58*, 81–94. [CrossRef]
6. Gosain, A.; DiPietro, L.A. Aging and Wound Healing. *World J. Surg.* **2004**, *28*, 321–326. [CrossRef] [PubMed]
7. Broughton, G.; Janis, J.E.; Attinger, C.E. The Basic Science of Wound Healing. *Plast. Reconstr. Surg.* **2006**, *117*, 12S–34S. [CrossRef] [PubMed]
8. Campos, A.C.L.; Groth, A.K.; Branco, A.B. Assessment and Nutritional Aspects of Wound Healing. *Curr. Opin. Clin. Nutr. Metab. Care* **2008**, *11*, 281–288. [CrossRef]
9. Sindrilaru, A.; Scharffetter-Kochanek, K. Disclosure of the Culprits: Macrophages-Versatile Regulators of Wound Healing. *Adv. Wound Care* **2013**, *2*, 357–368. [CrossRef]
10. Arnold, L.; Henry, A.; Poron, F.; Baba-Amer, Y.; van Rooijen, N.; Plonquet, A.; Gherardi, R.K.; Chazaud, B. Inflammatory Monocytes Recruited after Skeletal Muscle Injury Switch into Antiinflammatory Macrophages to Support Myogenesis. *J. Exp. Med.* **2007**, *204*, 1057–1069. [CrossRef]
11. Roy, S.; Khanna, S.; Nallu, K.; Hunt, T.K.; Sen, C.K. Dermal Wound Healing Is Subject to Redox Control. *Mol. Ther.* **2006**, *13*, 211–220. [CrossRef]
12. Bosurgi, L.; Cao, Y.G.; Cabeza-Cabrerizo, M.; Tucci, A.; Hughes, L.D.; Kong, Y.; Weinstein, J.S.; Licona-Limon, P.; Schmid, E.T.; Pelorosso, F.; et al. Macrophage Function in Tissue Repair and Remodeling Requires IL-4 or IL-13 with Apoptotic Cells. *Science* **2017**, *356*, 1072–1076. [CrossRef] [PubMed]
13. Duffield, J.S.; Forbes, S.J.; Constandinou, C.M.; Clay, S.; Partolina, M.; Vuthoori, S.; Wu, S.; Lang, R.; Iredale, J.P. Selective Depletion of Macrophages Reveals Distinct, Opposing Roles during Liver Injury and Repair. *J. Clin. Investig.* **2005**, *115*, 56–65. [CrossRef] [PubMed]
14. Lazarus, G.S.; Cooper, D.M.; Knighton, D.R.; Margolis, D.J.; Pecoraro, R.E.; Rodeheaver, G.; Robson, M.C. Definitions and Guidelines for Assessment of Wounds and Evaluation of Healing. *Arch. Dermatol.* **1994**, *130*, 489–493. [CrossRef] [PubMed]
15. Serra, R.; Gallelli, L.; Butrico, L.; Buffone, G.; Caliò, F.G.; De Caridi, G.; Massara, M.; Barbetta, A.; Amato, B.; Labonia, M.; et al. From Varices to Venous Ulceration: The Story of Chronic Venous Disease Described by Metalloproteinases. *Int. Wound J.* **2017**, *14*, 233–240. [CrossRef]
16. de Franciscis, S.; Gallelli, L.; Amato, B.; Butrico, L.; Rossi, A.; Buffone, G.; Caliò, F.G.; De Caridi, G.; Grande, R.; Serra, R. Plasma MMP and TIMP Evaluation in Patients with Deep Venous Thrombosis: Could They Have a Predictive Role in the Development of Post-Thrombotic Syndrome? *Int. Wound J.* **2016**, *13*, 1237–1245. [CrossRef] [PubMed]
17. Zhang, M.-Z.; Yao, B.; Yang, S.; Jiang, L.; Wang, S.; Fan, X.; Yin, H.; Wong, K.; Miyazawa, T.; Chen, J.; et al. CSF-1 Signaling Mediates Recovery from Acute Kidney Injury. *J. Clin. Investig.* **2012**, *122*, 4519–4532. [CrossRef]
18. Minutti, C.M.; Knipper, J.A.; Allen, J.E.; Zaiss, D.M.W. Tissue-Specific Contribution of Macrophages to Wound Healing. *Semin. Cell Dev. Biol.* **2017**, *61*, 3–11. [CrossRef]
19. Italiani, P.; Boraschi, D. From Monocytes to M1/M2 Macrophages: Phenotypical vs. Functional Differentiation. *Front. Immunol.* **2014**, *5*, 514. [CrossRef]
20. Barrientos, S.; Stojadinovic, O.; Golinko, M.S.; Brem, H.; Tomic-Canic, M. Growth Factors and Cytokines in Wound Healing. *Wound Repair Regen.* **2008**, *16*, 585–601. [CrossRef]
21. Zhang, Q.; Raoof, M.; Chen, Y.; Sumi, Y.; Sursal, T.; Junger, W.; Brohi, K.; Itagaki, K.; Hauser, C.J. Circulating Mitochondrial DAMPs Cause Inflammatory Responses to Injury. *Nature* **2010**, *464*, 104–107. [CrossRef]
22. Ramachandran, P.; Iredale, J.P.; Fallowfield, J.A. Resolution of Liver Fibrosis: Basic Mechanisms and Clinical Relevance. *Semin. Liver. Dis.* **2015**, *35*, 119–131. [CrossRef] [PubMed]
23. Rodero, M.P.; Khosrotehrani, K. Skin Wound Healing Modulation by Macrophages. *Int. J. Clin. Exp. Pathol* **2010**, *3*, 643–653. [PubMed]

24. Novak, M.L.; Koh, T.J. Phenotypic Transitions of Macrophages Orchestrate Tissue Repair. *Am. J. Pathol.* **2013**, *183*, 1352–1363. [CrossRef]

25. Lech, M.; Anders, H.-J. Macrophages and Fibrosis: How Resident and Infiltrating Mononuclear Phagocytes Orchestrate All Phases of Tissue Injury and Repair. *Biochim. Biophys. Acta* **2013**, *1832*, 989–997. [CrossRef] [PubMed]

26. Martin, K.E.; García, A.J. Macrophage Phenotypes in Tissue Repair and the Foreign Body Response: Implications for Biomaterial-Based Regenerative Medicine Strategies. *Acta Biomater.* **2021**, *133*, 4–16. [CrossRef]

27. Mosser, D.M.; Edwards, J.P. Exploring the Full Spectrum of Macrophage Activation. *Nat. Rev. Immunol.* **2008**, *8*, 958–969. [CrossRef] [PubMed]

28. Kim, S.Y.; Nair, M.G. Macrophages in Wound Healing: Activation and Plasticity. *Immunol. Cell Biol.* **2019**, *97*, 258–267. [CrossRef]

29. Sen, C.K.; Roy, S. Redox Signals in Wound Healing. *Biochim. Biophys. Acta* **2008**, *1780*, 1348–1361. [CrossRef]

30. Ojha, N.; Roy, S.; He, G.; Biswas, S.; Velayutham, M.; Khanna, S.; Kuppusamy, P.; Zweier, J.L.; Sen, C.K. Assessment of Wound-Site Redox Environment and the Significance of Rac2 in Cutaneous Healing. *Free Radic. Biol. Med.* **2008**, *44*, 682–691. [CrossRef]

31. Hattori, H.; Subramanian, K.K.; Sakai, J.; Jia, Y.; Li, Y.; Porter, T.F.; Loison, F.; Sarraj, B.; Kasorn, A.; Jo, H.; et al. Small-Molecule Screen Identifies Reactive Oxygen Species as Key Regulators of Neutrophil Chemotaxis. *Proc. Natl. Acad. Sci. USA* **2010**, *107*, 3546–3551. [CrossRef]

32. Schafer, M.; Werner, S. Oxidative Stress in Normal and Impaired Wound Repair. *Pharmacol. Res.* **2008**, *58*, 165–171. [CrossRef] [PubMed]

33. He, L.; He, T.; Farrar, S.; Ji, L.; Liu, T.; Ma, X. Antioxidants Maintain Cellular Redox Homeostasis by Elimination of Reactive Oxygen Species. *Cell. Physiol. Biochem.* **2017**, *44*, 532–553. [CrossRef]

34. Hakami, N.Y.; Dusting, G.J.; Chan, E.C.; Shah, M.H.; Peshavariya, H.M. Wound Healing After Alkali Burn Injury of the Cornea Involves Nox4-Type NADPH Oxidase. *Investig. Ophthalmol. Vis. Sci* **2020**, *61*, 20. [CrossRef]

35. Sen, C.K.; Khanna, S.; Babior, B.M.; Hunt, T.K.; Ellison, E.C.; Roy, S. Oxidant-Induced Vascular Endothelial Growth Factor Expression in Human Keratinocytes and Cutaneous Wound Healing. *J. Biol. Chem.* **2002**, *277*, 33284–33290. [CrossRef] [PubMed]

36. Hallberg, C.K.; Trocme, S.D.; Ansari, N.H. Acceleration of Corneal Wound Healing in Diabetic Rats by the Antioxidant Trolox. *Res. Commun. Mol. Pathol. Pharmacol.* **1996**, *93*, 3–12. [PubMed]

37. Shukla, A.; Rasik, A.M.; Patnaik, G.K. Depletion of Reduced Glutathione, Ascorbic Acid, Vitamin E and Antioxidant Defence Enzymes in a Healing Cutaneous Wound. *Free Radic. Res.* **1997**, *26*, 93–101. [CrossRef] [PubMed]

38. McDaniel, D.H.; Ash, K.; Lord, J.; Newman, J.; Zukowski, M. Accelerated Laser Resurfacing Wound Healing Using a Triad of Topical Antioxidants. *Dermatol. Surg.* **1998**, *24*, 661–664. [CrossRef] [PubMed]

39. Bilgen, F.; Ural, A.; Kurutas, E.B.; Bekerecioglu, M. The Effect of Oxidative Stress and Raftlin Levels on Wound Healing. *Int. Wound J.* **2019**, *16*, 1178–1184. [CrossRef]

40. Feng, C.; Yu, B.; Song, C.; Wang, J.; Zhang, L.; Ji, X.; Wang, Y.; Fang, Y.; Liao, Z.; Wei, M.; et al. Static Magnetic Fields Reduce Oxidative Stress to Improve Wound Healing and Alleviate Diabetic Complications. *Cells* **2022**, *11*, 443. [CrossRef]

41. Schilrreff, P.; Alexiev, U. Chronic Inflammation in Non-Healing Skin Wounds and Promising Natural Bioactive Compounds Treatment. *Int. J. Mol. Sci.* **2022**, *23*, 4928. [CrossRef]

42. Zhang, W.; Chen, L.; Xiong, Y.; Panayi, A.C.; Abududilibaier, A.; Hu, Y.; Yu, C.; Zhou, W.; Sun, Y.; Liu, M.; et al. Antioxidant Therapy and Antioxidant-Related Bionanomaterials in Diabetic Wound Healing. *Front. Bioeng. Biotechnol.* **2021**, *9*, 707479. [CrossRef] [PubMed]

43. Aviello, G.; Knaus, U.G. ROS in Gastrointestinal Inflammation: Rescue Or Sabotage? *Br. J. Pharmacol.* **2017**, *174*, 1704–1718. [CrossRef] [PubMed]

44. Rasik, A.M.; Shukla, A. Antioxidant Status in Delayed Healing Type of Wounds: Delayed Healing Wounds and Antioxidants. *Int. J. Exp. Pathol.* **2001**, *81*, 257–263. [CrossRef] [PubMed]

45. Dworzański, J.; Strycharz-Dudziak, M.; Kliszczewska, E.; Kiełczykowska, M.; Dworzańska, A.; Drop, B.; Polz-Dacewicz, M. Glutathione Peroxidase (GPx) and Superoxide Dismutase (SOD) Activity in Patients with Diabetes Mellitus Type 2 Infected with Epstein-Barr Virus. *PLoS ONE* **2020**, *15*, e0230374. [CrossRef]

46. Prasad, A.; Manoharan, R.R.; Sedlářová, M.; Pospíšil, P. Free Radical-Mediated Protein Radical Formation in Differentiating Monocytes. *Int. J. Mol. Sci* **2021**, *22*, 9963. [CrossRef]

47. Helfinger, V.; Palfi, K.; Weigert, A.; Schröder, K. The NADPH Oxidase Nox4 Controls Macrophage Polarization in an NFκB-Dependent Manner. *Oxid. Med. Cell. Longev.* **2019**, *2019*, 3264858. [CrossRef]

48. Lee, C.F.; Qiao, M.; Schröder, K.; Zhao, Q.; Asmis, R. Nox4 Is a Novel Inducible Source of Reactive Oxygen Species in Monocytes and Macrophages and Mediates Oxidized Low Density Lipoprotein-Induced Macrophage Death. *Circ. Res.* **2010**, *106*, 1489–1497. [CrossRef]

49. Xu, Q.; Choksi, S.; Qu, J.; Jang, J.; Choe, M.; Banfi, B.; Engelhardt, J.F.; Liu, Z.-G. NADPH Oxidases Are Essential for Macrophage Differentiation. *J. Biol. Chem.* **2016**, *291*, 20030–20041. [CrossRef]

50. Zhang, Y.; Choksi, S.; Chen, K.; Pobezinskaya, Y.; Linnoila, I.; Liu, Z.-G. ROS Play a Critical Role in the Differentiation of Alternatively Activated Macrophages and the Occurrence of Tumor-Associated Macrophages. *Cell Res.* **2013**, *23*, 898–914. [CrossRef]

51. Griess, B.; Mir, S.; Datta, K.; Teoh-Fitzgerald, M. Scavenging Reactive Oxygen Species Selectively Inhibits M2 Macrophage Polarization and Their Pro-Tumorigenic Function in Part, via Stat3 Suppression. *Free Radic. Biol. Med.* **2020**, *147*, 48–60. [CrossRef]

52. Tan, H.-Y.; Wang, N.; Li, S.; Hong, M.; Wang, X.; Feng, Y. The Reactive Oxygen Species in Macrophage Polarization: Reflecting Its Dual Role in Progression and Treatment of Human Diseases. *Oxidative Med. Cell. Longev.* **2016**, *2016*, 2795090. [CrossRef] [PubMed]

53. Nassif, R.M.; Chalhoub, E.; Chedid, P.; Hurtado-Nedelec, M.; Raya, E.; Dang, P.M.-C.; Marie, J.-C.; El-Benna, J. Metformin Inhibits ROS Production by Human M2 Macrophages via the Activation of AMPK. *Biomedicines* **2022**, *10*, 319. [CrossRef] [PubMed]

54. Sanmun, D.; Witasp, E.; Jitkaew, S.; Tyurina, Y.Y.; Kagan, V.E.; Åhlin, A.; Palmblad, J.; Fadeel, B. Involvement of a Functional NADPH Oxidase in Neutrophils and Macrophages during Programmed Cell Clearance: Implications for Chronic Granulomatous Disease. *Am. J. Physiol.-Cell Physiol.* **2009**, *297*, C621–C631. [CrossRef] [PubMed]

55. Brown, K.L.; Christenson, K.; Karlsson, A.; Dahlgren, C.; Bylund, J. Divergent Effects on Phagocytosis by Macrophage-Derived Oxygen Radicals. *J. Innate Immun.* **2009**, *1*, 592–598. [CrossRef]

56. Kumar, A.; Barrett, J.P.; Alvarez-Croda, D.-M.; Stoica, B.A.; Faden, A.I.; Loane, D.J. NOX2 Drives M1-like Microglial/Macrophage Activation and Neurodegeneration Following Experimental Traumatic Brain Injury. *Brain Behav. Immun.* **2016**, *58*, 291–309. [CrossRef]

57. Balce, D.R.; Li, B.; Allan, E.R.O.; Rybicka, J.M.; Krohn, R.M.; Yates, R.M. Alternative Activation of Macrophages by IL-4 Enhances the Proteolytic Capacity of Their Phagosomes through Synergistic Mechanisms. *Blood* **2011**, *118*, 4199–4208. [CrossRef]

58. Han, C.; Sheng, Y.; Wang, J.; Zhou, X.; Li, W.; Zhang, C.; Guo, L.; Yang, Y. NOX4 Promotes Mucosal Barrier Injury in Inflammatory Bowel Disease by Mediating Macrophages M1 Polarization through ROS. *Int. Immunopharmacol.* **2022**, *104*, 108361. [CrossRef]

59. Lokmic, Z.; Musyoka, J.; Hewitson, T.D.; Darby, I.A. Hypoxia and Hypoxia Signaling in Tissue Repair and Fibrosis. *Int. Rev. Cell Mol. Biol.* **2012**, *296*, 139–185. [CrossRef]

60. Smith, K.A.; Waypa, G.B.; Schumacker, P.T. Redox Signaling during Hypoxia in Mammalian Cells. *Redox. Biol.* **2017**, *13*, 228–234. [CrossRef]

61. Kim, J.; Tchernyshyov, I.; Semenza, G.L.; Dang, C.V. HIF-1-Mediated Expression of Pyruvate Dehydrogenase Kinase: A Metabolic Switch Required for Cellular Adaptation to Hypoxia. *Cell Metab.* **2006**, *3*, 177–185. [CrossRef]

62. Finkel, T. Signal Transduction by Mitochondrial Oxidants. *J. Biol. Chem.* **2012**, *287*, 4434–4440. [CrossRef] [PubMed]

63. Semenza, G.L. HIF-1 and Mechanisms of Hypoxia Sensing. *Curr. Opin. Cell Biol.* **2001**, *13*, 167–171. [CrossRef]

64. Chandel, N.S.; McClintock, D.S.; Feliciano, C.E.; Wood, T.M.; Melendez, J.A.; Rodriguez, A.M.; Schumacker, P.T. Reactive Oxygen Species Generated at Mitochondrial Complex III Stabilize Hypoxia-Inducible Factor-1α during Hypoxia. *J. Biol. Chem.* **2000**, *275*, 25130–25138. [CrossRef]

65. Wang, D.; Malo, D.; Hekimi, S. Elevated Mitochondrial Reactive Oxygen Species Generation Affects the Immune Response via Hypoxia-Inducible Factor-1alpha in Long-Lived Mclk1+/− Mouse Mutants. *J. Immunol.* **2010**, *184*, 582–590. [CrossRef] [PubMed]

66. Nishi, K.; Oda, T.; Takabuchi, S.; Oda, S.; Fukuda, K.; Adachi, T.; Semenza, G.L.; Shingu, K.; Hirota, K. LPS Induces Hypoxia-Inducible Factor 1 Activation in Macrophage-Differentiated Cells in a Reactive Oxygen Species-Dependent Manner. *Antioxid. Redox. Signal.* **2008**, *10*, 983–995. [CrossRef] [PubMed]

67. Li, C.; Wang, Y.; Li, Y.; Yu, Q.; Jin, X.; Wang, X.; Jia, A.; Hu, Y.; Han, L.; Wang, J.; et al. HIF1α-Dependent Glycolysis Promotes Macrophage Functional Activities in Protecting against Bacterial and Fungal Infection. *Sci. Rep.* **2018**, *8*, 3603. [CrossRef] [PubMed]

68. Cramer, T.; Yamanishi, Y.; Clausen, B.E.; Förster, I.; Pawlinski, R.; Mackman, N.; Haase, V.H.; Jaenisch, R.; Corr, M.; Nizet, V.; et al. HIF-1α Is Essential for Myeloid Cell-Mediated Inflammation. *Cell* **2003**, *112*, 645–657. [CrossRef]

69. Kerber, E.L.; Padberg, C.; Koll, N.; Schuetzhold, V.; Fandrey, J.; Winning, S. The Importance of Hypoxia-Inducible Factors (HIF-1 and HIF-2) for the Pathophysiology of Inflammatory Bowel Disease. *Int. J. Mol. Sci.* **2020**, *21*, 8551. [CrossRef]

70. Hayden, M.S.; Ghosh, S. Shared Principles in NF-KappaB Signaling. *Cell* **2008**, *132*, 344–362. [CrossRef]

71. Lawrence, T.; Fong, C. The Resolution of Inflammation: Anti-Inflammatory Roles for NF-KappaB. *Int. J. Biochem. Cell Biol.* **2010**, *42*, 519–523. [CrossRef]

72. Müller, C.W.; Harrison, S.C. The Structure of the NF-Kappa B P50:DNA-Complex: A Starting Point for Analyzing the Rel Family. *FEBS Lett.* **1995**, *369*, 113–117. [CrossRef]

73. Bonizzi, G.; Karin, M. The Two NF-KappaB Activation Pathways and Their Role in Innate and Adaptive Immunity. *Trends Immunol.* **2004**, *25*, 280–288. [CrossRef] [PubMed]

74. Roux, C.; Jafari, S.M.; Shinde, R.; Duncan, G.; Cescon, D.W.; Silvester, J.; Chu, M.F.; Hodgson, K.; Berger, T.; Wakeham, A.; et al. Reactive Oxygen Species Modulate Macrophage Immunosuppressive Phenotype through the Up-Regulation of PD-L1. *Proc. Natl. Acad. Sci. USA* **2019**, *116*, 4326–4335. [CrossRef] [PubMed]

75. Formentini, L.; Santacatterina, F.; Núñez de Arenas, C.; Stamatakis, K.; López-Martínez, D.; Logan, A.; Fresno, M.; Smits, R.; Murphy, M.P.; Cuezva, J.M. Mitochondrial ROS Production Protects the Intestine from Inflammation through Functional M2 Macrophage Polarization. *Cell Rep.* **2017**, *19*, 1202–1213. [CrossRef] [PubMed]

76. Luo, D.; Guo, Y.; Cheng, Y.; Zhao, J.; Wang, Y.; Rong, J. Natural Product Celastrol Suppressed Macrophage M1 Polarization against Inflammation in Diet-Induced Obese Mice via Regulating Nrf2/HO-1, MAP Kinase and NF-KB Pathways. *Aging* **2017**, *9*, 2069–2082. [CrossRef]

77. Ma, Q. Role of Nrf2 in Oxidative Stress and Toxicity. *Annu. Rev. Pharmacol. Toxicol.* **2013**, *53*, 401–426. [CrossRef]

78. Li, W.; Kong, A.-N. Molecular Mechanisms of Nrf2-Mediated Antioxidant Response. *Mol. Carcinog.* **2009**, *48*, 91–104. [CrossRef]

79. Canning, P.; Sorrell, F.J.; Bullock, A.N. Structural Basis of Keap1 Interactions with Nrf2. *Free Radic. Biol. Med.* **2015**, *88*, 101–107. [CrossRef]

80. Piotrowska, M.; Swierczynski, M.; Fichna, J.; Piechota-Polanczyk, A. The Nrf2 in the Pathophysiology of the Intestine: Molecular Mechanisms and Therapeutic Implications for Inflammatory Bowel Diseases. *Pharmacol. Res.* **2021**, *163*, 105243. [CrossRef]

81. Kovac, S.; Angelova, P.R.; Holmström, K.M.; Zhang, Y.; Dinkova-Kostova, A.T.; Abramov, A.Y. Nrf2 Regulates ROS Production by Mitochondria and NADPH Oxidase. *Biochim. Biophys. Acta (BBA)-Gen. Subj.* **2015**, *1850*, 794–801. [CrossRef]

82. Kobayashi, E.H.; Suzuki, T.; Funayama, R.; Nagashima, T.; Hayashi, M.; Sekine, H.; Tanaka, N.; Moriguchi, T.; Motohashi, H.; Nakayama, K.; et al. Nrf2 Suppresses Macrophage Inflammatory Response by Blocking Proinflammatory Cytokine Transcription. *Nat. Commun.* **2016**, *7*, 11624. [CrossRef] [PubMed]

83. Ding, S.; Li, C.; Cheng, N.; Cui, X.; Xu, X.; Zhou, G. Redox Regulation in Cancer Stem Cells. *Oxid. Med. Cell. Longev.* **2015**, *2015*, 750798. [CrossRef] [PubMed]

84. Li, M.; Yu, H.; Pan, H.; Zhou, X.; Ruan, Q.; Kong, D.; Chu, Z.; Li, H.; Huang, J.; Huang, X.; et al. Nrf2 Suppression Delays Diabetic Wound Healing Through Sustained Oxidative Stress and Inflammation. *Front. Pharmacol.* **2019**, *10*, 1099. [CrossRef] [PubMed]

85. Weis, N.; Weigert, A.; von Knethen, A.; Brüne, B. Heme Oxygenase-1 Contributes to an Alternative Macrophage Activation Profile Induced by Apoptotic Cell Supernatants. *MBoC* **2009**, *20*, 1280–1288. [CrossRef]

86. Khor, T.O.; Huang, M.-T.; Kwon, K.H.; Chan, J.Y.; Reddy, B.S.; Kong, A.-N. Nrf2-Deficient Mice Have an Increased Susceptibility to Dextran Sulfate Sodium-Induced Colitis. *Cancer Res.* **2006**, *66*, 11580–11584. [CrossRef]

87. Osburn, W.O.; Karim, B.; Dolan, P.M.; Liu, G.; Yamamoto, M.; Huso, D.L.; Kensler, T.W. Increased Colonic Inflammatory Injury and Formation of Aberrant Crypt Foci in Nrf2-Deficient Mice upon Dextran Sulfate Treatment. *Int. J. Cancer* **2007**, *121*, 1883–1891. [CrossRef]

88. Feng, R.; Morine, Y.; Ikemoto, T.; Imura, S.; Iwahashi, S.; Saito, Y.; Shimada, M. Nrf2 Activation Drive Macrophages Polarization and Cancer Cell Epithelial-Mesenchymal Transition during Interaction. *Cell Commun. Signal.* **2018**, *16*, 54. [CrossRef] [PubMed]

89. Ghosh, S.; Mukherjee, S.; Choudhury, S.; Gupta, P.; Adhikary, A.; Baral, R.; Chattopadhyay, S. Reactive Oxygen Species in the Tumor Niche Triggers Altered Activation of Macrophages and Immunosuppression: Role of Fluoxetine. *Cell Signal.* **2015**, *27*, 1398–1412. [CrossRef]

90. Zhuyan, J.; Chen, M.; Zhu, T.; Bao, X.; Zhen, T.; Xing, K.; Wang, Q.; Zhu, S. Critical Steps to Tumor Metastasis: Alterations of Tumor Microenvironment and Extracellular Matrix in the Formation of Pre-Metastatic and Metastatic Niche. *Cell Biosci.* **2020**, *10*, 89. [CrossRef]

91. Catalano, V.; Turdo, A.; Di Franco, S.; Dieli, F.; Todaro, M.; Stassi, G. Tumor and Its Microenvironment: A Synergistic Interplay. *Semin. Cancer Biol.* **2013**, *23*, 522–532. [CrossRef]

92. Kennel, K.B.; Greten, F.R. Immune Cell–Produced ROS and Their Impact on Tumor Growth and Metastasis. *Redox Biol.* **2021**, *42*, 101891. [CrossRef] [PubMed]

93. Schito, L.; Semenza, G.L. Hypoxia-Inducible Factors: Master Regulators of Cancer Progression. *Trends Cancer* **2016**, *2*, 758–770. [CrossRef] [PubMed]

94. Lotfi, R.; Eisenbacher, J.; Solgi, G.; Fuchs, K.; Yildiz, T.; Nienhaus, C.; Rojewski, M.T.; Schrezenmeier, H. Human Mesenchymal Stem Cells Respond to Native but Not Oxidized Damage Associated Molecular Pattern Molecules from Necrotic (Tumor) Material. *Eur. J. Immunol.* **2011**, *41*, 2021–2028. [CrossRef] [PubMed]

95. van Beijnum, J.R.; Nowak-Sliwinska, P.; van den Boezem, E.; Hautvast, P.; Buurman, W.A.; Griffioen, A.W. Tumor Angiogenesis Is Enforced by Autocrine Regulation of High-Mobility Group Box 1. *Oncogene* **2013**, *32*, 363–374. [CrossRef]

96. Patidar, A.; Selvaraj, S.; Sarode, A.; Chauhan, P.; Chattopadhyay, D.; Saha, B. DAMP-TLR-Cytokine Axis Dictates the Fate of Tumor. *Cytokine* **2018**, *104*, 114–123. [CrossRef]

97. Gilkes, D.M.; Semenza, G.L.; Wirtz, D. Hypoxia and the Extracellular Matrix: Drivers of Tumour Metastasis. *Nat. Rev. Cancer* **2014**, *14*, 430–439. [CrossRef]

98. Sakamoto, S.; Kyprianou, N. Targeting Anoikis Resistance in Prostate Cancer Metastasis. *Mol. Aspects. Med.* **2010**, *31*, 205–214. [CrossRef]

99. Alizadeh, A.M.; Shiri, S.; Farsinejad, S. Metastasis Review: From Bench to Bedside. *Tumour. Biol.* **2014**, *35*, 8483–8523. [CrossRef]

100. Wang, S.-S.; Jiang, J.; Liang, X.-H.; Tang, Y.-L. Links between Cancer Stem Cells and Epithelial-Mesenchymal Transition. *Onco Targets Ther.* **2015**, *8*, 2973–2980. [CrossRef]

101. Aggarwal, V.; Tuli, H.S.; Varol, A.; Thakral, F.; Yerer, M.B.; Sak, K.; Varol, M.; Jain, A.; Khan, M.A.; Sethi, G. Role of Reactive Oxygen Species in Cancer Progression: Molecular Mechanisms and Recent Advancements. *Biomolecules* **2019**, *9*, 735. [CrossRef]

102. Sorolla, M.A.; Hidalgo, I.; Sorolla, A.; Montal, R.; Pallisé, O.; Salud, A.; Parisi, E. Microenvironmental Reactive Oxygen Species in Colorectal Cancer: Involved Processes and Therapeutic Opportunities. *Cancers* **2021**, *13*, 5037. [CrossRef] [PubMed]

103. Bauer, G. Targeting Extracellular ROS Signaling of Tumor Cells. *Anticancer Res.* **2014**, *34*, 1467–1482. [PubMed]

104. Liou, G.-Y.; Storz, P. Reactive Oxygen Species in Cancer. *Free Radic. Res.* **2010**, *44*, 479–496. [CrossRef] [PubMed]

105. Redza-Dutordoir, M.; Averill-Bates, D.A. Activation of Apoptosis Signalling Pathways by Reactive Oxygen Species. *Biochim. Biophys. Acta* **2016**, *1863*, 2977–2992. [CrossRef] [PubMed]

106. Assi, M. The Differential Role of Reactive Oxygen Species in Early and Late Stages of Cancer. *Am. J. Physiol. Regul. Integr. Comp. Physiol.* **2017**, *313*, R646–R653. [CrossRef]

107. Spitz, D.R.; Sim, J.E.; Ridnour, L.A.; Galoforo, S.S.; Lee, Y.J. Glucose Deprivation-Induced Oxidative Stress in Human Tumor Cells. A Fundamental Defect in Metabolism? *Ann. N. Y. Acad. Sci.* **2000**, *899*, 349–362. [CrossRef]

108. Forsythe, J.A.; Jiang, B.H.; Iyer, N.V.; Agani, F.; Leung, S.W.; Koos, R.D.; Semenza, G.L. Activation of Vascular Endothelial Growth Factor Gene Transcription by Hypoxia-Inducible Factor 1. *Mol. Cell. Biol.* **1996**, *16*, 4604–4613. [CrossRef]

109. Han, X.; Sun, S.; Zhao, M.; Cheng, X.; Chen, G.; Lin, S.; Guan, Y.; Yu, X. Celastrol Stimulates Hypoxia-Inducible Factor-1 Activity in Tumor Cells by Initiating the ROS/Akt/P70S6K Signaling Pathway and Enhancing Hypoxia-Inducible Factor-1α Protein Synthesis. *PLoS ONE* **2014**, *9*, e112470. [CrossRef]

110. Karar, J.; Maity, A. PI3K/AKT/MTOR Pathway in Angiogenesis. *Front. Mol. Neurosci.* **2011**, *4*, 51. [CrossRef]

111. Xia, C.; Meng, Q.; Liu, L.-Z.; Rojanasakul, Y.; Wang, X.-R.; Jiang, B.-H. Reactive Oxygen Species Regulate Angiogenesis and Tumor Growth through Vascular Endothelial Growth Factor. *Cancer Res.* **2007**, *67*, 10823–10830. [CrossRef]

112. Liu, L.-Z.; Hu, X.-W.; Xia, C.; He, J.; Zhou, Q.; Shi, X.; Fang, J.; Jiang, B.-H. Reactive Oxygen Species Regulate Epidermal Growth Factor-Induced Vascular Endothelial Growth Factor and Hypoxia-Inducible Factor-1alpha Expression through Activation of AKT and P70S6K1 in Human Ovarian Cancer Cells. *Free Radic. Biol. Med.* **2006**, *41*, 1521–1533. [CrossRef] [PubMed]

113. West, X.Z.; Malinin, N.L.; Merkulova, A.A.; Tischenko, M.; Kerr, B.A.; Borden, E.C.; Podrez, E.A.; Salomon, R.G.; Byzova, T.V. Oxidative Stress Induces Angiogenesis by Activating TLR2 with Novel Endogenous Ligands. *Nature* **2010**, *467*, 972–976. [CrossRef]

114. Wartenberg, M.; Budde, P.; De Mareés, M.; Grünheck, F.; Tsang, S.Y.; Huang, Y.; Chen, Z.-Y.; Hescheler, J.; Sauer, H. Inhibition of Tumor-Induced Angiogenesis and Matrix-Metalloproteinase Expression in Confrontation Cultures of Embryoid Bodies and Tumor Spheroids by Plant Ingredients Used in Traditional Chinese Medicine. *Lab. Investig.* **2003**, *83*, 87–98. [CrossRef] [PubMed]

115. Boudreau, H.E.; Casterline, B.W.; Rada, B.; Korzeniowska, A.; Leto, T.L. Nox4 Involvement in TGF-Beta and SMAD3-Driven Induction of the Epithelial-to-Mesenchymal Transition and Migration of Breast Epithelial Cells. *Free Radic. Biol. Med.* **2012**, *53*, 1489–1499. [CrossRef] [PubMed]

116. Yazaki, K.; Matsuno, Y.; Yoshida, K.; Sherpa, M.; Nakajima, M.; Matsuyama, M.; Kiwamoto, T.; Morishima, Y.; Ishii, Y.; Hizawa, N. ROS-Nrf2 Pathway Mediates the Development of TGF-B1-Induced Epithelial-Mesenchymal Transition through the Activation of Notch Signaling. *Eur. J. Cell Biol.* **2021**, *100*, 151181. [CrossRef]

117. Padua, D.; Massagué, J. Roles of TGFbeta in Metastasis. *Cell Res.* **2009**, *19*, 89–102. [CrossRef]

118. Diebold, I.; Petry, A.; Hess, J.; Görlach, A. The NADPH Oxidase Subunit NOX4 Is a New Target Gene of the Hypoxia-Inducible Factor-1. *Mol. Biol. Cell* **2010**, *21*, 2087–2096. [CrossRef]

119. Zhang, B.; Liu, Z.; Hu, X. Inhibiting Cancer Metastasis via Targeting NAPDH Oxidase 4. *Biochem. Pharmacol.* **2013**, *86*, 253–266. [CrossRef]

120. Du, S.; Miao, J.; Zhu, Z.; Xu, E.; Shi, L.; Ai, S.; Wang, F.; Kang, X.; Chen, H.; Lu, X.; et al. NADPH Oxidase 4 Regulates Anoikis Resistance of Gastric Cancer Cells through the Generation of Reactive Oxygen Species and the Induction of EGFR. *Cell Death Dis.* **2018**, *9*, 948. [CrossRef]

121. Kim, H.; Sung, J.Y.; Park, E.-K.; Kho, S.; Koo, K.H.; Park, S.-Y.; Goh, S.-H.; Jeon, Y.K.; Oh, S.; Park, B.-K.; et al. Regulation of Anoikis Resistance by NADPH Oxidase 4 and Epidermal Growth Factor Receptor. *Br. J. Cancer* **2017**, *116*, 370–381. [CrossRef]

122. Binker, M.G.; Binker-Cosen, A.A.; Richards, D.; Oliver, B.; Cosen-Binker, L.I. EGF Promotes Invasion by PANC-1 Cells through Rac1/ROS-Dependent Secretion and Activation of MMP-2. *Biochem. Biophys. Res. Commun.* **2009**, *379*, 445–450. [CrossRef] [PubMed]

123. Tobar, N.; Villar, V.; Santibanez, J.F. ROS-NFkappaB Mediates TGF-Beta1-Induced Expression of Urokinase-Type Plasminogen Activator, Matrix Metalloproteinase-9 and Cell Invasion. *Mol. Cell. Biochem.* **2010**, *340*, 195–202. [CrossRef] [PubMed]

124. Basak, D.; Uddin, M.N.; Hancock, J. The Role of Oxidative Stress and Its Counteractive Utility in Colorectal Cancer (CRC). *Cancers* **2020**, *12*, 3336. [CrossRef] [PubMed]

125. Peiris-Pagès, M.; Martinez-Outschoorn, U.E.; Pestell, R.G.; Sotgia, F.; Lisanti, M.P. Cancer Stem Cell Metabolism. *Breast Cancer Res.* **2016**, *18*, 55. [CrossRef] [PubMed]

126. Dando, I.; Cordani, M.; Dalla Pozza, E.; Biondani, G.; Donadelli, M.; Palmieri, M. Antioxidant Mechanisms and ROS-Related MicroRNAs in Cancer Stem Cells. *Oxid. Med. Cell. Longev.* **2015**, *2015*, 425708. [CrossRef]

127. Sato, A.; Okada, M.; Shibuya, K.; Watanabe, E.; Seino, S.; Narita, Y.; Shibui, S.; Kayama, T.; Kitanaka, C. Pivotal Role for ROS Activation of P38 MAPK in the Control of Differentiation and Tumor-Initiating Capacity of Glioma-Initiating Cells. *Stem Cell Res.* **2014**, *12*, 119–131. [CrossRef]

128. Cao, Y.; Fang, Y.; Cai, J.; Li, X.; Xu, F.; Yuan, N.; Zhang, S.; Wang, J. ROS Functions as an Upstream Trigger for Autophagy to Drive Hematopoietic Stem Cell Differentiation. *Hematology* **2016**, *21*, 613–618. [CrossRef]

129. Singel, K.L.; Segal, B.H. Neutrophils in the Tumor Microenvironment: Trying to Heal the Wound That Cannot Heal. *Immunol. Rev.* **2016**, *273*, 329–343. [CrossRef]

130. Mantovani, A.; Sica, A. Macrophages, Innate Immunity and Cancer: Balance, Tolerance, and Diversity. *Curr. Opin. Immunol.* **2010**, *22*, 231–237. [CrossRef]

131. Kotsafti, A.; Scarpa, M.; Castagliuolo, I.; Scarpa, M. Reactive Oxygen Species and Antitumor Immunity-From Surveillance to Evasion. *Cancers* **2020**, *12*, 1748. [CrossRef]

132. Wang, H.; Tian, T.; Zhang, J. Tumor-Associated Macrophages (TAMs) in Colorectal Cancer (CRC): From Mechanism to Therapy and Prognosis. *Int. J. Mol. Sci* **2021**, *22*, 8470. [CrossRef] [PubMed]

133. Ryder, M.; Ghossein, R.A.; Ricarte-Filho, J.C.M.; Knauf, J.A.; Fagin, J.A. Increased Density of Tumor-Associated Macrophages Is Associated with Decreased Survival in Advanced Thyroid Cancer. *Endocr. Relat. Cancer* **2008**, *15*, 1069–1074. [CrossRef] [PubMed]
134. Sørensen, M.D.; Dahlrot, R.H.; Boldt, H.B.; Hansen, S.; Kristensen, B.W. Tumour-Associated Microglia/Macrophages Predict Poor Prognosis in High-Grade Gliomas and Correlate with an Aggressive Tumour Subtype. *Neuropathol. Appl. Neurobiol.* **2018**, *44*, 185–206. [CrossRef] [PubMed]
135. Jeong, H.; Hwang, I.; Kang, S.H.; Shin, H.C.; Kwon, S.Y. Tumor-Associated Macrophages as Potential Prognostic Biomarkers of Invasive Breast Cancer. *J. Breast Cancer* **2019**, *22*, 38–51. [CrossRef] [PubMed]
136. Zhou, Y.; Que, K.-T.; Zhang, Z.; Yi, Z.J.; Zhao, P.X.; You, Y.; Gong, J.-P.; Liu, Z.-J. Iron Overloaded Polarizes Macrophage to Proinflammation Phenotype through ROS/Acetyl-P53 Pathway. *Cancer Med.* **2018**, *7*, 4012–4022. [CrossRef]
137. Wu, Q.; Allouch, A.; Paoletti, A.; Leteur, C.; Mirjolet, C.; Martins, I.; Voisin, L.; Law, F.; Dakhli, H.; Mintet, E.; et al. NOX2-Dependent ATM Kinase Activation Dictates pro-Inflammatory Macrophage Phenotype and Improves Effectiveness to Radiation Therapy. *Cell Death Differ.* **2017**, *24*, 1632–1644. [CrossRef]
138. Mantovani, A.; Locati, M. Tumor-Associated Macrophages as a Paradigm of Macrophage Plasticity, Diversity, and Polarization: Lessons and Open Questions. *Arterioscler. Thromb. Vasc. Biol.* **2013**, *33*, 1478–1483. [CrossRef]
139. Chen, L.-M.; Tseng, H.-Y.; Chen, Y.-A.; Al Haq, A.T.; Hwang, P.-A.; Hsu, H.-L. Oligo-Fucoidan Prevents M2 Macrophage Differentiation and HCT116 Tumor Progression. *Cancers* **2020**, *12*, 421. [CrossRef]
140. Salpeter, S.J.; Pozniak, Y.; Merquiol, E.; Ben-Nun, Y.; Geiger, T.; Blum, G. A Novel Cysteine Cathepsin Inhibitor Yields Macrophage Cell Death and Mammary Tumor Regression. *Oncogene* **2015**, *34*, 6066–6078. [CrossRef]
141. Riabov, V.; Gudima, A.; Wang, N.; Mickley, A.; Orekhov, A.; Kzhyshkowska, J. Role of Tumor Associated Macrophages in Tumor Angiogenesis and Lymphangiogenesis. *Front. Physiol.* **2014**, *5*, 75. [CrossRef]
142. Luput, L.; Licarete, E.; Sesarman, A.; Patras, L.; Alupei, M.C.; Banciu, M. Tumor-Associated Macrophages Favor C26 Murine Colon Carcinoma Cell Proliferation in an Oxidative Stress-Dependent Manner. *Oncol. Rep.* **2017**, *37*, 2472–2480. [CrossRef] [PubMed]
143. Kalluri, R. The Biology and Function of Fibroblasts in Cancer. *Nat. Rev. Cancer* **2016**, *16*, 582–598. [CrossRef] [PubMed]
144. Chan, J.S.K.; Tan, M.J.; Sng, M.K.; Teo, Z.; Phua, T.; Choo, C.C.; Li, L.; Zhu, P.; Tan, N.S. Cancer-Associated Fibroblasts Enact Field Cancerization by Promoting Extratumoral Oxidative Stress. *Cell Death Dis.* **2017**, *8*, e2562. [CrossRef] [PubMed]
145. Onfroy-Roy, L.; Hamel, D.; Malaquin, L.; Ferrand, A. Colon Fibroblasts and Inflammation: Sparring Partners in Colorectal Cancer Initiation? *Cancers* **2021**, *13*, 1749. [CrossRef] [PubMed]
146. Martinez-Outschoorn, U.E.; Balliet, R.M.; Rivadeneira, D.B.; Chiavarina, B.; Pavlides, S.; Wang, C.; Whitaker-Menezes, D.; Daumer, K.M.; Lin, Z.; Witkiewicz, A.K.; et al. Oxidative Stress in Cancer Associated Fibroblasts Drives Tumor-Stroma Co-Evolution: A New Paradigm for Understanding Tumor Metabolism, the Field Effect and Genomic Instability in Cancer Cells. *Cell Cycle* **2010**, *9*, 3256–3276. [CrossRef]
147. Policastro, L.L.; Ibañez, I.L.; Notcovich, C.; Duran, H.A.; Podhajcer, O.L. The Tumor Microenvironment: Characterization, Redox Considerations, and Novel Approaches for Reactive Oxygen Species-Targeted Gene Therapy. *Antioxid. Redox Signal.* **2013**, *19*, 854–895. [CrossRef]
148. Giannoni, E.; Bianchini, F.; Calorini, L.; Chiarugi, P. Cancer Associated Fibroblasts Exploit Reactive Oxygen Species through a Proinflammatory Signature Leading to Epithelial Mesenchymal Transition and Stemness. *Antioxid. Redox Signal.* **2011**, *14*, 2361–2371. [CrossRef]
149. Erez, N.; Truitt, M.; Olson, P.; Arron, S.T.; Hanahan, D. Cancer-Associated Fibroblasts Are Activated in Incipient Neoplasia to Orchestrate Tumor-Promoting Inflammation in an NF-KappaB-Dependent Manner. *Cancer Cell* **2010**, *17*, 135–147. [CrossRef]
150. Zhao, H.; Huang, J.; Li, Y.; Lv, X.; Zhou, H.; Wang, H.; Xu, Y.; Wang, C.; Wang, J.; Liu, Z. ROS-Scavenging Hydrogel to Promote Healing of Bacteria Infected Diabetic Wounds. *Biomaterials* **2020**, *258*, 120286. [CrossRef]
151. Konaté, M.M.; Antony, S.; Doroshow, J.H. Inhibiting the Activity of NADPH Oxidase in Cancer. *Antioxid. Redox Signal.* **2020**, *33*, 435–454. [CrossRef]
152. Hadi, T.; Douhard, R.; Dias, A.M.M.; Wendremaire, M.; Pezzè, M.; Bardou, M.; Sagot, P.; Garrido, C.; Lirussi, F. Beta3 Adrenergic Receptor Stimulation in Human Macrophages Inhibits NADPHoxidase Activity and Induces Catalase Expression via PPARγ Activation. *Biochim. Biophys. Acta* **2017**, *1864*, 1769–1784. [CrossRef] [PubMed]
153. Carvalho, M.T.B.; Araújo-Filho, H.G.; Barreto, A.S.; Quintans-Júnior, L.J.; Quintans, J.S.S.; Barreto, R.S.S. Wound Healing Properties of Flavonoids: A Systematic Review Highlighting the Mechanisms of Action. *Phytomedicine* **2021**, *90*, 153636. [CrossRef] [PubMed]
154. Magrì, A.; Germano, G.; Lorenzato, A.; Lamba, S.; Chilà, R.; Montone, M.; Amodio, V.; Ceruti, T.; Sassi, F.; Arena, S.; et al. High-Dose Vitamin C Enhances Cancer Immunotherapy. *Sci. Transl. Med.* **2020**, *12*, eaay8707. [CrossRef] [PubMed]
155. Luchtel, R.A.; Bhagat, T.; Pradhan, K.; Jacobs, W.R.; Levine, M.; Verma, A.; Shenoy, N. High-Dose Ascorbic Acid Synergizes with Anti-PD1 in a Lymphoma Mouse Model. *Proc. Natl. Acad. Sci. USA* **2020**, *117*, 1666–1677. [CrossRef] [PubMed]
156. Dunnill, C.; Patton, T.; Brennan, J.; Barrett, J.; Dryden, M.; Cooke, J.; Leaper, D.; Georgopoulos, N.T. Reactive Oxygen Species (ROS) and Wound Healing: The Functional Role of ROS and Emerging ROS-Modulating Technologies for Augmentation of the Healing Process. *Int. Wound J.* **2017**, *14*, 89–96. [CrossRef]

MDPI

St. Alban-Anlage 66

4052 Basel

Switzerland

Tel. +41 61 683 77 34

Fax +41 61 302 89 18

www.mdpi.com

Biomedicines Editorial Office

E-mail: biomedicines@mdpi.com

www.mdpi.com/journal/biomedicines